Carlos Arturo Pomala

Image Processing—Continuous to Discrete Volume I

Geometric, Transform, and Statistical Methods

Edward R. Dougherty
Fairleigh Dickinson University
Charles R. Giardina
City University of New York

Prentice-Hall, Inc.
Englewood Cliffs, NJ 07632

Library of Congress Cataloging-in-Publication Data

Dougherty, Edward R.
  Image processing.

  Bibliography: v. 1, p.
  Includes index.
  1. Image processing—Mathematics.   I. Giardina,
Charles Robert.   II. Title.
TA1632.D67 1987   621.36′7   86-20453
ISBN 0-13-453283-X (v. 1)

Editorial/production supervision and
  interior design: Linda Zuk, WordCrafters Editorial Services, Inc.
Cover design: Edsal Enterprises
Manufacturing buyer: Ed O'Dougherty

© 1987 by Prentice-Hall, Inc.
A division of Simon & Schuster
Englewood Cliffs, New Jersey 07632

*All rights reserved. No part of this book may be
reproduced, in any form or by an means,
without permission in writing from the publisher.*

Printed in the United States of America

10 9 8 7 6 5 4 3 2 1

ISBN 0-13-453283-X   025

Prentice-Hall International (UK) Limited, *London*
Prentice-Hall of Australia Pty. Limited, *Sydney*
Prentice-Hall Canada Inc., *Toronto*
Prentice-Hall Hispanoamericana, S.A., *Mexico*
Prentice-Hall of India Private Limited, *New Delhi*
Prentice-Hall of Japan, Inc., *Tokyo*
Prentice-Hall of Southeast Asia Pte. Ltd., *Singapore*
Editora Prentice-Hall do Brasil, Ltda., *Rio de Janeiro*

To our children—
Russell and John
               E.D.

Barbara, Steven,
Michael, and Joey
               C.G.

Special thanks to Barbara
for helping with the typing,
and to Steven and Russell
for their artistic contributions

Contents

PREFACE ix

**1
INTRODUCTION 1**
1.1 Image Recognition, 1
1.2 Image Processing Transformations, 8

**2
BOUND MATRIX AND BLOCK DIAGRAM IMAGE REPRESENTATION 11**
2.1 Digital Image Representation, 11
2.2 Bound Matrices, 13
2.3* Bound Matrices as Equivalence Classes, 17
2.4 Arithmetic Operations on Digital Images, 19
2.5 Geometric Operations on Digital Images, 23
2.6 Database-Type Image Operations, 26
2.7 Structural Transformations, 28
2.8 Thresholding, 30
2.9 Vector-Type Operations, 35
2.10 Moving-Average Filters, 40
2.11 Image Vectors, 48
2.12 Digital Image Gradients, 52
2.13 Edge Detection by Image Gradients, 62
2.14 Edge Detection by Compass Gradients, 67
2.15 Best-Fit Plane, 69
2.16 Gray Level Histogram, 75
2.17 Co-Occurrence Matrix, 79
2.18 Best Quadric Fit, 81
Exercises, 87

3
MORPHOLOGICAL IMAGE PROCESSING 91
3.1 Minkowski Algebra, 91
3.2 Opening and Closing, 105
3.3 Convex Sets, 113
3.4 Granulometries, 120
3.5 Image Functionals, 123
3.6 Integral Geometry and Image Functionals, 131
3.7 Digital Minkowski Algebra, 139
3.8 Digital Openings and Closings, 144
3.9 Digital Size and Shape Distributions, 146
3.10* Increasing τ-Mappings, 152
3.11* Algebraic Openings and Closings of Images, 159
3.12* Algebraic Granulometric Characterization, 163
3.13 Sampling-Error Bounds for Morphological Operations, 171
Exercises, 178

4
TRANSFORM TECHNIQUES IN IMAGE PROCESSING 183
4.1 Analog (Continuous) Images, 183
4.2 Fourier Transforms of Integrable Signals (One Dimension), 187
4.3 The Fourier Transform of Integrable Images, 198
4.4* The Plancherel Theorem (Fourier Transform on L_2), 203
4.5 Sine and Cosine Transforms, 206
4.6 Mellin Transform of Signals, 208
4.7 Hilbert Transform, 213
4.8 Radon Transform, 217
4.9* Inversion of the Radon Transform, 226
4.10* Image Creation From Projections via the Fourier Transform, 232
4.11* Convolutional Method of Image Creation from Projections, 234
4.12 Hough Transform, 236
Exercises, 238

5
PROJECTION METHODS IN IMAGE PROCESSING 241
5.1 Projections in Euclidean Space, 241
5.2 Inner-Product Spaces, 245
5.3 Orthonormal Systems, 248
5.4 Projections in Hilbert Space, 253
5.5 Trigonometric Fourier Series of Square Integrable Signals, 257
5.6 Trigonometric Fourier Series of Integrable Signals, 260
5.7 Walsh Expansions, 266
5.8 Complete Orthonormal Systems for Expansion of Images, 272
5.9 Matrix Image Transforms, 275

Contents vii

5.10 Chain Codes for Boundary Representation of Images, 277
5.11 Fourier Series Representation of Chain-Encoded,
 Line-Drawing-Type Images, 280
5.12 Normalization of Fourier Series Representations, 289
5.13 Walsh Function Feature Representations, 292
Exercises, 302

6
PROBABILITY FOR IMAGE PROCESSING 305

6.1 Probability Space, 305
6.2 Conditional Probabilities, 309
6.3 Random Variables and Probability Distribution Functions, 311
6.4 The Probability Density Function, 315
6.5 Mathematical Expectation and Parameters, 320
6.6 Some Important Probability Density Functions, 327
6.7 Functions of a Single Random Variable, 328
6.8 Random Vectors and Joint Distributions, 335
6.9 Marginal and Conditional Distributions, 340
6.10 Expected Values for Random Vectors, 346
6.11 Functions of Two or More Random Variables, 351
6.12 Multivariate Gaussian Density, 360
6.13 Moment Transformation for Feature Representation, 362
Exercises, 367

7
INFERENTIAL STATISTICAL TECHNIQUES IN IMAGE PROCESSING 371

7.1 Estimators and Estimates, 371
7.2 Estimation Classification Criteria, 376
7.3 Estimation Procedures, 383
7.4 Matrix Estimation Techniques, 392
7.5 Kalman-Bucy Filtering, 402
7.6 Least-Squares Image Restoration, 407
7.7 Detection, 410
7.8 Hotelling Transform, 419
Exercises, 428

Appendix
DELTA FUNCTIONS 431

BIBLIOGRAPHY 437

INDEX 447

Preface

This series of image processing texts is intended to provide a comprehensive and serious exposition of image processing methodologies for the professional image engineer. It presents, at the graduate level, fundamental mathematical methods utilized in image analysis. Topics are explored in depth with many fully-worked-out, relevant examples.

Chapter 1 provides a general statement of the problems that constitute the domain of image processing. It also presents a characterization of the four levels which we believe form a conceptually useful partition of the modeling, mathematics, and processing that enter into the solution of any intelligent recognizer system.

The first half of Chapter 2 introduces the bound matrix data structure for image representation, which, for machine-implementable algorithms, provides a unified and consistent framework for algorithm specification. The second half of the chapter provides a concise account of some fundamental topics concerning gray-level processing. These include thresholding, linear filtering, and gradient-type edge detection. The topics emphasized are of a geometric nature and are introduced in a structured manner through the use of image vectors.

One of the most topical subject matters within the field of image processing is the mathematical morphology pioneered by G. Matheron. In Chapter 3 we provide a rigorous and somewhat comprehensive account of the operations that underlie the morphological method. These include Minkowski algebra, morphological feature parameter generation, and size distributions. In order to rectify the lack of any systematic body of detailed proofs in the current literature, we have provided these proofs in the current volume. However, the presentation is augmented by many drawings and examples, and the exposition can be read without digesting the proofs. (In no other chapter have proofs been given; rather, theorems have been footnoted to quality references.) The last three sections of the chapter give the Matheron representation theorems for different classes of morphological filters. Due to our desire to provide an in-depth account of the underlying morphological principles, no gray-level morphology has been discussed herein. Due to the umbra transformation between two-dimensional gray-scale images and three-dimensional constant images, once the theory presented here is digested, the jump to gray-level theory should present no great difficulty.

Chapter 4 presents the standard material on the Fourier transform, in terms of precisely stated theorems on clearly defined function classes. A multitude of examples are provided. The Plancherel theorem for square-integrable signals and images is explained in detail. The Radon transform, which has proved so beneficial in medical imaging, is defined and illustrated. Inversion of the Radon transform is given in terms of Riesz potentials. The potential theoretic inversion methodology is then related to Hilbert, Fourier, and convolutional image creation. Geometric intuition is emphasized throughout.

Many compression and feature generation techniques may be classified under the general theory of projections. In Chapter 5, this theory is presented in the context of Hilbert space, as is done customarily in applied mathematics. As in Chapter 4, geometric motivation is emphasized. Specific expansions for signals and images are given in terms of the trigonometric system and the Walsh system. Applications involving chain code representation, the matrix image transform, feature normalization, error bounds for feature recognition, and epsilon-nets are given.

Chapter 6 introduces the basics of probability theory. While the chapter is self-contained, it is not intended as a treatise on probability; rather, it provides a solid foundation for readers who are interested in studying those aspects of image processing which require probability and statistics. It will prepare students for the wide-ranging articles in the literature, and it will provide a core body of knowledge to facilitate a smooth transition into the stochastic process methodologies to be explicated in Volume 2. Many examples involving images faithfully illustrate the definitions and properties discussed. Included are the introduction of random fields and an operational calculus, employing delta-functions, which leads to a unification of the discrete and continuous cases. A wide variety of the standard densities are employed in the examples.

The final chapter of the current volume concerns inferential statistical techniques used in image processing. Estimation techniques include least-squares estimation, maximum likelihood filters, Baysian estimation, and Kalman–Bucy filtering. Detection (hypothesis testing) is presented in the Bayesian context, and the Neymann–Pearson criteria are given. The chapter concludes with a detailed account of the Hotelling transform. Throughout the chapter, examples supplement the theory.

Because of the large amount of material presented and the diversity of that material, the book has been written in a partitioned format. After Sections 2.1 through 2.8, the following units are essentially independent:

1. Sections 2.9 through 2.18
2. Chapter 3

3. Chapter 4
4. Chapter 5
5. Chapters 6 and 7

Depending upon the level, courses can be implemented around Chapters 2 and 3 (geometric and morphological processing); Sections 2.1 through 2.8, Chapter 4, and Chapter 5 (continuous and discrete transforms); or Sections 2.1 through 2.8, Chapter 6, and Chapter 7 (probability and statistics).

Finally, some sections have been starred to indicate that they require a higher degree of mathematical sophistication than the text in general. These sections may be skipped without loss of continuity. However, some of these sections—in particular those on the representation theorems of Matheron and on inversion of the Radon transform—contain material of paramount importance to the future of image analysis. It is our belief that a serious attempt should be made to at least gain familiarity with the content of these sections, even if full understanding of the mathematical subtleties is not attained.

1

Introduction

1.1 IMAGE RECOGNITION

At the outset it should be recognized that image processing is a discipline in and of itself and not merely a particular subject within the domain of computer science, electrical engineering, or some other established branch of science. While it is certainly true that image processing makes use of automated computation and, in a real-time sense, is dependent upon machine implementation, the analysis of computational problems relevant to image processing remains properly a part of the whole. Moreover, though image processing tends to draw the attention of electrical engineers due to its physical dependency upon electronic technology, its application extends deeply into biological science, materials science, medicine, physics, chemistry, photography, robotics, quality control, and numerous other disciplines of contemporary technology. Most assuredly, it, too, is a discipline in its own right.

Perhaps the most widely known aspect of image processing is its relationship to computer vision, robotics, and artificial intelligence. In a sense, all three of these topics can be grouped under the last, and we shall assume such a categorization. Given a mathematical representation of an image, we wish to extract information from that image which can be employed in a decision procedure. The image itself may result from any number of different kinds of sensors. These sensors may utilize an optical technique, such as in a camera, they may sense heat, as with infrared (IR) sensors, they may sense illumination, as in synthetic aperture radar (SAR), or they might involve X rays, as in tomographical medical imaging. Although each particular sensing technique may possess its own requirements insofar as imaging methodology is concerned, the scope of that methodology is not constrained by the demands of any specialized technology. The purpose of image processing continues to be the extraction of information upon which intelligence-type procedures can be implemented. Its own subject matter is determined by its purpose.

To begin with, let us consider a specific approach to the construction of a computer vision system. Since an imitation of human intelligence is a stated goal, it is most natural to view the automated vision system as some sort of replication of our own human vision system. Indeed, the attitude we adopt in framing mathematical categories relating to image intelligence naturally results from reflection on those categories which occur naturally within ourselves. Image understanding, as we simulate it in logico-mathematical systems, is bound to employ categories that are defined in terms of shape, texture, orientation, edge demarcation, area, length, and relative position, to name a few.

The human vision system is not passive. The brain takes the raw sensory data and transforms it into percepts which correspond to mental conceptual categories. The "real" image, if such a thing can even be given meaning, is not a subject of mental analysis. It is the percepts formed by the brain which form the material upon which conceptual analysis is performed. Any data that does not conform to the requirements of perceptual organization cannot be processed.

Even when data is organized into perceptions, judgments made by the brain are sometimes internally contradictory. For instance, consider the images in Figure 1.1. It is well known that most people conclude, on first viewing them, that line L_1 is longer than line L_2; however, measurement soon "proves" otherwise. Here we are presented with two contradictory judgments. Though we tend to make the decision that L_1 and L_2 have the same length, we do so in the realization that we are choosing one decision process over another. Even a simple categorical parameter such as length presents perceptual anomalies.

In sum, the sequence of steps running from the initial data collection to the final decision procedure is both long and strewn with pitfalls. The arsenal of mathematical tools necessary for traversing each path is substantial. Moreover, there are many paths, each one leading from a different source to different decision requirements. In Section 1.2 we will present a breakdown of the major stages of processing that might be present in any data-to-decision chain. The methods presented in this text are image processing methods precisely because each is useful for performing operations required in one or more links of that chain, or for performing some peripheral tasks associated with the chain.

Figure 1.2 provides a coarse breakdown of the stages involved in an image recognizer system. It is certainly not exhaustive of all possibilities, nor are the particular blocks necessarily distinct. Moreover, any individual system may only employ some subcollection of the delineated stages.

The raw data is gathered by an imaging sensor or collection of sensors. The

Figure 1.1 Two Lines of Equal Length

Sec. 1.1   Image Recognition

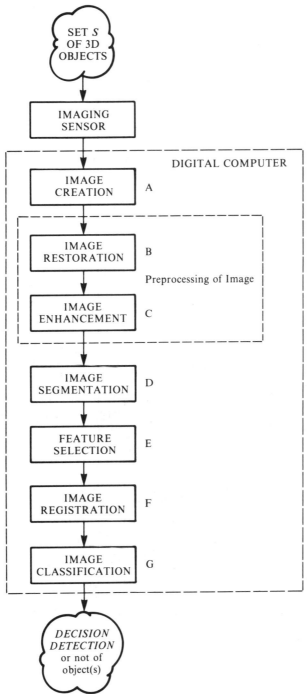

Figure 1.2   Some Stages for Image Recognition Systems

resulting image, which must be created from the data (Figure 1.2, block A), may be Euclidean (continuous) or digital. In the first case, it is represented by some continuously changing level of gray, as in an everyday photograph. If digital processing is to be performed, the image must be digitized: the continuously changing levels of gray, which are represented (in the two-dimensional case) by a function $F(x, y)$ of two real variables, must be approximated by a function $f(i, j)$ of two integer variables. This approximation will no doubt lose some of the quantitative characteristics of the Euclidean image, since $f(i, j)$ is a coarse replica of $F(x, y)$. The digitized image $f(i, j)$ is said to be a *sampled version* of $F(x, y)$. At once we are confronted by the troublesome sampling problem: the digital image is defined only on a grid within the domain of the image; so how fine a grid is necessary in order not to lose a significant quantity of information? The answer depends upon both the methodology that is eventually to be applied to the digital image and the fineness of the ultimate decision process. Figure 1.3 gives a digitization of a signal $f(t)$ (a function of one real variable).

A problem closely related to digitization is quantization. Given that the range values of both $F(x, y)$ and the digitized version $f(i, j)$ are real numbers, and given that we wish to utilize a discrete range for computer implementation of image processing algorithms, what scale do we choose for the representation of the gray values $f(i, j)$? Any choice of a uniform discrete scale will require the alteration of actual gray values at the grid points (i, j). Not only must the direct mathematical representation obtained from the sensor be sampled on a grid, but also the gray values at the sample points must be altered. Figure 1.4 gives a quantization of a signal.

Once the incoming data has been formed into an image, it will most likely have to be restored or cleaned up (Figure 1.2, block B). Restoration involves the

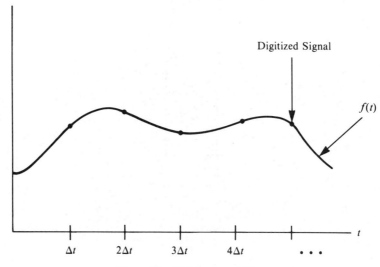

Figure 1.3 Digitization of Signal

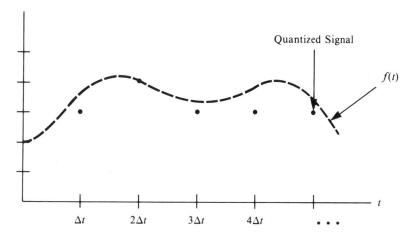

Figure 1.4 Quantization of Digitized Signal

filtering out of noise, the correction of sensor bias, and perhaps the alteration of the originally created image into one that is more compatible with what is expected. It is at this stage that image processing closely parallels the perception formation inherent in human mental perceptual processes.

Consider the image in Figure 1.5. At first glance, it appears to be a circle. However, even if we were to grant that the circular part were perfect, there is a tiny hole in it. In effect, the brain restores the image to one to which its perceptual categories are most responsive. It has filtered out the noise (the hole) and has interpreted the actual image to be one that it "expects." If the observer happens to have astigmatism, corrective lenses could be used to correct "sensor" bias.

From the perspective of intelligence, the restoration stage is crucial. A computer vision system cannot be passive. More than that, it cannot consist solely of an inference system and a knowledge base. It must be endowed with heuristics. Its design must be categorical so that it can constitute images that are compatible

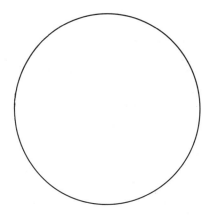

Figure 1.5 Circle with Hole

with its purpose, and its algorithms must be adaptive so that they can be upgraded through the application of statistical techniques. Restoration must be "smart." Each of the major areas of this text, dealing with geometric, transform, and statistical considerations, plays a substantial role in image restoration.

An image may be enhanced prior to further processing (Figure 1.2, block C). That is, features that are of special interest to the problem at hand might be emphasized so that they are more distinctive. For instance, edges of objects within the image might be singled out or darkened. Or the background might be eliminated or whitened. Indeed, as is common, the entire image may be transformed into one that is black and white. Many of the operations of Chapters 2 and 3 are useful for enhancement.

Sometimes it is profitable to segment an image into mutually exclusive pieces that are more amenable to later processing and recognition (Figure 1.2, block D). Curves or shapes of interest might then be located.

In terms of intelligence processing, block E of Figure 1.2 is paramount. Here features of the image are selected, and it is upon these features, and not on the image or enhanced image itself, that decision processing operates. Features are parameters that can be obtained geometrically, statistically, or through the use of transform techniques. The selected features represent those aspects of the image that we believe to be useful for characterization. The methodology for feature parameter construction is as extensive as that of the mathematical apparatus that can be brought to bear on the image. The features may involve parameters such as boundary length or area from geometry, Fourier coefficients from projection theory, moments from statistics, size distributions from morphology, or some combination of these together with a host of others. Indeed, it is common to consider a feature vector, that is, a collection of feature parameters, rather than simply a single feature.

Since the sensor itself does not represent an absolute frame of reference, its relevant aspects, such as velocity and attitude, must be registered (Figure 1.2, block F). Thus, just as a human observer must take him- or herself into account, so too must the attributes of the sensor be accounted for.

Once a feature vector has been judiciously selected and the parameters computed, the vector can be compared to some archetypal vector in the knowledge base to see whether the parameters are sufficiently close to make an identification based upon the features under consideration (Figure 1.2, block G). A categorical decision must be made as to what features are to be utilized and what norming operation is to be employed in the comparison. These decisions must be made prior to any actual feature calculation. Should there be no archetypal vector forms in the knowledge base corresponding to the observational feature vector under consideration, no decision can be made. On the other hand, if there are vector forms present in the knowledge base which correspond to the given feature vector, then classification can take place. That is, based upon the feature vector of the observed image and a comparison of it with vectors in the knowledge base, the image can be classified. Notice that this classification is inherently uncertain. It

Figure 1.6 Image with Various Processing Styles

is highly unlikely that a perfect match will be obtained and, even if it were, some uncertainty would still remain. Nevertheless, we must depend upon the features we have chosen. Based upon them, a detection decision must be made which, in the end, may be no decision at all.

Figures 1.6a through 1.6d give a visual indication of the manner in which an original image, Figure 1.6a, can be processed to give varying outputs.

1.2 IMAGE PROCESSING TRANSFORMATIONS

The preceding analysis of a typical image recognizer system leads us to consider four levels of transformations that appear in image processing:

Level 0   Image representation
Level 1   Image-to-image transformations
Level 2   Image-to-parameter transformations
Level 3   Parameter-to-decision transformations

We will consider each level in turn.

Level 0 consists of digitization and quantization transformations. Suppose $F(x, y)$ is a real-valued function of two real variables. Then a digitizing operation $F \to f$, where f is a real-valued function on a square grid of lattice points in R^2, the Euclidean plane, is a level-0 transformation. Similarly, a quantization of the range of f to yield a function f_0 whose range is some equally spaced discrete subset of R would also be a level-0 transformation.

One might also wish to invert a digitization, that is, construct a Euclidean image from one that is digital. A reconstruction of this type goes to the heart of the sampling problem: what is the relationship between the original Euclidean image and the reconstructed image if no intermediate processing has occurred? More generally, suppose that the Euclidean image is processed by means of transformations which output only Euclidean images, and that a digitization of the original image is processed by digital analogs of the Euclidean processes. If a Euclidean image is constructed from the final output of the digital processing, what is its relationship to the final output of the Euclidean processing? Figure 1.7 provides a schematic of this more general digitization problem. Suffice it to say that the problem is one of the most profound in image processing. Its complexity can be seen in the results of Section 3.13. A much easier problem is addressed in Section 2.18, where the local least-squares best fit of a quadratic surface to a digital image is presented.

Level-1 transformations are those which take images as inputs and yield an image output that is of the same type as the input images, i.e., continuous goes to continuous and digital goes to digital. Although there might be auxiliary real-number or vector inputs, the operations are essentially image to image. It is at

Sec. 1.2   Image Processing Transformations

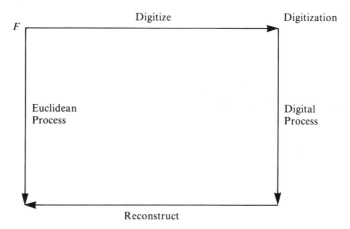

Figure 1.7 Commutative Diagram for Digitization

level 1 that most processing takes place. Restoration, enhancement, segmentation, and some types of feature selection (see Figure 1.2) are level-1 operations. It is these operations that are most commonly associated with image processing in the everyday world. However, while it is possible to use entire images in the classification problem, proceeding in such a manner is not likely to be fruitful, at least in a real-time situation. Accordingly, we must concern ourselves with level-2 operations.

Level-2 operations involve the production of feature parameters. Consequently, these operations are image to parameter. Since it is the feature parameters that are utilized in the eventual intelligence processing, the mathematical properties satisfied by the parameters are of supreme importance. Not only must we know how they behave in and of themselves, but it is also useful to know the extent to which a specific set of parameters characterizes a particular problem. Further, a couple of important questions present themselves, viz., Is there some number of parameters that jointly prove sufficient for identification given some criterion that must be met? and Is there some parameter or set of parameters that is best for the task relative to some measure of goodness? Some of these questions are addressed in Section 3.5 for the geometric parameters employed in morphological analysis. General goodness questions regarding the estimation of statistical parameters are discussed in Section 7.2.

Before proceeding to the level-3 operations, we should mention that the distinction between the preceding levels is not absolute. For instance, if f is a digital image and v is a feature vector, we could look upon v as an image having only a single row of pixels. (See Section 2.16.) Indeed, we can view a real number as an image consisting of a single pixel. But such mathematical equivalences are not the point. It is not the mathematical structural changes that are most fundamental, but rather, the classes of problems and the corresponding classes of operations required to solve those problems.

A typical case in point involves the area of compression: given a certain amount of information encoded in a certain amount of symbols, is it possible to transform the symbol set to some other symbol set of lesser cardinality without losing any information or, more realistically, losing only an insubstantial amount of information? Compression involves levels 0, 1, and 2. In fact, since one might wish to reduce the number of feature parameters in a feature vector for reasons of storage, computation time, or data transmission, compression often involves vector-to-vector transformations. In this volume our main interest in compression involves the image-to-image compression and image-to-parameter compression engendered by projections (Chapter 5). In point of fact, the delineation of the levels is only to help clarify a sometimes confusing mass of transformations which are included in image processing methodology.

Once the processing at levels 0, 1, and 2 is complete, the logical operations of level 3 take place. These involve decision and detection techniques. Except for statistical detection (Section 7.7), decision operations have not been included in this volume. Although they are certainly pertinent to image processing, they tend to belong to the area of mathematics that concerns the intersection of image processing and artificial intelligence. Consequently, our purpose has only been to lay the groundwork for intelligence. In this respect, the Bayesian estimation in Section 7.3 and the criterion of sufficiency in Section 7.2 are especially relevant.

In conclusion, each of the four levels has a branch of mathematics that tends to be associated with it. Level 0 involves the theory of approximation. Level 1 tends to be geometric in nature since images, as objects of human sensation, appear to possess geometric qualities. While gradient analysis and morphology make direct use of geometric qualities, we should not lose sight of the fact that most transform techniques also are geometric in character. (See Section 4.9 on inversion of the Radon transform and Section 5.1 on Euclidean projections.) Level 2 tends to be measure-theoretic in that measurements of one sort or another are being taken. This is especially apparent in the parameters which are integral geometric, probabilistic, and statistical.

Finally, level 3 is associated with logic, where by the term 'logic', we do not restrict ourselves to the standard propositional or first-order predicate calculi. Indeed, if we wish to model the inherent uncertainty of our decision procedures, there is no doubt that logics that employ possibility-type distributions must eventually be employed. Moreover, decision algorithms must be adaptive. They must be capable of autonomous reconfiguration when presented with unexpected information. Put precisely, the algorithmic decision structure must be multitiered so that various levels can operate on other levels. The explication of adaptive and fuzzy logics would take us far afield, however, so we shall make no further mention of these notions in this, the first, volume.

2

Bound Matrix and Block Diagram Image Representation

2.1 DIGITAL IMAGE REPRESENTATION

From a classical mathematical point of view, an *image* is a real-valued function defined on some finite subset of the set of integral lattice points $Z \times Z$. Though this definition appears to be rather straightforward, for the sake of algorithm specification and program implementation it is useful to introduce the concept of a bound matrix. Bound matrices will serve as vehicles for the representation of digital images.

By computational necessity, every digital image has gray values on only a finite set of pixels. Consequently, there are always an infinite number of lattice points on which no gray value is defined, and each digital image is considered to be of the form $f: A \to R$, where A is a finite subset of $Z \times Z$ and R is the set of real numbers. Unfortunately, this seemingly simple characterization leads to operational difficulties. Different digital images have different domains. Therefore, when images are pieced together or segmented, or when operations are performed on only part of an image, there often arises a specification problem relating to the domain of definition of the image under consideration. Many of the problems can be avoided by choosing a slightly altered image representation in which all digital images are assumed to be defined on the entire grid $Z \times Z$.

Before proceeding to a formal presentation of bound matrices, an example might be helpful. Consider the image f of Figure 2.1. Each real number on the grid denotes a gray value corresponding to the pixel on which it is positioned. For example, $f(2, 1) = 7$ and $f(2, 2) = -2$. While one often thinks of digital images as being defined on rectangular frames, in general there is no reason for making such an assumption. In particular, the image f under consideration certainly does not possess a rectangular domain, although it might have been made by joining two smaller images, each of which did possess a rectangular domain.

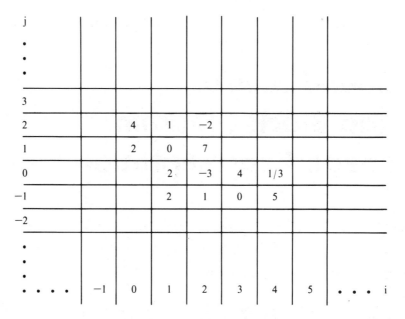

Figure 2.1 Image Given in x-y Plane

It is our intention to present an image representation scheme that allows each image to be depicted as a type of matrix, and hence in rectangular fashion. This can be accomplished by associating with each image a so-called *bound matrix*. Intuitively, this is achieved by laying a matrix over the lattice representation of the image and "filling in" the missing gray values with stars (*). In the case of Figure 2.1, the following are possible bound matrix representations of f:

$$A = \begin{pmatrix} 4 & 1 & -2 & * & * \\ 2 & 0 & 7 & * & * \\ * & 2 & -3 & 4 & \frac{1}{3} \\ * & 2 & 1 & 0 & 5 \end{pmatrix}_{0,2}$$

and

$$B = \begin{pmatrix} * & * & * & * & * & * \\ 4 & 1 & -2 & * & * & * \\ 2 & 0 & 7 & * & * & * \\ * & 2 & -3 & 4 & \frac{1}{3} & * \\ * & 2 & 1 & 0 & 5 & * \end{pmatrix}_{0,3}$$

In each of A and B, the subscript pair denotes the (i, j) position in the lattice of the uppermost, leftmost entry in the bound matrix. For instance, the uppermost, leftmost entry of B is a star, and that star occupies the (0,3) position in the lattice, which in terms of the image f is a point outside the domain. Not only does the bound matrix give the gray values of the image, it also gives the positions of those values on the lattice $Z \times Z$. More will be said with regard to this positional representation, though it should be apparent that the positioning is in essence accomplished geometrically by overlaying the bound matrix in a manner determined by the subscript pair.

The example of f, together with two different bound matrix representations, illustrates the point that the bound matrix representation of an image is not unique; indeed, there are an infinite number of such representations. Nonetheless, each one contains the same amount of information about the original image. Because of this, we usually choose the *minimal* bound matrix associated with f, that is, the one that has the minimal number of entries (real numbers or stars) among all bound matrix representations of f. In the preceding example, A is the minimal bound matrix associated with f.

We mentioned previously that we would like to treat a digital image as though it were defined on the whole of the lattice $Z \times Z$. This can be accomplished by making the assumption that each bound matrix has the value $*$ at each pixel outside its actual matrix frame. Then, even though the "actual" domain of the image is finite, its bound matrix representations cover the entire grid structure and, as a result, an image can be manipulated as though it were defined everywhere. One need only define all image operations appropriately. It should be noted that the question of infinity does not arise with regard to storage since each bound matrix can be represented with a finite amount of information; the assumption regarding stars outside the given frame makes no demand for further data representation.

2.2 BOUND MATRICES

Consider the array-type structure consisting of m by n entities given by

$$\begin{pmatrix} a_{11} & a_{12} & \cdots & a_{1n} \\ a_{21} & a_{22} & \cdots & a_{2n} \\ \vdots & \vdots & & \vdots \\ a_{m1} & a_{m2} & \cdots & a_{mn} \end{pmatrix}_{r,t}$$

where

1. each a_{pq} is a real number or a star $(*)$
2. $1 \leq p \leq m$, $1 \leq q \leq n$
3. r and t are integers.

Such a data structure is called a *bound matrix*, or an *m by n bound matrix*, and is denoted by $(a_{pq})_{rt}$.

Every bound matrix can be associated with a digital image f. This is accomplished by "locating the bound matrix in the lattice." The location on $Z \times Z$ of the entry a_{11}, which may be a star, is (r, t). The location of the entry a_{pq} in

$Z \times Z$ is $(q + r - 1, t + 1 - p)$. The image associated with the bound matrix $(a_{pq})_{rt}$ is the image whose domain consists of those lattice points for which the bound matrix has a real number as an entry. Since the gray value at such a pixel is given by that real entry, it follows that

$$f(r, t) = \begin{cases} a_{11} & \text{if } a_{11} \text{ is a real number} \\ \text{undefined} & \text{if } a_{11} = * \end{cases}$$

In general,

$$f(i, j) = \begin{cases} a_{t+1-j, i+1-r} & \text{if } a_{t+1-j, i+1-r} \text{ is a real number} \\ \text{undefined} & \text{if } a_{t+1-j, i+1-r} = * \end{cases}$$

In accordance with our desire to treat an image as though it were defined on the whole of $Z \times Z$, the image will be given the value $*$ wherever it is undefined whenever we are in the bound matrix context. Thus, the preceding representation reduces simply to

$$f(i, j) = a_{t+1-j, i+1-r}$$

The foregoing relationships can best be seen by "overlaying" the bound matrix on the pixel grid structure ("hanging" the matrix), as illustrated in Figure 2.2. Note that, from that figure, we have the further relation

$$f(i, j) = a_{m+s-j, i+1-r}$$

Also, the equivalent identities

$$a_{pq} = f(q + r - 1, m + s - p) = f(q + r - 1, t + 1 - p)$$

hold if we make the assumption that $f(i, j) = *$ for any pixel (i, j) not in the actual domain of f, i.e., if we make the assumption that f is defined on the entire lattice $Z \times Z$ and has the value $*$ at any pixel not in the actual domain.

The preceding discussion explained what is meant by associating an image with a given bound matrix. Conversely, given an image f, there are many bound matrices associated with it. Accordingly, we say that an m by n bound matrix $(a_{pq})_{rt}$ "represents" a digital image $f: D \to R$ if the gray values of f lie in $(a_{pq})_{rt}$ in the proper place when $(a_{pq})_{rt}$ is positioned on the lattice, and the value of any other entry a_{pq} is $*$. More rigorously, $(a_{pq})_{rt}$ *represents the digital image* $f: D \to R$ if (1) for every (i, j) in D there corresponds the element $a_{t+1-j, i+1-r}$ in $(a_{pq})_{rt}$, where $a_{t+1-j, i+1-r} = f(i, j)$; and (2) all other entries a_{pq} in the bound matrix $(a_{pq})_{rt}$ have the value $*$.

The bound matrices A and B of Section 2.1 both represent the image f of Figure 2.1. Conversely, one can say that f is the digital image associated with A and is also the image associated with B. Due to these associations, we will often write $f = A$ and $f = B$. Though technically the structure f and the structure A are different, for all intents and purposes they can be construed to be the same and will therefore be identified. Furthermore, by assuming that all entries outside

Sec. 2.2 Bound Matrices

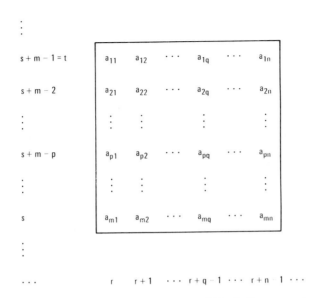

Figure 2.2 Overlay on x-y Plane of Bound Matrix Representation

the frame of the bound matrix have the value $*$, we can write $A = B$, although as bound matrices the two differ structurally, not only from a strict mathematical perspective, but also in the manner in which they are stored in memory. To illustrate these matters, the image f of Figure 2.1 satisfies both

$$f = \begin{pmatrix} 4 & 1 & -2 & * & * \\ 2 & 0 & 7 & * & * \\ * & 2 & -3 & 4 & \frac{1}{3} \\ * & 2 & 1 & 0 & 5 \end{pmatrix}_{0,2}$$

This is the value at the location $(0, 2)$. Thus, by definition, $a_{11} = f(0, 2) = 4$.

and

$$f = \begin{pmatrix} * & * & * & * & * & * \\ 4 & 1 & -2 & * & * & * \\ 2 & 0 & 7 & * & * & * \\ * & 2 & -3 & 4 & \frac{1}{3} & * \\ * & 2 & 1 & 0 & 5 & * \end{pmatrix}_{0,3}$$

This is the value for the pixel located three units to the right and two units down from $(0, 3)$. Thus, $a_{34} = f(3, 1) = *$.

It is important to know how to go from a "picture" representation of an image to a bound matrix representation and vice versa. While algorithms of the form

$$f(i, j) = a_{t+1-j, i+1-r}$$

are necessary for computer implementation and mechanization purposes, changes of representation can most simply be accomplished geometrically by overlaying one structure on the other. There is no need to memorize the formulas for relating values using the matrix notation a_{pq} with the function notation $f(i, j)$. The complexity of the relations arises from the fact that the notation used in representing tuples in matrices is backwardly rotated relative to the x-y coordinate labeling system.

At times it is convenient to represent the position of a bound matrix by noting the location of the origin pixel (0, 0) in the matrix rather than by locating the uppermost, leftmost entry of the bound matrix in the lattice. This alternative representation can be achieved simply by circling the bound matrix entry that corresponds to the origin. For instance, the image f of Figure 2.1 can be represented by

$$f = \begin{pmatrix} 4 & 1 & -2 & * & * \\ 2 & 0 & 7 & * & * \\ \circledast & 2 & -3 & 4 & \frac{1}{3} \\ * & 2 & 1 & 0 & 5 \end{pmatrix}$$

In general, a digital image $f: D \to R$ with nonempty, finite domain D containing d elements requires $3d$ pieces of data for storage: for each of the d pixels, the i and j locations must be registered along with the gray value $f(i, j)$. This same image, when represented by an m by n bound matrix $(a_{pq})_{rt}$, can be specified by using $(mn + 4)$ pieces of data. The data are the positive integers m and n, followed by the mn entries a_{pq} of the matrix, and then by the absolute "location" of the a_{11} entry, namely, the coordinates r and t. The gray values of the other pixels are found by either computation or geometry.

As mentioned earlier, each image (or bound matrix) has associated with it a minimal bound matrix, that is, the bound matrix representation that has the least number of entries. It is obvious that the minimal bound matrix representing f is specified with the least amount of data among all bound matrices representing f. When the image f has a domain that is nearly rectangular in shape, a common occurrence, the minimal bound matrix representation provides a data compression close to one-third. Instead of specifying the location of each gray value of f, the location of the element a_{11} of the minimal bound matrix is specified. Then, all mn elements of that matrix are easily found by using relative-address-type techniques.

A measure of the amount of information in an m by n bound matrix $(a_{pq})_{rt}$ is given by the *information density* u, which is defined as the ratio of the number s of nonstarred values to the total number mn of entries, i.e., $u = s/mn$. When $u = 1$, there are no stars, and the image "occupies the whole matrix." In this case the bound matrix is said to be *saturated*. Since a minimal bound matrix represents a digital image with the least amount of data, it possesses the maximum information density among all representations. For the image f of Figure 2.1, A is the minimal bound matrix and $u(A) = \frac{7}{10}$. For the same image, $u(B) = \frac{7}{15}$.

2.3 BOUND MATRICES AS EQUIVALENCE CLASSES

In this section a mathematically rigorous description of the identification between digital images and bound matrices will be presented. Understanding the exposition requires a familiarity with one-to-one, onto mappings, as well as with equivalence relations and equivalence classes. Those who are not familiar with these concepts may skip the section without any loss of continuity.

By definition, a digital image is a function from some finite, nonempty subset D in $Z \times Z$, the integral lattice, into the set of real numbers R. Let I denote the class of all such functions, and let Q denote the set of all functions $g: Z \times Z \to R \cup \{*\}$ such that the set of all $g(i, j)$ with $g(i, j) \neq *$ is finite. Then there exists a natural identification between the class of images I and $Q - \{z\}$, where z is the element in Q defined by $z(i, j) = *$ for all (i, j) in $Z \times Z$. This identification is given by the isomorphism (one-to-one, onto mapping)

$$\Psi: I \to Q - \{z\}$$

defined by

$$[\Psi(f)](i, j) = \begin{cases} f(i, j) & \text{if } f(i, j) \text{ is defined} \\ * & \text{otherwise} \end{cases}$$

Under the identification, each element of $Q - \{z\}$ is also an image. The extra element z in Q is called the *null image*. Note that Ψ operates on an image by "filling in all valueless pixels with stars."

Now let B denote the class of all bound matrices as defined at the outset of Section 2.2. Define the mapping

$$\Phi: B \to Q$$

by

$$[\Phi((a_{pq})_{rt})](i, j) = \begin{cases} a_{t+1-j, i+1-r} & \text{for } 1 \leq t + 1 - j \leq m \\ & \text{and } 1 \leq i + 1 - r \leq n, \\ & \text{where } (a_{pq})_{rt} \text{ is } m \text{ by } n \\ * & \text{otherwise} \end{cases}$$

The mapping Φ is simply a rigorous specification of the identification between a bound matrix and an image which was given both algebraically and geometrically in Section 2.2. To see that the mapping Φ is onto, let g be any element of Q. Let $(c_{pq})_{rt}$ be the bound matrix defined as follows:

1. Let r be the minimum integer value such that there exists a j with $g(r, j) \neq *$, and let t be the maximum integer value such that there exists an i with $g(i, t) \neq *$.
2. Let m be the maximum integer value such that there exists an i with

$g(i, t + 1 - m) \neq *$, and let n be the maximum integer value such that there exists a j with $g(r - 1 + n, j) \neq *$.

3. $c_{pq} = g(q + r - 1, t + 1 - p)$ for $1 \leq p \leq m$, $1 \leq q \leq n$.

Then

$$[\Phi((c_{pq})_{rt})](i, j) = c_{t+1-j, i+1-r}$$
$$= g(i + 1 - r + r - 1, t + 1 - t - 1 + j) = g(i, j)$$

if $1 \leq t + 1 - j \leq m$ and $1 \leq i + 1 - r \leq n$, i.e.,

if $1 \leq p \leq m$ and $1 \leq q \leq n$

Otherwise, $[\Phi((c_{pq})_{rt})](i, j)$ is equal to $*$. Since this means precisely that

$$\Phi((c_{pq})_{rt}) = g$$

the mapping Φ is onto. For purposes of intuition, it should be noted that the bound matrix $(c_{pq})_{rt}$ is the minimal bound matrix for g.

We now define an equivalence relation $\sim$ on the class of bound matrices B. We define $(a_{pq})_{rt} \sim (b_{pq})_{xy}$, where $(a_{pq})_{rt}$ is m by n and $(b_{pq})_{xy}$ is m' by n', if and only if

$$\Phi((a_{pq})_{rt}) = \Phi((b_{pq})_{xy})$$

Stated simply, two bound matrices are equivalent if and only if they result in the same element of Q when operated on by Φ. It should be noted that $\sim$ is reflexive, symmetric, and transitive, the transitivity following from the fact if A_1, A_2, and A_3 are bound matrices with $A_1 \sim A_2$ and $A_2 \sim A_3$, then $\Phi(A_1) = \Phi(A_2) = \Phi(A_3)$, and hence $A_1 \sim A_3$. Now let $\overline{B}$ be the set of equivalence classes in B under the relation $\sim$. Then Φ naturally induces the mapping

$$\overline{\Phi}: \overline{B} \to Q$$

by

$$\overline{\Phi}([A]) = \Phi(A),$$

where $[A]$ denotes the equivalence class of the bound matrix A. $\overline{\Phi}$ is well defined since $A_1 \sim A_2$ implies that

$$\overline{\Phi}([A_1]) = \Phi(A_1) = \Phi(A_2) = \overline{\Phi}([A_2])$$

Moreover, $\overline{\Phi}$ is onto since Φ is onto and $\overline{\Phi}$ is one-to-one by construction; indeed,

$$\overline{\Phi}([A_1]) = \overline{\Phi}([A_2])$$

implies that $\Phi(A_1) = \Phi(A_2)$, which in turn implies that $[A_1] = [A_2]$.

The preceding remarks show that it is the equivalence classes within the

Sec. 2.4   Arithmetic Operations on Digital Images

collection of bound matrices B which are actually identified with the images of Q. More precisely, we have the mappings

$$I \xrightarrow{\Psi} Q - \{z\} \subset Q \xrightarrow{\overline{\Phi}^{-1}} \overline{B}$$

where Ψ and $\overline{\Phi}$ are both one-to-one and onto.

2.4 ARITHMETIC OPERATIONS ON DIGITAL IMAGES

Arithmetic operations can be performed upon images just as they can be performed upon real numbers, vectors, and matrices. For instance, two images can be added. The addition is performed "pointwise," the values of corresponding pixels being added numerically. Formally, ADD is a binary operation defined by

$$[\text{ADD}(f, g)](i, j) = \begin{cases} f(i, j) + g(i, j) & \text{if } f(i, j) \neq * \text{ and } g(i, j) \neq * \\ * & \text{if } f(i, j) = * \text{ or } g(i, j) = * \end{cases}$$

where f and g are the input images, and where the bound matrix representation of the images is assumed (as it will be whenever digital images are being discussed). It should be noted that in terms of actual image domains, $\text{ADD}(f, g)$ is undefined whenever f or g is undefined. Consequently, the domain of $\text{ADD}(f, g)$ is equal to the intersection of the two input domains.

Example 2.1

Consider the images

$$f = \begin{pmatrix} 0 & 4 & * & * \\ * & 0 & -1 & 3 \end{pmatrix}_{0,1}$$

and

$$g = \begin{pmatrix} 2 & 2 & -1 \\ 7 & 3 & 3 \\ 4 & 0 & * \end{pmatrix}_{0,2}$$

Image addition yields

$$\text{ADD}(f, g) = \begin{pmatrix} 7 & 7 \\ * & 0 \end{pmatrix}_{0,1}$$

Note that the domain of the output image is equal to the intersection of the domains of the input images.

In order to facilitate the implementation of image processing algorithms, block diagrams are used extensively in digital image processing. They generally

take the form

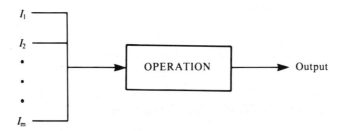

where $I_1, I_2, \ldots, I_m$ are the inputs. The inputs may be of different types—for example, images, real numbers, vectors, sets, or any other structure that appears in image processing. Moreover, the order of the inputs may be consequential; therefore, block diagrams will always be drawn with the proper input order specified vertically.

For the operator ADD, a binary operation with two image inputs, the block diagram is given by

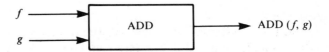

Since image addition is performed pointwise, and since real-number addition is commutative, ADD is commutative, i.e.,

$$\text{ADD}(f, g) = \text{ADD}(g, f)$$

Consequently, the order in which f and g are input in the block diagram is irrelevant.

Besides adding two images, one can multiply two images. This operation, which is also defined pointwise, is given by

$$[\text{MULT}(f, g)](i, j) = \begin{cases} f(i, j) \cdot g(i, j) & \text{if } f(i, j) \neq * \text{ and } g(i, j) \neq * \\ * & \text{if } f(i, j) = * \text{ or } g(i, j) = * \end{cases}$$

As in the case of the addition operator, the domain of MULT(f, g) is equal to the intersection of the two input domains. Moreover, MULT is commutative and has a block diagram similar to ADD.

Example 2.2

Utilizing the images f and g of Example 2.1,

$$\text{MULT}(f, g) = \begin{pmatrix} 0 & 12 \\ * & 0 \end{pmatrix}_{0,1}$$

Sec. 2.4 Arithmetic Operations on Digital Images

Two other elementary binary operations are defined on pairs of digital images: the maximum operator MAX and the minimum operator MIN. MAX compares two images pointwise and returns the maximum value at each pixel, and MIN behaves likewise except that it returns the minimum value. The two are respectively defined by

$$[MAX(f, g)](i, j) = \begin{cases} \max\{f(i,j), g(i,j)\} & \text{if } f(i,j) \neq * \text{ and } g(i,j) \neq * \\ * & \text{otherwise} \end{cases}$$

and

$$[MIN(f, g)](i, j) = \begin{cases} \min\{f(i,j), g(i,j)\} & \text{if } f(i,j) \neq * \text{ and } g(i,j) \neq * \\ * & \text{otherwise} \end{cases}$$

Both MAX and MIN are commutative. The block diagram for MAX is

and that for MIN is similar. Once again, notice that the output is defined only on the intersection of the two input domains.

Example 2.3

Using the images f and g of Example 2.1,

$$MAX(f, g) = \begin{pmatrix} 7 & 4 \\ * & 0 \end{pmatrix}_{0,1}$$

and

$$MIN(f, g) = \begin{pmatrix} 0 & 3 \\ * & 0 \end{pmatrix}_{0,1}$$

The operators thus far introduced have been binary operators whose inputs are of the same type, viz., images. The next operation is a form of multiplication which is also binary; however, its inputs are of different types. It is called SCALAR, for scalar multiplication, and is defined by

$$[SCALAR(t; f)](i, j) = \begin{cases} t \cdot f(i,j) & \text{if } f(i,j) \neq * \\ * & \text{if } f(i,j) = * \end{cases}$$

where t is a real number and f is a digital image. Simply stated, SCALAR($t; f$) is found by multiplying each gray value of f by t. Scalar multiplication in image processing is analogous to the operation of the same name in vector algebra. It

has the block diagram

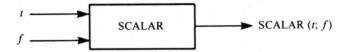

Note the semicolon between t and f in the notation for scalar multiplication. This is used to indicate that t and f are different sorts of entities.

Example 2.4

Let $t = 4$ and g be the second image given in Example 2.1. Then

$$\text{SCALAR}(4; g) = \begin{pmatrix} 8 & 8 & -4 \\ 28 & 12 & 12 \\ 16 & 0 & * \end{pmatrix}_{0,2}$$

A unary operator is one that requires only one input. The simplest image-to-image unary transformation is subtraction, or negation. It is defined pointwise by

$$[\text{SUB}(f)](i, j) = \begin{cases} -f(i, j) & \text{if } f(i, j) \neq * \\ * & \text{if } f(i, j) = * \end{cases}$$

Subtraction has a one-input block diagram:

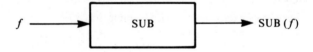

Note that $\text{SUB}(f) = \text{SCALAR}(-1; f)$.

Example 2.5

Letting f be the image given in Example 2.1,

$$\text{SUB}(f) = \begin{pmatrix} 0 & -4 & * & * \\ * & 0 & 1 & -3 \end{pmatrix}_{0,1}$$

A unary-type division, or reciprocation, is also defined on images. This operation is given by

$$[\text{DIV}(f)](i, j) = \begin{cases} 1/f(i, j) & \text{if } f(i, j) \neq * \text{ and } f(i, j) \neq 0 \\ * & \text{otherwise} \end{cases}$$

In other words, one divided by zero is, as usual, undefined.

Example 2.6

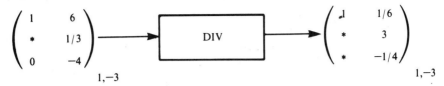

Before leaving this section, some comments on notation might be appropriate. Why do we use the prefix acronym notation ADD(f, g) instead of infix notation, say, of the form ($f + g$)? There are two paramount reasons for doing so. First, there are many binary operations that are repeatedly used in image processing. For instance, there are at least three addition-type operations and five multiplication-type operations that will be introduced in this text. It would be very difficult to arrive at a convenient and readable infix symbology that could accommodate these addition and multiplication operations, let alone all the others. Second, from a programming point of view, operators such as ADD and MAX are procedures, and ADD(f, g) represents a procedure call. Thus, this notation is convenient in that, as will be seen shortly, a digital image processing algorithm, when implemented on a computer, consists of a flow of such procedures.

2.5 GEOMETRIC OPERATIONS ON DIGITAL IMAGES

The operations introduced in the previous section operate on the gray levels of a digital image; they deal with light to dark intensity. The operations to be discussed in this section are spatial in nature; they alter an image's position and orientation in the lattice.

One of the most useful image operators is the translation operator TRAN, a ternary operator having two integer inputs i and j and an image input f. Intuitively, TRAN(f; i, j) is an image that is the "same as f except that it is moved over i pixels to the right and j pixels up." If (in bound matrix form)

$$f = \begin{pmatrix} a_{11} & a_{12} & \cdots & a_{1n} \\ \vdots & \vdots & & \vdots \\ a_{m1} & a_{m2} & \cdots & a_{mn} \end{pmatrix}_{r,t}$$

then

$$\text{TRAN}(f; i, j) = \begin{pmatrix} a_{11} & a_{12} & \cdots & a_{1n} \\ \vdots & \vdots & & \vdots \\ a_{m1} & a_{m2} & \cdots & a_{mn} \end{pmatrix}_{r+i,t+j}$$

Notice that none of the values a_{pq} are changed; only the location of the gray values change by a translation. The block diagram for TRAN is given by

$$f = (a_{pq})_{rt} \longrightarrow$$
$$i \longrightarrow \boxed{\text{TRAN}} \longrightarrow \text{TRAN}(f; i, j) = (a_{pq})_{r+i, t+j}$$
$$j \longrightarrow$$

Whereas the preceding definition of TRAN is given in terms of the bound matrix representation, the operator can also be defined pointwise by

$$[\text{TRAN}(f; i, j)](u, v) = f(u - i, v - j)$$

Example 2.7

Let

$$f = \begin{pmatrix} 1 & 8 \\ 2 & * \end{pmatrix}_{1,2}$$

Then

$$\text{TRAN}(f; 2, -1) = \begin{pmatrix} 1 & 8 \\ 2 & * \end{pmatrix}_{3,1}$$

The action of TRAN is perhaps best seen by employing the origin specification notation:

$$\begin{pmatrix} * & 1 & 8 \\ * & 2 & * \\ \circledast & * & * \end{pmatrix} \longrightarrow \boxed{\text{TRAN}} \longrightarrow \begin{pmatrix} * & * & * & 1 & 8 \\ \circledast & * & * & 2 & * \end{pmatrix}$$

Images can be rotated 90°, 180°, 270°, and 360° = 0°. The counterclockwise 90° rotation operator NINETY is defined pointwise by

$$[\text{NINETY}(f)](i, j) = f(j, -i)$$

For example, if $f(4, 2) = 5$, then $[\text{NINETY}(f)](-2, 4) = 5$. The effect of applying NINETY to an image is to rotate it 90°, with the lattice origin $(0, 0)$ being the pivot.

Example 2.8

Let

$$f = \begin{pmatrix} 4 & 0 \\ * & 2 \\ 1 & 3 \end{pmatrix}_{3,4} = \begin{pmatrix} * & * & * & 4 & 0 \\ * & * & * & * & 2 \\ * & * & * & 1 & 3 \\ * & * & * & * & * \\ \circledast & * & * & * & * \end{pmatrix} \quad \text{Rotate } f \text{ 90°}$$
with origin as pivot.

origin

Sec. 2.5   Geometric Operations on Digital Images   **25**

Then

$$f \longrightarrow \boxed{\text{NINETY}} \longrightarrow \begin{pmatrix} 0 & 2 & 3 \\ 4 & * & 1 \end{pmatrix}_{-4,4} = \begin{pmatrix} 0 & 2 & 3 & * & * \\ 4 & * & 1 & * & * \\ * & * & * & * & * \\ * & * & * & * & * \\ * & * & * & * & \circledast \end{pmatrix}$$

In general, if f is given by an m by n minimal bound matrix with first row, first column value at (r, t), then NINETY(f) is given by an n by m minimal bound matrix with first row, first column value at $(-t, r + n - 1)$.

In addition to the 90° rotation operator, there are the 180° and 270° rotation operators, specified as NINETY^2 and NINETY^3, respectively. The first is obtained by two successive applications of NINETY, and the second by three successive applications. NINETY^2 is of particular importance. It rotates the domain D 180° around the origin, which means that (i, j) is an element of D if and only if $(-i, -j)$ is an element in the domain of NINETY$^2(f)$.

Images can also be flipped. There is a horizontal, a vertical, and two distinct diagonal flips (45° and 135°). Only the 135° flip, denoted FLIP, will be discussed in detail. The remaining ones are obtained by using FLIP in conjunction with NINETY. Pointwise, FLIP is defined by

$$[\text{FLIP}(f)](i, j) = f(-j, -i)$$

Example 2.9

$$\begin{pmatrix} 2 & 3 \\ 0 & * \\ 1 & 5 \end{pmatrix}_{2,3} \longrightarrow \boxed{\text{FLIP}} \longrightarrow \begin{pmatrix} 2 & 0 & 1 \\ 3 & * & 5 \end{pmatrix}_{-3,-2}$$

or, in origin specification notation,

$$\begin{pmatrix} * & * & 2 & 3 \\ * & * & 0 & * \\ * & * & 1 & 5 \\ \circledast & * & * & * \end{pmatrix} \longrightarrow \boxed{\text{FLIP}} \longrightarrow \begin{pmatrix} * & * & * & \circledast \\ * & * & * & * \\ 2 & 0 & 1 & * \\ 3 & * & 5 & * \end{pmatrix}$$

There are several points of note regarding image flipping. First, in the subscript notation, the coordinates of the upper left pixel have "flipped." Second, in matrix terms, FLIP outputs the transpose of the input. Third, the geometric effect of FLIP in the bound matrix representation is to rigidly fasten the image to the 135° line through the origin and then to twist this line 180° on its own axis.

To see how other flipping operations can be generated using FLIP in conjunction with NINETY, consider a flip around the x-axis. This flip is obtained by applying FLIP and NINETY in succession.

Example 2.10

Suppose NINETY is applied to the output of FLIP in Example 2.9. The result is the image

$$\begin{pmatrix} * & * & * & * \\ * & * & 1 & 5 \\ * & * & 0 & * \\ * & * & 2 & 3 \end{pmatrix}$$

which is precisely the original image of the example flipped about the horizontal axis.

2.6 DATABASE-TYPE IMAGE OPERATIONS

The image processing operations of selection and extension are similar to their counterparts found in database algebras. The first of these, SELECT, extracts part of an image from a given image. Intuitively, it leaves entries within a specified "window" unchanged and it removes all other gray values from the image, leaving stars in their places. SELECT requires five inputs, including the image, the size of the window, and the location of the window. Given an image f, the window size m by n where $m, n \geq 1$, and the location (r', t') of the first row, first column position of the output image, SELECT is defined by

$$[\text{SELECT}(f; m, n, r', t')](i, j) = \begin{cases} f(i, j) & \text{for } (i, j) \in W \\ * & \text{otherwise} \end{cases}$$

where W (the window) is the set of pixels

$$W = \{(i, j): r' \leq i < r' + n, \ t' - m < j \leq t'\}$$

SELECT has the block diagram

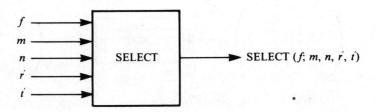

Example 2.11

Let

$$f = \begin{pmatrix} 3 & * & 4 & 5 \\ -1 & 3 & 0 & 2 \\ 3 & 2 & * & -4 \end{pmatrix}_{2,1}$$

Sec. 2.6   Database-Type Image Operations   27

Then

$$\text{SELECT}(f; 4, 3, 4, 0) = \begin{pmatrix} 0 & 2 & * \\ * & -4 & * \\ * & * & * \\ * & * & * \end{pmatrix}_{4,0}$$

Thus, the bound matrix output generated by SELECT in this example is 4 by 3 and has its a_{11} entry located at $(4, 0)$.

The other database-type operator to be presently introduced is the extension operator EXTEND. Given two input images f and g, in that order,

$$[\text{EXTEND}(f, g)](i, j) = \begin{cases} f(i, j) & \text{if } f(i, j) \neq * \\ g(i, j) & \text{otherwise} \end{cases}$$

Intuitively, EXTEND(f, g) is obtained by "adjoining" to f that part of g which does not intersect f. The image f is called *dominant* and the image g is called *subordinate*. EXTEND is not commutative, so f must appear above g whenever the block diagram

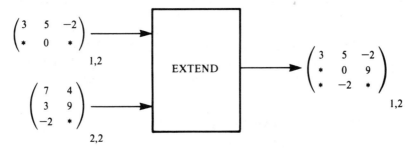

is employed.

Example 2.12

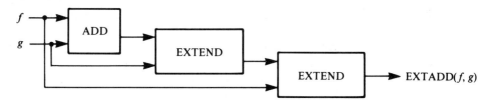

Using only the image processing operators thus far defined, one can construct many "higher-level" operations by composition. For instance, the following block diagram gives the specification of a new addition operator called *extended addition*:

Letting D_f and D_g be the respective domains of f and g,

$$\text{EXTADD}(f, g) = \begin{cases} \text{ADD}(f, g) & \text{on } D_f \cap D_g \\ f & \text{on } D_f - D_g \\ g & \text{on } D_g - D_f \end{cases}$$

In other words, EXTADD adds the images pointwise on their common domain of definition and leaves them unaltered elsewhere. One should pay careful attention to how the higher level operator EXTADD has been constructed from the lower level operators ADD and EXTEND. This block diagram methodology will be used extensively in the sequel. In this instance it yields

$$\text{EXTADD}(f, g) = \text{EXTEND}[\text{EXTEND}(\text{ADD}(f, g), g), f]$$

By construction, EXTADD is commutative.

Example 2.13

Let f and g be the two images respectively given in Example 2.12. Then

$$\text{EXTADD}(f, g) = \begin{pmatrix} 3 & 12 & 2 \\ * & 3 & 9 \\ * & -2 & * \end{pmatrix}_{1,2}$$

Notice that the domain of EXTADD(f, g) is the union of the two input domains, whereas the domain of ADD(f, g) would be the intersection of the two input domains.

Many other useful operations will be derived through the utilization of more primitive transformations. For example, in addition to EXTADD, there are extended arithmetic binary operations for multiplication, maximum, and minimum, viz., EXTMULT, EXTMAX, and EXTMIN, respectively. These operations are derived in a manner similar to the derivation of EXTADD and, like that operator, each performs the appropriate arithmetic operation on the intersection of the two input domains and leaves the inputs as they were elsewhere.

2.7 STRUCTURAL TRANSFORMATIONS

To this point, all of the image processing transformations introduced have been of the *image-to-image* variety in that at least one of the inputs has been an image and the output has been an image. Although in some cases there are "auxiliary" inputs of a different sort, it would not be inappropriate to categorize these operators as belonging to the class of level-1 type transformations as described in Chapter 1. In this section, three operators that are not of the image-to-image variety will be discussed.

An image (bound matrix) is a data structure consisting of a two-dimensional array together with four integers. Through the use of relative address-type techniques, the structure can be changed to a three-dimensional array or, equivalently,

Sec. 2.7   Structural Transformations   29

a two-dimensional array consisting of absolute locations (i, j) together with a one-dimensional array consisting of real-valued gray levels $f(i, j)$. Now imagine two stacks called DOMAIN and RANGE, the first consisting of n ordered pairs (i_k, j_k), $k = 1, 2, \ldots, n$, and the second consisting of n real numbers y_k, $k = 1, 2, \ldots, n$. Together the stacks implicitly contain an image $f(i_k, j_k) = y_k$, for if they were popped simultaneously, the corresponding words would form a location together with its gray value. One could go so far as to say that an image is nothing more than a pair of stacks, (DOMAIN, RANGE). Yet when the stacks are separated, each can be operated upon as a data structure independently of the other. Once such operations are completed, the stacks can again be considered as a pair, and a new image is "created."

To formalize the preceding remarks, we define the binary operator CREATE, which takes an array of ordered pairs of integers and an array consisting of real numbers or stars (*) and outputs an image (or bound matrix). By definition, if

$$D = [(i_1, j_1), (i_2, j_2), \ldots, (i_n, j_n)]$$

and

$$R = [y_1, y_2, \ldots, y_n]$$

then

$$[CREATE(D, R)](i, j)] = \begin{cases} y_k & \text{if } (i, j) = (i_k, j_k) \\ * & \text{otherwise} \end{cases}$$

Except for the addressing techniques required to change a three-dimensional array to a bound matrix, the operator CREATE is purely "structural" in that it simply alters structure. Nevertheless, from the perspectives of logic, low-level programming, and architecture, it needs to be specified. It is important to remember that the two input stacks are not to be considered an image until "joined" by CREATE. This point is crucial since the image results not only from the contents of the stacks, but also from the ordering within the stacks. As long as the stacks remain independent, each can have its ordering permuted by some nonimage-type operation. The output of CREATE depends upon those orderings.

Example 2.14

Consider the arrays

$$D = [(1, 1), (1, 3), (3, 2), (3, 3), (4, 1)]$$
$$R = [0, *, 3, 4, -2]$$

Then

$$CREATE(D, R) = \begin{pmatrix} * & * & 4 & * \\ * & * & 3 & * \\ 0 & * & * & -2 \end{pmatrix}_{1,3}$$

Just as CREATE is used to form an image by joining two arrays, the operators DOMAIN and RANGE can be employed to decompose an image into two arrays, neither of which is an image. DOMAIN takes an image input and yields an array of ordered pairs which make up the domain of the image. RANGE takes an image and yields an array consisting of the gray values of the image. Each "reads" from the top down and from left to right. This last specification assures that DOMAIN and RANGE work in a compatible manner in the sense that the exact image can be reconstructed by CREATE so long as the arrays DOMAIN(f) and RANGE(f) have not been transformed in any manner. Thus,

$$\text{CREATE}[\text{DOMAIN}(f), \text{RANGE}(f)] = f$$

In terms of the classification scheme introduced in Chapter 1, DOMAIN and RANGE are both level-2 operators. Note that if we identify an array of length one with a real number, RANGE can be used to take a 1 by 1 image to a real number. Indeed,

$$\text{RANGE}[(s)_{r,t}] = [s] = s$$

where the last equality results from the aforementioned identification.

Example 2.15

Let f be the first image given in Example 2.12. Then

$$\text{DOMAIN}(f) = [(1, 2), (2, 2), (2, 1), (3, 2)]$$

and

$$\text{RANGE}(f) = [3, 5, 0, -2]$$

As an illustration of the usage of the structural transformations, consider the problem of taking an image f and defining a new image that has the same domain as f but has the single gray value t on that domain. (Such an image is known as a *constant* image.) The operation CONST to produce this image is defined as

$$\text{CONST}(f; t) = \text{CREATE}[\text{DOMAIN}(f), T]$$

where $T = [t, t, \ldots, t]$ is an array with n elements, where n is the cardinality (number of points) of the domain of f. This latter number, CARD(f), is simply the size of the array DOMAIN(f).

2.8 THRESHOLDING

So far, we have discussed numerous low-level operators that are used for the construction of algorithms in image processing. In this section, we introduce a slightly higher level operator that is often used to extract a figure or a feature of particular interest from within an image. The threshold operator THRESH pro-

Sec. 2.8 Thresholding

duces a "binary" image having only two gray values, 0 (white) and 1 (black). It takes two inputs, an image f and a real number t, and returns an image that has the value 1 wherever the gray level of the input image f exceeds or is equal to t, and has the value 0 wherever the gray level of f is less than t. Pixels outside the actual domain of f are not affected, i.e., star-valued pixels are invariant under THRESH. Rigorously, THRESH is defined by

$$[\text{THRESH}(f; t)](i, j) = \begin{cases} 1 & \text{if } f(i, j) \geq t \\ 0 & \text{if } f(i, j) < t \\ * & \text{if } f(i, j) = * \end{cases}$$

As an illustration of the operation of THRESH, consider the image f given in Figure 2.3 and sketched in Figure 2.4. In that image, the gray level quantization runs from 0 (white) to 8 (black). The result of thresholding with threshold value $t = 7$ is given in Figure 2.5, where the dashes (–) are used to represent zeros in order to render the thresholded image more discernible to the eye. The letter H is clearly distinguishable in the thresholded image. Yet notice the two zero (dashed) pixels at the upper left of the H in THRESH(f; 7). These originally had gray value 6, a fairly dark value, but were whitened due to the choice of $t = 7$. Overall, it can be seen that the left side of the image f is lighter than the right side. This seeming distortion might be due to sensor bias, reflected light, or garbled data transmission, to name a few possibilities. Similar to the loss of two pixels on the left of the H are six extra 1-valued (black) pixels on the right. Whereas the lost pixels on the left could be recovered by thresholding at $t = 6$, all but one on the right could be removed by thresholding at $t = 8$. However, the input $t = 7$ appears to be best: the lower threshold value will create excess distortion by

$$\begin{pmatrix}
* & 0 & 0 & 0 & 0 & * & 0 & 0 & 0 & 0 & 1 & 2 & 3 & * & 3 & 3 \\
0 & 0 & 0 & 0 & 0 & 1 & 1 & 6 & 0 & 0 & 2 & 3 & 3 & 3 & 3 & 4 \\
0 & 0 & 0 & 2 & 1 & 1 & 0 & 0 & * & 4 & 4 & 6 & 5 & 5 & 5 & 4 \\
0 & 0 & 2 & 7 & 7 & 7 & 5 & 0 & 2 & 5 & 8 & 8 & 8 & 6 & 5 & 5 \\
0 & 0 & 2 & 6 & 7 & 7 & 2 & 0 & 4 & 6 & 8 & 8 & 8 & 6 & 5 & 3 \\
0 & 0 & 2 & 6 & 7 & 8 & 2 & 0 & 3 & 5 & 8 & 8 & 8 & 7 & 6 & 5 \\
0 & 1 & 2 & 7 & 8 & 7 & 7 & 7 & 8 & 8 & 8 & 8 & 8 & 6 & 6 & 4 \\
0 & 0 & 4 & 7 & 7 & 7 & 7 & 7 & 7 & 8 & 8 & 8 & 8 & 7 & 5 & 6 \\
0 & 2 & 2 & 7 & 7 & 7 & 3 & 4 & 4 & 7 & 8 & 8 & 8 & 6 & 4 & 3 \\
1 & 1 & 4 & 7 & 8 & 8 & 2 & 0 & 1 & * & 8 & 8 & 8 & 5 & 6 & 7 \\
1 & 0 & 2 & 7 & 7 & 8 & 2 & 2 & 1 & 5 & 8 & 8 & 8 & 6 & 6 & 5 \\
0 & 1 & 3 & 3 & 4 & 2 & 2 & 0 & 1 & 4 & 6 & 8 & 7 & 5 & 4 & 3 \\
0 & 0 & 2 & 1 & 1 & 2 & 0 & 0 & 0 & 3 & 4 & 4 & 6 & 6 & 5 & 3 \\
* & 0 & 0 & 0 & 0 & 0 & 0 & 0 & 2 & 2 & 3 & 2 & 3 & 5 & 3 & 3
\end{pmatrix}$$

Figure 2.3 Image f with Gray Levels 0 Through 8

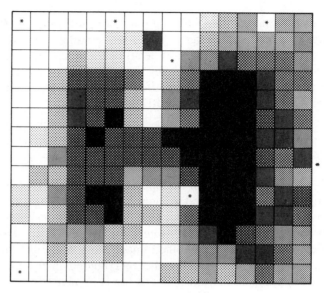

Figure 2.4 Sketch of Image f

the inclusion of too many pixels on the right, while the higher input will result in too few pixels on the left. (It would be a useful exercise to compute both THRESH(f; 6) and THRESH(f; 8).)

Notice in the preceding example that no value of t will result in the output of a perfect H. While there are many reasons for such inexactness, some of which have already been pointed out, the phenomenon of fuzzy demarcation is inherent

Figure 2.5 Thresholded Version of Image f at Level $t = 7$

Sec. 2.8 Thresholding

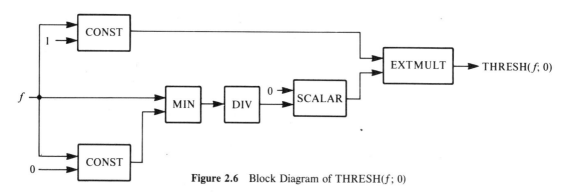

Figure 2.6 Block Diagram of THRESH(f; 0)

to image processing. Very often gray values appear extreme, in the sense that an intelligent observer would likely conclude that they do not correctly reflect the darkness of the actual image under consideration. Such gray value readings are referred to as *noise*. Noise is a fact of life in image analysis, and its removal is a central problem to the discipline.

A specification of THRESH(f; 0) in terms of lower level operators is given in Figure 2.6. As will be our custom, it is given in block diagram form. Representations of higher level operators in terms of lower level operators is important to image processing because it allows for a structured approach to algorithm development and a ready format for programming implementation.

Rather than threshold to a binary image, it is sometimes desirable to threshold in such a manner as to leave unaltered that part of the image above the threshold value and to whiten the remaining part of the image. This operation is known as *truncation* and is defined by

$$[\text{TRUNC}(f; t)](i, j) = \begin{cases} f(i, j) & \text{if } f(i, j) \geq t \\ 0 & \text{if } f(i, j) < t \\ * & \text{if } f(i, j) = * \end{cases}$$

TRUNC is easily implemented by using THRESH; in fact,

$$\text{TRUNC}(f; t) = \text{MULT}[f, \text{THRESH}(f; t)]$$

Example 2.16

Let

$$f = \begin{pmatrix} 6 & 4 & 6 & 0 & 1 & * \\ 5 & 6 & 7 & 0 & 0 & 2 \\ 2 & 1 & 1 & 0 & 0 & 4 \end{pmatrix}_{2,3}$$

Then

$$\text{TRUNC}(f; 5) = \begin{pmatrix} 6 & - & 6 & - & - & * \\ 5 & 6 & 7 & - & - & - \\ - & - & - & - & - & - \end{pmatrix}_{2,3}$$

Several variants of the basic thresholding operator are utilized in image processing, all involving variations on the ⩽ sign. One of these involves identifying the solution to the equation $f(i, j) = t$. It gives a one at each pixel where the equation is satisfied, zero at each pixel where it is not, and leaves star-valued pixels unaltered. This operator is defined by

$$[\text{EQUAL}(f; t)](i, j) = \begin{cases} 1 & \text{if } f(i, j) = t \\ 0 & \text{if } f(i, j) \neq t \text{ and } f(i, j) \neq * \\ * & \text{if } f(i, j) = * \end{cases}$$

Another threshold-type operator is GREATER, which is given by

$$[\text{GREATER}(f; t)](i, j) = \begin{cases} 1 & \text{if } f(i, j) > t \\ 0 & \text{if } f(i, j) \leq t \\ * & \text{if } f(i, j) = * \end{cases}$$

Other variations can be defined, each with its own block diagram. For example, THRESH has the block diagram

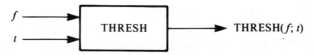

Example 2.17

Let

$$f = \begin{pmatrix} 1 & 2 & 1 & 0 & 0 & 1 & 2 \\ 2 & 1 & 7 & 8 & 5 & 7 & * \\ 1 & 2 & 7 & 12 & 13 & 8 & 1 \\ 0 & 1 & 8 & 12 & 12 & 7 & 2 \\ 2 & 1 & 7 & 6 & 6 & 8 & 0 \\ 1 & 0 & 0 & 3 & 3 & 1 & 1 \end{pmatrix}_{2,6}$$

Then

$$\text{GREATER}(f; 5) = \begin{pmatrix} - & - & - & - & - & - & - \\ - & - & 1 & 1 & - & 1 & * \\ - & - & 1 & 1 & 1 & 1 & - \\ - & - & 1 & 1 & 1 & 1 & - \\ - & - & 1 & 1 & 1 & 1 & - \\ - & - & - & - & - & - & - \end{pmatrix}_{2,6}$$

Using the symbols ⩾ and ⩽, we can define the ternary operator BETWEEN in an obvious manner, and we have

$$\text{BETWEEN}(f; 5, 11) = \begin{pmatrix} - & - & - & - & - & - & - \\ - & - & 1 & 1 & 1 & 1 & * \\ - & - & 1 & - & - & 1 & - \\ - & - & 1 & - & - & 1 & - \\ - & - & 1 & 1 & 1 & 1 & - \\ - & - & - & - & - & - & - \end{pmatrix}_{2,6}$$

Thus, GREATER has helped locate a square (except for one pixel) in the image and BETWEEN has located the edge of the square.

There are many ways of employing thresholding. Often these involve some sort of preprocessing prior to applying THRESH.

2.9 VECTOR-TYPE OPERATIONS

In this section, several often-employed image processing operations will be introduced. Each of these is a two-dimensional analog of an operation in vector algebra and a discrete, two-dimensional analog of an operation in calculus.

Given a digital image f, it is often necessary to find the sum of all the gray values. Accordingly, the gray value summation operator PIXSUM is defined by

$$\text{PIXSUM}(f) = \sum_{(i,j) \in D_f} f(i, j)$$

where D_f is the domain of f. In terms of lower level operators, PIXSUM is specified by

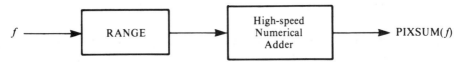

The array of gray values output by RANGE is simply sent through some high-speed adder to obtain PIXSUM(f). In the event that f is the null image, i.e., the image consisting of only stars, PIXSUM(f) is defined to be simply a $*$. Consequently, PIXSUM is a unary operator taking an image input and outputting an element of $R \cup \{*\}$, where R denotes the set of real numbers.

Once the gray level summation for an image has been obtained, the average gray level may be determined by dividing PIXSUM by the cardinality of the image. That is, the gray level average operator AVER is defined by

$$\text{AVER}(f) = \frac{\text{PIXSUM}(f)}{\text{CARD}(f)}$$

By convention, if CARD(f) = 0 (f is the null image), then AVER(f) is defined to be a star. Both AVER and PIXSUM will be called level-2 transformations, even though they might output a star rather than a real number.

The image processing operator DOT is the two-dimensional analog of the usual vector dot product, the only essential complication being the handling of star-valued pixels. DOT is defined by

$$\text{DOT}(f, g) = \begin{cases} \sum_{(i,j) \in D} f(i,j) \times g(i,j) & \text{if } f \text{ and } g \text{ have the common domain } D \\ * & \text{if } f \text{ and } g \text{ have different domains} \end{cases}$$

As is the case in vector algebra, the gray level values of corresponding pixels are multiplied and then the sum of those products is taken. If any terms in the summation have a $*$ as a factor, then the dot product is defined to be $*$.

Example 2.18

Consider the images f, g, and h defined respectively by

$$f = \begin{pmatrix} 2 & 3 & 1 \\ -1 & 0 & * \end{pmatrix}_{3,1}, \quad g = \begin{pmatrix} 4 & 2 & 0 \\ 3 & 3 & * \end{pmatrix}_{3,1}, \quad h = \begin{pmatrix} 2 & 1 & * \\ 2 & -5 & 3 \end{pmatrix}_{3,1}$$

Then PIXSUM$(f) = 5$, AVER$(f) = 1$, DOT$(f, g) = 11$, and DOT$(f, h) = *$. Note that DOT is a binary operator with two image inputs and an output that is an element of $R \cup \{*\}$.

Some basic properties of the image dot product are analogous to those of the dot product of vector algebra. We have

1. DOT(f, g) = DOT(g, f)
2. DOT$[f, $ ADD$(g, h)]$ = DOT(f, g) + DOT(f, h) if $D_f = D_g = D_h$
3. DOT$[$SCALAR$(t; f), g]$ = $t \times$ DOT(f, g)

The preceding properties are, respectively, commutativity, distributivity, and associativity. In associativity, the convention is adopted that the multiplication of a $*$ by either a real number or a $*$ is again a $*$. The proofs of the three properties are identical to the vector algebra proofs, except that one must pay attention to stars.

The implementation of DOT in terms of lower level operators is given by the block diagram of Figure 2.7. Although a detailed explanation of such block diagrams will not always be provided, in this instance one will be provided since the block diagram for DOT is the first one of any significant complexity presented in the text. Referring to Figure 2.7, first suppose that f and g have a common domain. Then applying the sequence EXTEND, NINETY^2, and DOMAIN simply gives their common domain rotated by 180°. The parallel translations of f and g by that rotated domain result in two sets of images, one having each gray value of f situated at the origin and the other having each gray value of g situated at the origin. SELECT then picks out singleton images situated at the origin, MULT multiplies common pairs, ADD sums the products, and RANGE outputs the resulting real number.

Suppose on the other hand that f and g do not have common domains. Then applying the sequence EXTEND, NINETY^2, and DOMAIN gives the 180° rotation of a set of pixels that is strictly larger than at least one of the original domains, and hence, the parallel translations and selections yield at least one image that is null. Consequently, at least one MULT yields a null image, and ADD does likewise. The output of RANGE is then $*$.

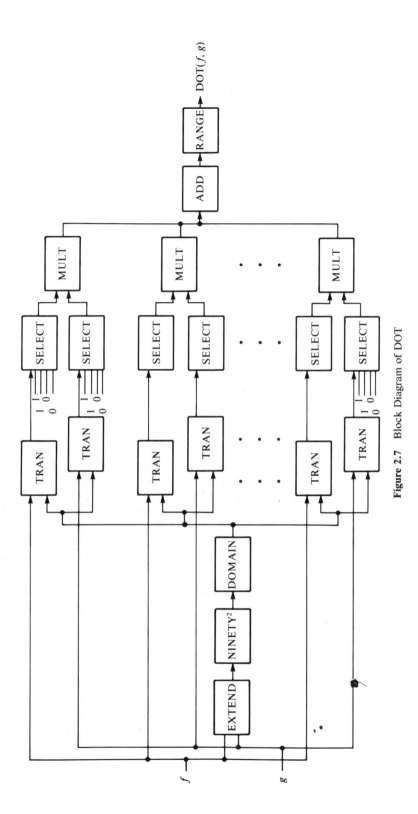

Figure 2.7 Block Diagram of DOT

In a fashion similar to that for Euclidean vectors, a norming operation is defined on digital images by

$$\text{NORM}(f) = \left[\sum_{(i,j) \in D_f} [f(i,j)]^2 \right]^{1/2}$$

where D_f is the domain of f. NORM(f) gives a quantitative measure of the absolute gray values. If it happens, as is most common, that the gray values run from zero upward, then NORM(f) gives an overall quantification of the darkness of the image.

As is the case with the usual Euclidean vector norm,

$$\text{NORM}(f) = \sqrt{\text{DOT}(f, f)}$$

Other properties of NORM that are analogs of the Euclidean vector norm are:

1. $\text{NORM}(f) \geq 0$
2. $\text{NORM}[\text{SCALAR}(t;f)] = |t| \times \text{NORM}(f)$
3. $\text{NORM}[\text{SCALAR}(1/\text{NORM}(f);f)] = 1$, if $\text{NORM}(f) \neq 0$
4. $|\text{DOT}(f, g)| \leq \text{NORM}(f) \times \text{NORM}(g)$ if $D_f = D_g$
5. $\text{NORM}[\text{EXTADD}(f, g)] \leq \text{NORM}(f) + \text{NORM}(g)$

The preceding properties are straightforward generalizations of the usual Euclidian properties, except that one must be attentive to the stars. For example, property 4 is the digital image form of the Cauchy-Schwarz inequality, with the proviso that the domains coincide. This restriction results from the fact that $\text{DOT}(f, g) = *$ for $D_f \neq D_g$. Given this condition, the proof is exactly the same as in the Euclidean vector case.

Property 5 is the triangle inequality, except that EXTADD is employed instead of ADD. In this instance the Euclidean proof does not immediately generalize. It needs a slight alteration; indeed,

$(\text{NORM}[\text{EXTADD}(f, g)])^2$

$$= \sum_{(i,j) \text{ in } D_f - D_g} f(i,j)^2 + \sum_{(i,j) \text{ in } D_g - D_f} g(i,j)^2 + \sum_{(i,j) \text{ in } D_f \cap D_g} [f(i,j) + g(i,j)]^2$$

Let the first two summations be denoted by S_1 and S_2, respectively. The last sum is simply $(\text{NORM}[\text{ADD}(f, g)])^2$, where $\text{ADD}(f, g)$ is an image that is defined precisely on $D_f \cap D_g$. Now let f' denote the image that equals f on $D_f \cap D_g$ and is $*$ elsewhere, and let g' denote the image that equals g on $D_f \cap D_g$ and is

∗ elsewhere. Then $\text{ADD}(f, g) = \text{ADD}(f', g')$, and

$$\begin{aligned}
(\text{NORM}[\text{ADD}(f, g)])^2 &= \text{DOT}[\text{ADD}(f, g), \text{ADD}(f, g)] \\
&= \text{DOT}[\text{ADD}(f', g'), \text{ADD}(f', g')] \\
&= \text{DOT}(f', f') + 2\,\text{DOT}(f', g') + \text{DOT}(g', g') \\
&\leq \text{NORM}(f')^2 + 2\,\text{NORM}(f')\text{NORM}(g') + \text{NORM}(g')^2 \\
&= [\text{NORM}(f') + \text{NORM}(g')]^2
\end{aligned}$$

where the Cauchy-Schwarz inequality was used on $\text{DOT}(f', g')$ to obtain the inequality. Next, let S_3 denote the summation of $f(i, j)^2$ over $D_f \cap D_g$ and S_4 denote the summation of $g(i, j)^2$ over the same set of pixels. Then

$$\begin{aligned}
(\text{NORM}[\text{EXTADD}(f, g)])^2 &= S_1 + S_2 + \text{NORM}[\text{ADD}(f, g)]^2 \\
&\leq S_1 + S_2 + [\text{NORM}(f') + \text{NORM}(g')]^2 \\
&= S_1 + S_2 + [S_3^{1/2} + S_4^{1/2}]^2 \\
&= S_1 + S_2 + S_3 + S_4 + 2 S_3^{1/2} S_4^{1/2} \\
&\leq S_1 + S_3 + S_2 + S_4 + 2[S_1 + S_3]^{1/2}[S_2 + S_4]^{1/2} \\
&= [(S_1 + S_3)^{1/2} + (S_2 + S_4)^{1/2}]^2 \\
&= [\text{NORM}(f) + \text{NORM}(g)]^2
\end{aligned}$$

Finally, taking square roots gives property 5, the triangle inequality for extended addition. Notice that, as an immediate consequence, there is a triangle inequality for ADD, since

$$\text{NORM}[\text{ADD}(f, g)] \leq \text{NORM}[\text{EXTADD}(f, g)]$$

because the summation for $\text{NORM}[\text{EXTADD}(f, g)]$ includes all the terms in the summation for $\text{NORM}[\text{ADD}(f, g)]$.

The last operator to be defined in this section is REST, which restricts some *primary* input image f to the domain of a *secondary* input image g. Rigorously,

$$[\text{REST}(f, g)](i, j) = \begin{cases} f(i, j) & \text{if } g(i, j) \neq * \\ * & \text{if } g(i, j) = * \end{cases}$$

REST is related to the selection operator in that it picks out a portion of some input image f. Notice that REST is certainly not commutative. Its specification

is given by

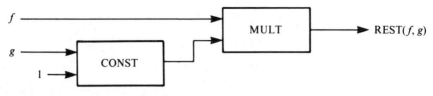

Example 2.19

Let f and h be as in Example 2.18. Then

$$\text{REST}(f, h) = \begin{pmatrix} 2 & 3 \\ -1 & 0 \end{pmatrix}_{3,1}$$

2.10 MOVING-AVERAGE FILTERS

Among the most common methodologies employed in image processing, as well as signal processing, is the application of a *moving-average filter*. There are two types of such filters: the *space-invariant* and the *space-variant* varieties.

Given a pixel (i, j), a *neighborhood* of (i, j) is a set of pixels containing (i, j). A particularly important collection of neighborhoods is the set of *square* neighborhoods. The square neighborhood containing (i, j) is given by

$$\text{SQUARE}(i, j) = \{(i + 1, j - 1), (i, j - 1), (i - 1, j - 1), (i - 1, j),$$
$$(i - 1, j + 1), (i, j + 1), (i + 1, j + 1), (i + 1, j), (i, j)\}$$

Thus, SQUARE(i, j) consists of (i, j) together with the pixels immediately above it, immediately below it, at its sides, and at its corners. The pixels above, below, and at the sides of (i, j) are called *strong* neighbors of (i, j), and the ones at the corners of (i, j) are called *weak* neighbors of (i, j).

Associated with the notion of a neighborhood is that of a *mask*. A mask is nothing but a digital image containing the origin. Given a neighborhood N of a pixel (i, j), a mask of the same shape can be placed over N in order to weight the pixels of N. The mask (image) M is translated so that its center (origin pixel) is located at (i, j). The intent here is to create a weighting scheme that will be utilized in the definition of certain image operations. If the gray values of the mask are all nonnegative, and if PIXSUM$(M) = 1$, then M is called an *averaging mask*.

Example 2.20

Let mask

$$M = \begin{pmatrix} \frac{1}{32} & \frac{3}{32} & \frac{1}{32} \\ \frac{3}{32} & \frac{1}{2} & \frac{3}{32} \\ \frac{1}{32} & \frac{3}{32} & \frac{1}{32} \end{pmatrix}$$

Sec. 2.10 Moving-Average Filters

Then TRAN($M;i, j$) is situated over SQUARE(i, j) and is centered at (i, j). If we treat the output of DOMAIN as simply a set rather than an array, then

$$\text{DOMAIN}[\text{TRAN}(M;i, j)] = \text{SQUARE}(i, j)$$

Note that M is an averaging mask.

Given a mask M and an arbitrary digital image f, we define

$$[\text{FILTER}(f;M)](i, j) = \sum_{(u,v) \text{ in } N} [\text{TRAN}(M;i, j)](u, v) \times f(u, v)$$

where N is the neighborhood corresponding to the domain of the translated mask and where the stipulation is made that if any $f(u, v)$ in the summation are star valued then $[\text{FILTER}(f;M)](i, j) = *$. Notice that the operator FILTER takes two input images, the second one of which must be a mask. If the mask input is fixed, FILTER($\cdot$, M) can be viewed as a single input operator. For a fixed mask M, FILTER($\cdot$, M) is called a *space-invariant moving-average filter*—even though M might not be an averaging mask.

Example 2.21

Let M be the mask of Example 2.20, and let

$$f = \begin{pmatrix} 0 & 0 & 0 & 0 & 0 & 0 & 0 \\ * & 0 & 0 & 0 & 0 & 0 & 0 \\ 0 & 0 & 8 & 8 & 8 & 0 & 0 \\ 0 & 0 & 8 & 8 & 8 & 0 & 0 \\ 0 & 0 & 8 & 8 & 8 & 0 & 6 \\ 0 & 0 & 0 & 6 & 0 & 6 & 0 \\ 0 & 0 & 6 & 0 & 6 & 0 & 6 \end{pmatrix}_{-3,3}$$

Then FILTER($f;M$) is given by

$$\begin{pmatrix} * & * & * & * & * & * & * \\ * & * & 1 & 1\tfrac{1}{4} & 1 & \tfrac{1}{4} & * \\ * & * & 5\tfrac{3}{4} & 6\tfrac{3}{4} & 5\tfrac{3}{4} & 1 & * \\ * & 1\tfrac{1}{4} & 6\tfrac{3}{4} & 8 & 6\tfrac{3}{4} & 1\tfrac{7}{16} & * \\ * & 1 & 5\tfrac{15}{16} & 7\tfrac{5}{16} & 6\tfrac{1}{8} & 2\tfrac{1}{8} & * \\ * & \tfrac{7}{16} & 2\tfrac{1}{8} & 4\tfrac{5}{8} & 2\tfrac{11}{16} & 3\tfrac{13}{16} & * \\ * & * & * & * & * & * & * \end{pmatrix}_{-3,3}$$

Several comments are in order concerning the preceding example. The outer columns and rows of the image have been lost due to the stipulation regarding stars in the summation defining FILTER. There is also a loss of gray values at

$(-2, 1)$ and $(-2, 2)$ for a similar reason. Moreover, there has been an overall "smoothing" of the image. The original image contained a dark square surrounded by mostly white pixels, with a few scattered darker pixels interspersed among the white ones. The application of FILTER has reduced substantially the gray values of these apparently noisy pixels, but it has done so with a concomitant loss in the clarity of the demarcation between the square and its background. Whereas thresholding the image f with threshold value $t = 7$ would have produced a black-and-white image with the square black, thresholding the filtered image with threshold value $t = 7$ or $t = 6$ would not produce a figure recognizable as a square. Only six pixels would be black in THRESH[FILTER($f;M$);6].

The aforementioned smoothing characteristics typify the action of filtering a digital image with an averaging mask. At each pixel of the output, the gray value is a weighted average of the surrounding pixels, those determined by the domain of the mask. The net effect is a flattening of the input image f. Because of this, we shall write SMOOTH($f;M$) whenever M is an averaging mask.

Filters such as SMOOTH are called *low-pass* filters in that they attenuate high frequencies (rapidly fluctuating gray values). Low-pass filters transmit with relatively little alteration any input image that fluctuates slowly with respect to pixel changes, but they level out any input image that fluctuates rapidly with respect to gray value changes. In other words, rapidly oscillating highs and lows are mediated. Consequently, high-frequency salt-and-pepper noise is attenuated. An unwanted side effect is the blurring of contrast within the image: the distinguishability of features which results from substantial gray value variation in surrounding pixels is reduced.

It is possible to quantify the preceding comments regarding the reduction in gray level fluctuation resulting from SMOOTH. First, note that according to the definition of FILTER,

$$[\text{SMOOTH}(f;M)](i, j) = \sum_{(u,v) \text{ in } N(i,j)} M(u - i, v - j) \times f(u, v)$$

$$= \sum_{(u,v) \text{ in } D_M} M(u, v) \times f(u + i, v + j)$$

where $N(i, j)$ is the domain of TRAN($M;i, j$), D_M is the domain of M, and the usual stipulation is made regarding terms where the value of f is $*$.

Now let us examine the gray level fluctuation between two pixels of the output of SMOOTH. Let $g = \text{SMOOTH}(f;M)$, and let

$$Q[f;i, j, i', j'] = \max\{|f(u + i, v + j) - f(u + i', v + j')| : (u, v) \in D_M\}$$

Then Q measures, in absolute value, the maximum gray level variation of f between corresponding pixels in the neighborhoods $N(i, j)$ and $N(i', j')$. Using Q,

Sec. 2.10 Moving-Average Filters

we obtain

$$|g(i,j) - g(i',j')| = \left| \sum_{(u,v) \text{ in } D_M} M(u,v) \right.$$
$$\left. \times [f(u+i, v+j) - f(u+i', v+j')] \right|$$
$$\leq \sum_{(u,v) \text{ in } D_M} M(u,v)$$
$$\times |f(u+i, v+j) - f(u+i', v+j')|$$
$$\leq \sum_{(u,v) \text{ in } D_M} M(u,v) \times Q[f;i,j,i',j']$$
$$= Q[f;i,j,i',j'] \times \left[\sum_{(u,v) \text{ in } D_M} M(u,v) \right]$$
$$= Q[f;i,j,i',j']$$

where the last equality follows from the fact that the sum of the gray values of M is equal to 1. In other words, gray level variation in the smoothed image is dominated by the maximum of certain gray level variations in the original image, that maximum being taken over a domain whose size depends upon the averaging mask M.

So far we have been discussing low-pass filters. Other filters known as *high-pass* filters also exist. High-pass filters have little effect upon a rapidly fluctuating input image, but accentuate changes in a slowly fluctuating image. Thus, whereas a low-pass filter reduces high-frequency noise at the cost of blurring contrast, high-pass filters sharpen contrast at the cost of increasing noise.

Example 2.22

Consider the mask

$$M = \begin{pmatrix} 0 & -1 & 0 \\ -1 & \circledR{8} & -1 \\ 0 & -1 & 0 \end{pmatrix}$$

and the image

$$f = \begin{pmatrix} 0 & 0 & 0 & 0 & 0 & 0 & 0 \\ 0 & 1 & 1 & 1 & 1 & 1 & 0 \\ 0 & 1 & 2 & 2 & 2 & 1 & 0 \\ 0 & 1 & 2 & 2 & 2 & 1 & 0 \\ 0 & 1 & 2 & 2 & 2 & 1 & 0 \\ 0 & 1 & 1 & 1 & 1 & 1 & 0 \\ 0 & 0 & 0 & 0 & 0 & 0 & 0 \end{pmatrix}_{-3,3}$$

Then

$$\text{FILTER}(f;M) = \begin{pmatrix} 6 & 4 & 4 & 4 & 6 \\ 4 & 10 & 9 & 10 & 4 \\ 4 & 9 & 8 & 9 & 4 \\ 4 & 10 & 9 & 10 & 4 \\ 6 & 4 & 4 & 4 & 6 \end{pmatrix}_{-2,2}$$

Assuming a gray scale quantization running from 0 to 31, the square of gray value 2 in the middle of f would be difficult to distinguish from its surroundings. Filtering with the mask M has greatly increased the contrast between the square and its background. However, it has done so at the cost of introducing both seemingly noisy pixels at the corners of the square and gray level variation into the square itself. That is to say, easier recognition of the square has been accomplished, but the penalty has been some loss of uniformity and the introduction of some noise. The full impact of the dilemma can be seen in Example 2.23, where the same image has been filtered with a different high-pass filter. In that example, the square has been effectively lost.

Example 2.23

Let

$$N = \begin{pmatrix} 0 & -2 & 0 \\ -2 & \fbox{10} & -2 \\ 0 & -2 & 0 \end{pmatrix}$$

and let f be as in Example 2.22. Then

$$\text{FILTER}(f;N) = \begin{pmatrix} 6 & 2 & 2 & 2 & 6 \\ 2 & 8 & 6 & 8 & 2 \\ 2 & 6 & 4 & 6 & 2 \\ 2 & 8 & 6 & 8 & 2 \\ 6 & 2 & 2 & 2 & 6 \end{pmatrix}_{-2,2}$$

Thresholding with $t = 6$ yields

$$\text{THRESH}[\text{FILTER}(f;N);6] = \begin{pmatrix} 1 & - & - & - & 1 \\ - & 1 & 1 & 1 & - \\ - & 1 & - & 1 & - \\ - & 1 & 1 & 1 & - \\ 1 & - & - & - & 1 \end{pmatrix}_{-2,2}$$

Sec. 2.10   Moving-Average Filters   45

while thresholding with $t = 7$ yields

$$\text{THRESH}[\text{FILTER}(f;N);7] = \begin{pmatrix} - & - & - & - & - \\ - & 1 & - & 1 & - \\ - & - & - & - & - \\ - & 1 & - & 1 & - \\ - & - & - & - & - \end{pmatrix}_{-2,2}$$

The preceding example has once again shown the problematic nature of space-invariant moving-average filters. The situation can be somewhat improved by using a space-variant moving-average filter. Instead of employing a single mask and translating it to each pixel in the image, a space-variant filter utilizes a collection of masks, each one to be used at a specific pixel or set of pixels within the image domain. Consequently, the output image is defined by

$$g(i, j) = \sum_{(u,v) \text{ in } D_{M(i,j)}} [M(i, j)](u, v) \times f(u + i, v + j)$$

where $M(i, j)$ is the mask to be used in the computation of the output image value at pixel (i, j). The choice of which mask to use at a particular pixel (i, j) usually depends upon some prior experience. For example, if the sensor equipment is such that one portion of the image is usually noisy while the rest is not, one need only smooth on the noisy portion while leaving the rest of the image alone. In this instance there would only be two filtering masks used in the process. More involved situations are of course possible.

Example 2.24

In the image f of Example 2.21, it appears that the noise is constrained to the lower right corner of the image. Suppose the image engineer has seen this phenomenon occur repeatedly and has decided to smooth only the twelve pixels in the bottom three rows and the rightmost four columns. The image has been segmented into two regions, one comprising the pixels in the lower right corner to be filtered with the mask M and the other consisting of the remaining pixels to be filtered with the mask $I = (①)$. Therefore, the collection of masks is $\{I, M\}$. The output of the filter will agree with the input f at all pixels except those in the lower right corner, where it will agree with $\text{FILTER}(f;M)$. Though the filter in this example is quite trivial, it nonetheless illustrates the possible advantage of space-variant filters over those which are space invariant. Unfortunately, the choice of which masks to use where is highly problematic.

Before leaving this section, we wish to give an implementation of the operator FILTER in terms of lower level operations. This is done in Figure 2.8. In Example 2.25, a "walk-through" of the implementation is given for a specific

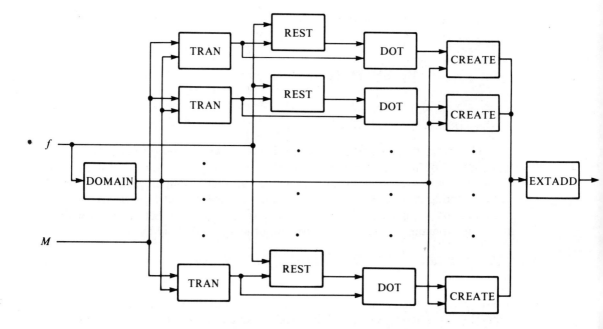

Figure 2.8 Block Diagram of FILTER

image and mask in which the gray value of FILTER$(f;M)$ is computed for a single lattice point.

Example 2.25

Let

$$f = \begin{pmatrix} 2 & 4 & -1 & 1 \\ 0 & 1 & 3 & 9 \\ 1 & 5 & 0 & -3 \end{pmatrix}_{0,2}$$

and let the mask M be given by

$$M = \begin{pmatrix} 1 & 2 & 1 \\ 2 & ④ & 2 \\ 1 & 2 & 1 \end{pmatrix}$$

A walk-through will now be given for the computation of FILTER$(f;M)$ according to the implementation of Figure 2.8. In the figure it can be seen that the algorithm works in a parallel fashion, one branch for each pixel in the domain of the image f. The following walk-through will follow only the domain pixel (2, 1).

Sec. 2.10 Moving-Average Filters 47

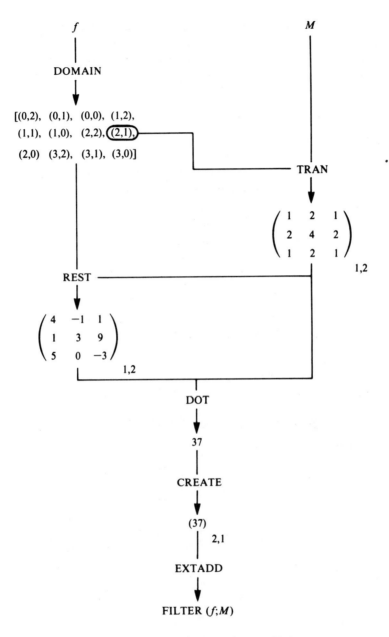

(Note that [FILTER($f;M$)](2,1) = 37.)

In both Figure 2.8 and Example 2.25 the opportunity for the exploitation of parallel design is evident. One of the major problems of image processing is the real-time constraint. While bound matrix representations of images might introduce some notational intricacies, they also expose the structural parallelism inherent in many imaging algorithms, thereby facilitating efficient design.

2.11 IMAGE VECTORS

There are times when one must consider an ordered collection of images $(f_1, f_2, \ldots, f_n)$. Such a collection is known as an *image vector* and is denoted by $\mathbf{f} = (f_1, f_2, \ldots, f_n)$. Like Euclidean vectors, image vectors can be "added componentwise" and "scalar multiplied." Indeed, we define

$$\mathbf{ADD}(\mathbf{f}, \mathbf{g}) = (\text{ADD}(f_1, g_1), \text{ADD}(f_2, g_2), \ldots, \text{ADD}(f_n, g_n))$$

and

$$\mathbf{SCALAR}(f; \mathbf{g}) = (\text{MULT}(f, g_1), \text{MULT}(f, g_2), \ldots, \text{MULT}(f, g_n))$$

where $\mathbf{f}$ and $\mathbf{g}$ are image vectors with n component images and f is simply an image. Using extended image addition and extended image multiplication, one can also define

$$\mathbf{EXTADD}(\mathbf{f}, \mathbf{g}) = (\text{EXTADD}(f_1, g_1), \ldots, \text{EXTADD}(f_n, g_n))$$

and

$$\mathbf{EXTSCALAR}(f; \mathbf{g}) = (\text{EXTMULT}(f, g_1), \ldots, \text{EXTMULT}(f, g_n))$$

Example 2.26

Let

$$\mathbf{f} = \left(\begin{pmatrix} 2 & 3 \\ * & 0 \end{pmatrix}, \begin{pmatrix} 1 & * \\ 2 & 1 \end{pmatrix}, \begin{pmatrix} 2 & -1 \\ 3 & 4 \end{pmatrix} \right)$$

$$\mathbf{g} = \left(\begin{pmatrix} * & 2 \\ \textcircled{1} & 0 \end{pmatrix}, \begin{pmatrix} * & 1 \\ \textcircled{6} & 0 \end{pmatrix}, \begin{pmatrix} * & 0 \\ \textcircled{-2} & 5 \end{pmatrix} \right)$$

and

$$f = \begin{pmatrix} 3 & 2 \\ * & 1 \end{pmatrix}$$

Then

$$\mathbf{ADD}(\mathbf{f}, \mathbf{g}) = \left(\begin{pmatrix} * & 5 \\ * & 0 \end{pmatrix}, \begin{pmatrix} * & * \\ \textcircled{8} & 1 \end{pmatrix}, \begin{pmatrix} * & -1 \\ \textcircled{1} & 9 \end{pmatrix} \right)$$

$$\mathbf{EXTADD}(\mathbf{f}, \mathbf{g}) = \left(\begin{pmatrix} 2 & 5 \\ \textcircled{1} & 0 \end{pmatrix}, \begin{pmatrix} 1 & 1 \\ \textcircled{8} & 1 \end{pmatrix}, \begin{pmatrix} 2 & -1 \\ \textcircled{1} & 9 \end{pmatrix} \right)$$

Sec. 2.11   Image Vectors   49

$$\text{SCALAR}(f; \mathbf{g}) = \left(\begin{pmatrix} * & 4 \\ * & 0 \end{pmatrix}, \begin{pmatrix} * & 2 \\ * & 0 \end{pmatrix}, \begin{pmatrix} * & 0 \\ * & 5 \end{pmatrix} \right)$$

$$\text{EXTSCALAR}(f; \mathbf{g}) = \left(\begin{pmatrix} 3 & 4 \\ \boxed{1} & 0 \end{pmatrix}, \begin{pmatrix} 3 & 2 \\ \boxed{6} & 0 \end{pmatrix}, \begin{pmatrix} 3 & 0 \\ \boxed{-2} & 5 \end{pmatrix} \right)$$

Suppose $\mathbf{f} = (f_1, f_2, \ldots, f_n)$ is an image vector. If we consider a fixed pixel (i, j), then the array

$$\mathbf{f}(i, j) = (f_1(i, j), f_2(i, j), \ldots, f_n(i, j))$$

is like an n-dimensional Euclidean vector except that some of the components might be star valued. For instance, in the previous example, $\mathbf{f}(1,1) = (3, *, -1)$. Should all component images have the same domain, say D, then $\mathbf{f}(i, j)$ is actually a Euclidean n-vector for every (i, j) in D. For instance, in the previous example each component vector of $\mathbf{g}$ had the domain $\{(0, 0), (1, 1), (1, 0)\}$, so that $\mathbf{g}$ evaluated at any lattice point in that domain is a Euclidean 3-vector.

There are three common ways of defining the norm, or magnitude, of a vector $\mathbf{V} = (x_1, x_2, \ldots, x_n)$ consisting of real- or complex-valued components, namely the l_∞-norm,

$$\| \mathbf{V} \|_\infty = \max\{| x_1 |, | x_2 |, \ldots, | x_n |\}$$

the l_1-norm,

$$\| \mathbf{V} \|_1 = \sum_{k=1}^{n} | x_k |$$

and the l_2-norm

$$\| \mathbf{V} \|_2 = \left[\sum_{k=1}^{n} | x_k |^2 \right]^{1/2}$$

These norms lead directly to three magnitude operators defined on image vectors, each such operation yielding an image h where the gray value $h(i, j)$ is one of the above norms computed on the vector $\mathbf{f}(i, j) = (f_1(i, j), f_2(i, j), \ldots, f_n(i, j))$. Though the entire procedure appears straightforward, a slight complication does arise due to the possibility of stars in some components of $\mathbf{f}(i, j)$. This problem is handled by the use of EXTMAX and EXTADD in defining the image vector magnitude operators. We define on image vectors the pixelwise magnitude operators MAG0, MAG1, and MAG2 by the respective block diagrams in Figure 2.9. Several comments are in order concerning the figure. First, the operator ABS takes a single input image and outputs an image that has, at each pixel, the absolute value of the gray value at the corresponding input pixel; the operator SQUARE takes a single input image and outputs an image that has, at each pixel, the square of the gray value of the corresponding input pixel; and the operator SQROOT takes a single input image and outputs an image that has, at each pixel, the square

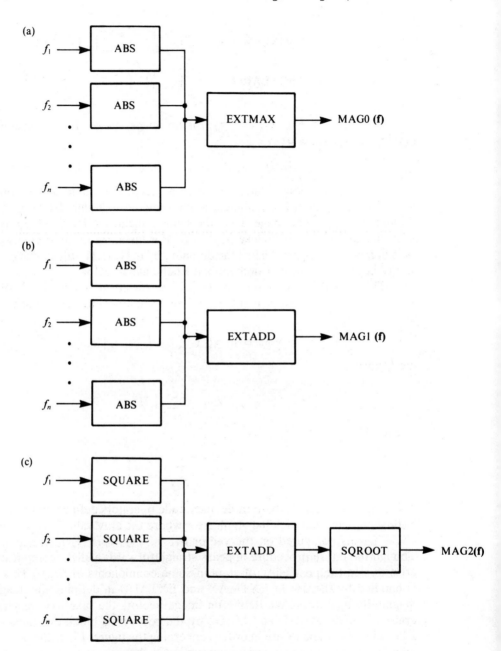

Figure 2.9 Block Diagrams of Magnitude Operations

Sec. 2.11   Image Vectors   51

root of the gray value of the corresponding input pixel. Second, in each case the domain of the output image is the union of the domains of the component images of the input image vector; thus, letting those domains be $D_1, D_2, \ldots, D_n$, respectively, the domain D_M of the output magnitude image is given by

$$D_M = \bigcup_{k=1}^{n} D_k$$

Finally, note that if all the component images are defined at pixel (i, j), i.e., if

$$(i, j) \in \bigcap_{k=1}^{n} D_k$$

then the gray value at (i, j) for each of the three image magnitude vectors is given by the appropriate original norm defined on the Euclidean n-vector $\mathbf{f}(i, j)$; that is,

$$[\text{MAG0}(\mathbf{f})](i, j) = \| \mathbf{f}(i, j) \|_\infty$$
$$[\text{MAG1}(\mathbf{f})](i, j) = \| \mathbf{f}(i, j) \|_1$$
$$[\text{MAG2}(\mathbf{f})](i, j) = \| \mathbf{f}(i, j) \|_2$$

Example 2.27

Let $\mathbf{f}$ be the image vector given in Example 2.26. Then

$$\text{MAG0}(\mathbf{f}) = \begin{pmatrix} 2 & 3 \\ ③ & 4 \end{pmatrix}$$

$$\text{MAG1}(\mathbf{f}) = \begin{pmatrix} 5 & 4 \\ ⑤ & 5 \end{pmatrix}$$

$$\text{MAG2}(\mathbf{f}) = \begin{pmatrix} 3 & \sqrt{10} \\ \sqrt{13} & \sqrt{17} \end{pmatrix}$$

Suppose $(i, j) \in D_M$, the union of the input component image domains. Then there exists at least one k for which $f_k(i, j) \neq *$. We define a new image vector $\mathbf{F}$ such that each component image has the domain D_M. $\mathbf{F}$ is given componentwise by $\mathbf{F} = (F_1, F_2, \ldots, F_n)$, where

$$F_p(i, j) = \begin{cases} f_p(i, j) & \text{if } f_p(i, j) \neq * \\ 0 & \text{if } f_p(i, j) = * \text{ and } (i, j) \in D_M \\ * & \text{if } (i, j) \notin D_M \end{cases}$$

Intuitively, F_p is found by replacing all stars in D_M with zeros. What is of consequence, however, is the fact that each component of $\mathbf{F}$ has the same domain

and

$$\mathrm{MAG0}(\mathbf{f}) = \mathrm{MAG0}(\mathbf{F})$$
$$\mathrm{MAG1}(\mathbf{f}) = \mathrm{MAG1}(\mathbf{F})$$
$$\mathrm{MAG2}(\mathbf{f}) = \mathrm{MAG2}(\mathbf{F})$$

These equalities are useful since the concurrence of the domains of the component images $F_1, F_2, \ldots, F_n$ implies that

$$[\mathrm{MAG0}(\mathbf{F})](i, j) = \| \mathbf{F}(i, j) \|_\infty$$
$$[\mathrm{MAG1}(\mathbf{F})](i, j) = \| \mathbf{F}(i, j) \|_1$$

and

$$[\mathrm{MAG2}(\mathbf{F})](i, j) = \| \mathbf{F}(i, j) \|_2$$

Consequently, certain properties of the l_∞-norm, the l_1-norm, and the l_2-norm generalize to the image vector magnitude operators.

Example 2.28

Let **f** and **g** be the respective image vectors given in Example 2.26. Then

$$\mathbf{F} = \left(\begin{pmatrix} 2 & 3 \\ \textcircled{0} & 0 \end{pmatrix}, \begin{pmatrix} 1 & 0 \\ \textcircled{2} & 1 \end{pmatrix}, \begin{pmatrix} 2 & -1 \\ \textcircled{3} & 4 \end{pmatrix} \right)$$

and $\mathbf{G} = \mathbf{g}$.

2.12 DIGITAL IMAGE GRADIENTS

For $f(x, y)$ a real-valued function of two real variables, the instantaneous rates of change in the x-direction and the y-direction are given by the partial derivatives $\partial f/\partial x$ and $\partial f/\partial y$, respectively. The notations $D_x f$ and $D_y f$, as well as f_x and f_y, are also employed. For digital images, the independent variables are integer valued rather than real valued, as in the case of the usual partial derivatives. Therefore, the notion of a derivative, or instantaneous rate of change, is no longer relevant. Instead, rates of change must be measured discretely over integral distances. In this section we study various image difference operators and the image gradients that result from them.

The simplest digital image difference operators are the partial difference operators DX and DY. We use the term *difference operator* without qualification to refer to these. They are respectively defined by

$$[\mathrm{DX}(f)](i, j) = \begin{cases} f(i, j) - f(i - 1, j) & \text{if } f(i, j) \neq * \text{ and} \\ & f(i - 1, j) \neq * \\ * & \text{otherwise} \end{cases}$$

and

$$[DY(f)](i,j) = \begin{cases} f(i,j) - f(i,j-1) & \text{if } f(i,j) \neq * \text{ and} \\ & f(i,j-1) \neq * \\ * & \text{otherwise} \end{cases}$$

The operators DX and DY are used to measure the digital rate of change in the horizontal and the vertical directions, respectively.

Example 2.29

Let

$$f = \begin{pmatrix} 1 & 2 & 3 & 5 \\ 3 & 3 & 4 & 5 \\ 3 & 5 & 4 & 6 \\ 4 & 5 & 6 & 7 \end{pmatrix}_{-1,2}$$

Then

$$DX(f) = \begin{pmatrix} * & 1 & 1 & 2 \\ * & 0 & 1 & 1 \\ * & 2 & -1 & 2 \\ * & 1 & 1 & 1 \end{pmatrix}_{-1,2}$$

and

$$DY(f) = \begin{pmatrix} -2 & -1 & -1 & 0 \\ 0 & -2 & 0 & -1 \\ -1 & 0 & -2 & -1 \\ * & * & * & * \end{pmatrix}_{-1,2}$$

Notice how the leftmost column in the bound matrix is lost in the application of DX and the bottom row is lost in the application of DY. Notice also that the increasing darkness of the image as one looks from left to right is reflected in the mostly positive values for DX(f), while the increasing lightness as one looks from bottom to top is reflected in the fact that there are no positive values in the bound matrix for DY(f).

Just as in the case of the usual partial derivatives, DX and DY measure change in only one direction each. Therefore, as in calculus, a gradient is introduced to rectify the situation. In calculus, the gradient ∇ is an operator that yields a vector-valued function (f_x, f_y). In the theory of digital images, the digital gradient GRAD takes one image as an input and outputs an image vector. More precisely, GRAD is defined to be the image vector

$$GRAD(f) = (DX(f), DY(f))$$

The digital gradient is sensitive to change in any direction. At a particular pixel (i, j),

$$[GRAD(f)](i,j) = ([DX(f)](i,j), [DY(f)](i,j))$$

which is a Euclidean 2-vector if both partial difference operators are defined at (i, j).

Since GRAD(f) is an image vector, we can apply each of the three image vector magnitude operators to it. This is most crucial since in image processing it is often the magnitude of a particular gradient which is important. Applying each of the three magnitude operators in turn, we obtain three new operators, each one taking an image as input and outputting an image that has at each pixel a value equal to the appropriate gradient magnitude for f at that pixel. These new gradient magnitude operators are

$$\text{GRADMAG0}(f) = \text{MAG0}[\text{GRAD}(f)]$$

$$\text{GRADMAG1}(f) = \text{MAG1}[\text{GRAD}(f)]$$

$$\text{GRADMAG2}(f) = \text{MAG2}[\text{GRAD}(f)]$$

GRADMAG0 measures the magnitude of the gradient relative to the l_∞-norm, GRADMAG1 measures it relative to the l_1-norm, and GRADMAG2 measures it relative to the l_2-norm.

Example 2.30

Let f be the image given in Example 2.29. Then

$$\text{GRAD}(f) = \left(\begin{pmatrix} * & 1 & 1 & 2 \\ * & 0 & 1 & 1 \\ * & 2 & -1 & 2 \\ * & 1 & 1 & 1 \end{pmatrix}_{-1,2}, \begin{pmatrix} -2 & -1 & -1 & 0 \\ 0 & -2 & 0 & -1 \\ -1 & 0 & -2 & -1 \\ * & * & * & * \end{pmatrix}_{-1,2} \right)$$

The reason DX(f) and DY(f) have not been put in minimal bound matrix form is to aid in the computation of the different gradient magnitudes. Now,

$$\text{GRADMAG0}(f) = \begin{pmatrix} 2 & 1 & 1 & 2 \\ 0 & 2 & 1 & 1 \\ 1 & 2 & 2 & 2 \\ * & 1 & 1 & 1 \end{pmatrix}_{-1,2}$$

$$\text{GRADMAG1}(f) = \begin{pmatrix} 2 & 2 & 2 & 2 \\ 0 & 2 & 1 & 2 \\ 1 & 2 & 3 & 3 \\ * & 1 & 1 & 1 \end{pmatrix}_{-1,2}$$

$$\text{GRADMAG2}(f) = \begin{pmatrix} 2 & \sqrt{2} & \sqrt{2} & 2 \\ 0 & 2 & 1 & \sqrt{2} \\ 1 & 2 & \sqrt{5} & \sqrt{5} \\ * & 1 & 1 & 1 \end{pmatrix}_{-1,2}$$

Two points relating to the preceding example should be mentioned. First, notice the role played by both EXTMAX and EXTADD in the respective definitions of the magnitude operators. If $[\text{DX}(f)](i, j) \neq *$ and $[\text{D}\dot{Y}(f)](i, j) = *$,

Sec. 2.12   Digital Image Gradients

then at (i, j) all the gradient magnitudes are equal to the absolute value of $DX(f)$ at (i, j). A corresponding remark holds if $DX(f)$ is star valued at (i, j) and $DY(f)$ is not. Second, at any pixel where both $DX(f)$ and $DY(f)$ are real valued,

$$[GRADMAG0(f)](i, j) = \| [GRAD(f)](i, j) \|_\infty$$

$$[GRADMAG1(f)](i, j) = \| [GRAD(f)](i, j) \|_1$$

$$[GRADMAG2(f)](i, j) = \| [GRAD(f)](i, j) \|_2$$

Of particular interest in image processing is the fact that both DX and DY can be viewed as moving-average filters. Indeed, consider the masks

$$D1 = (-1 \ \ ①)$$

and

$$D2 = \begin{pmatrix} ① \\ -1 \end{pmatrix}$$

Then

$$[FILTER(f;D1)](i,j) = [(-1) \times f(i - 1, j)] + [1 \times f(i, j)]$$

$$= f(i, j) - f(i - 1, j)$$

$$= [DX(f)](i, j)$$

Likewise,

$$[FILTER(f;D2)](i,j) = [DY(f)](i, j)$$

Each of the preceding filters is referred to as a *gradient filter*.

An obvious weakness of DX and DY is that they both measure only "one-sided" change. For instance, $[DX(f)](i, j)$ measures the change of gray value from the pixel to the left of (i, j) to (i, j) itself. In order for a difference operator to be "balanced," it should measure change symmetrically. This can be accomplished by averaging the change of gray values between $(i - 1, j)$ and (i, j) and between (i, j) and $(i + 1, j)$. In doing this, we obtain the *symmetric difference operators*

$$[SYMDX(f)](i, j) = \frac{1}{2} [f(i + 1, j) - f(i, j) + f(i, j) - f(i - 1, j)]$$

$$= \frac{1}{2} [f(i + 1, j) - f(i - 1, j)]$$

and

$$[SYMDY(f)](i, j) = \frac{1}{2} [f(i, j + 1) - f(i, j - 1)]$$

If any of three quantities $f(i - 1, j)$, $f(i, j)$, or $f(i + 1, j)$ is star valued, then so is $[SYMDX(f)](i, j)$. An analogous stipulation holds for $[SYMDY(f)](i, j)$.

The symmetric difference operators measure change "across" a pixel. Moreover, like the original difference operators, they can each be seen as filters. Indeed,

$$\text{SYMDX}(f) = \text{FILTER}\,[f;(-\tfrac{1}{2}\;\textcircled{0}\;\tfrac{1}{2})]$$

and

$$\text{SYMDY}(f) = \text{FILTER}\left[f;\begin{pmatrix}\tfrac{1}{2}\\ \textcircled{0}\\ -\tfrac{1}{2}\end{pmatrix}\right]$$

For the sake of mask simplicity, it is common practice in image processing to employ gradient masks with only integer entries. Consequently, for the symmetric difference operators, we define the masks

$$\text{SYM1} = (-1\;\;\textcircled{0}\;\;1)$$

and

$$\text{SYM2} = \begin{pmatrix}1\\ \textcircled{0}\\ -1\end{pmatrix}$$

We can then express SYMDX and SYMDY as scalar multiples of filters with the masks SYM1 and SYM2, respectively:

$$\text{SYMDX}(f) = \text{SCALAR}[\tfrac{1}{2};\,\text{FILTER}(f;\text{SYM1})]$$

$$\text{SYMDY}(f) = \text{SCALAR}[\tfrac{1}{2};\,\text{FILTER}(f;\text{SYM2})]$$

It should be recognized that the weighting factor 1/2 results from the fact that the operators derive from an average rate of change over two discrete intervals. Other weighting factors shall arise subsequently. Furthermore, if one is interested only in the relative magnitudes of SYMDX and SYMDY among the pixels of a given image, then the weighting factor can be disregarded since it operates uniformly across the image. This latter approach will be utilized in Section 2.13 on edge detection.

Given the symmetric difference operators SYMDX and SYMDY, we define the *symmetric gradient* as the image vector

$$\text{SYMGRAD}(f) = (\text{SYMDX}(f),\,\text{SYMDY}(f))$$

Like GRAD(f), SYMGRAD(f) is sensitive to change in all directions.

Since SYMGRAD(f) is an image 2-vector, each of the three magnitude operators MAG0, MAG1, and MAG2 can be applied to it. The three resulting operators are

$$\text{SYMMAG0}(f) = \text{MAG0}[\text{SYMGRAD}(f)]$$
$$\text{SYMMAG1}(f) = \text{MAG1}[\text{SYMGRAD}(f)]$$
$$\text{SYMMAG2}(f) = \text{MAG2}[\text{SYMGRAD}(f)]$$

Example 2.31

Consider the image f of Example 2.29. Then

$$\text{SYMGRAD}(f) = \left(\begin{pmatrix} * & 1 & \frac{3}{2} \\ * & \frac{1}{2} & 1 \\ * & \frac{1}{2} & \frac{1}{2} \\ * & 1 & 1 \end{pmatrix}_{-1,2} , \begin{pmatrix} * & * & * & * \\ -1 & -\frac{3}{2} & -\frac{1}{2} & -\frac{1}{2} \\ -\frac{1}{2} & -1 & -1 & -1 \end{pmatrix}_{-1,2} \right)$$

and

$$\text{SYMMAG1}(f) = \begin{pmatrix} * & 1 & \frac{3}{2} & * \\ 1 & 2 & \frac{3}{2} & \frac{1}{2} \\ \frac{1}{2} & \frac{3}{2} & \frac{3}{2} & 1 \\ * & 1 & 1 & * \end{pmatrix}_{-1,2}$$

Three other commonly employed image gradient operators shall now be introduced. Each arises from an averaging of gray level differences, each can be viewed as a scalar multiple of a moving-average filter, and each leads to three gradient magnitude operators, one each in the l_∞, the l_1, and the l_2 sense.

The symmetric difference operators average the gray value differences on the right and on the left of the pixel, and on the top and on the bottom of the pixel, respectively. For intance,

$$[\text{SYMDX}(f)](i, j) = \frac{1}{2}([\text{DX}(f)](i + 1, j) + [\text{DX}(f)](i, j))$$

The *Prewitt* difference operators go one step further in that they average gray level change over six discrete intervals. Thus, we have, in the x-direction,

$$[\text{PREWDX}(f)](i, j) = \frac{1}{6}([\text{DX}(f)](i + 1, j + 1) + [\text{DX}(f)](i, j + 1)$$
$$+ [\text{DX}(f)](i + 1, j) + [\text{DX}(f)](i, j)$$
$$+ [\text{DX}(f)](i + 1, j - 1) + [\text{DX}(f)](i, j - 1))$$
$$= \frac{1}{6}(f(i + 1, j + 1) + f(i + 1, j) + f(i + 1, j - 1)$$
$$- f(i - 1, j + 1) - f(i - 1, j) - f(i - 1, j - 1))$$

where it is assumed that all the gray values in SQUARE(i, j) are defined, or else [PREWDX(f)]$(i, j) = *$. If PREW1 is the *Prewitt mask*

$$PREW1 = \begin{pmatrix} -1 & 0 & 1 \\ -1 & ⓪ & 1 \\ -1 & 0 & 1 \end{pmatrix}$$

then

$$PREWDX(f) = SCALAR[\tfrac{1}{6};FILTER(f;PREW1)]$$

In an analogous fashion, if

$$PREW2 = \begin{pmatrix} 1 & 1 & 1 \\ 0 & ⓪ & 0 \\ -1 & -1 & -1 \end{pmatrix}$$

then the Prewitt difference operator in the y-direction is

$$PREWDY(f) = SCALAR[\tfrac{1}{6};FILTER(f;PREW2)]$$

In terms of gray values,

$$[PREWDY(f)](i, j) = \tfrac{1}{6}[f(i-1, j+1) + f(i, j+1) + f(i+1, j+1)$$
$$- f(i-1, j-1) - f(i, j-1) - f(i+1, j-1)]$$

Using the Prewitt difference operators, we define the *Prewitt gradient* operator to be the image vector

$$PREWGRAD(f) = (PREWDX(f), PREWDY(f))$$

The resulting gradient magnitude operators are given by

$$PREWMAG0(f) = MAG0[PREWGRAD(f)]$$
$$PREWMAG1(f) = MAG1[PREWGRAD(f)]$$
$$PREWMAG2(f) = MAG2[PREWGRAD(f)]$$

Example 2.32

Let f be the image given in Example 2.29. Then

$$PREWGRAD(f) = \left(\begin{pmatrix} \tfrac{2}{3} & 1 \\ \tfrac{2}{3} & \tfrac{5}{6} \end{pmatrix}_{0,1}, \begin{pmatrix} -1 & -\tfrac{5}{6} \\ -\tfrac{5}{6} & -1 \end{pmatrix}_{0,1} \right)$$

and

$$PREWMAG2(f) = \begin{pmatrix} \tfrac{\sqrt{13}}{3} & \tfrac{\sqrt{61}}{6} \\ \tfrac{\sqrt{41}}{6} & \tfrac{\sqrt{61}}{6} \end{pmatrix}_{0,1}$$

Sec. 2.12   Digital Image Gradients   **59**

The *Sobel* gradient is obtained by using a weighted average of the same gray value differences employed in the construction of the Prewitt gradient. Here, however, the intent is to "spread" the measure of change about SQUARE(i, j) and give twice as much weight to strong neighbors of (i, j). The Sobel masks are given by

$$\text{SOB1} = \begin{pmatrix} -1 & 0 & 1 \\ -2 & \textcircled{0} & 2 \\ -1 & 0 & 1 \end{pmatrix}$$

and

$$\text{SOB2} = \begin{pmatrix} 1 & 2 & 1 \\ 0 & \textcircled{0} & 0 \\ -1 & -2 & -1 \end{pmatrix}$$

The Sobel difference operators in the x- and y-directions are given respectively by

$$\text{SOBDX}(f) = \text{SCALAR}[\tfrac{1}{8}; \text{FILTER}(f; \text{SOB1})]$$

and

$$\text{SOBDY}(f) = \text{SCALAR}[\tfrac{1}{8}; \text{FILTER}(f; \text{SOB2})]$$

The Sobel gradient is given by the image vector

$$\text{SOBGRAD}(f) = (\text{SOBDX}(f), \text{SOBDY}(f))$$

Three corresponding magnitude operators, SOBMAG0, SOBMAG1, and SOBMAG2, are defined by respectively applying MAG0, MAG1, and MAG2 to SOBGRAD(f).

Example 2.33

Let f be the image given in Example 2.29. Then

$$\text{SOBGRAD}(f) = \left(\begin{pmatrix} \tfrac{5}{8} & 1 \\ \tfrac{5}{8} & \tfrac{3}{4} \end{pmatrix}_{0,1}, \begin{pmatrix} -\tfrac{9}{8} & -\tfrac{3}{4} \\ -\tfrac{7}{8} & -1 \end{pmatrix}_{0,1} \right)$$

and

$$\text{SOBMAG0}(f) = \begin{pmatrix} \tfrac{9}{8} & 1 \\ \tfrac{7}{8} & 1 \end{pmatrix}_{0,1}$$

In the definition of the Sobel gradient, the weighting factor 2 was applied to the strong neighbors. Other weighting factors could have been applied instead; of course, the resulting image gradients would then not have been called Sobel gradients. If the weighting factor is arbitrary, say λ, then the resulting gradient is given by the image vector

$$\left(\text{FILTER}\left(f;\begin{pmatrix}-1 & 0 & 1\\ -\lambda & \textcircled{0} & \lambda\\ -1 & 0 & 1\end{pmatrix}\right), \text{FILTER}\left(f;\begin{pmatrix}1 & \lambda & 1\\ 0 & \textcircled{0} & 0\\ -1 & -\lambda & -1\end{pmatrix}\right)\right)$$

scalar multiplied by $1/(4 + 2\lambda)$.

Another commonly employed image gradient is the *Roberts* gradient. This gradient is different from the others thus far introduced in that it is not composed of component images that measure change in the vertical and horizontal directions; rather, the respective components of the Roberts gradient measure change in the $-45°$ direction and the $45°$ direction. The Roberts gradient, which is generated by the Roberts masks

$$\text{ROB1} = \begin{pmatrix}-1 & 0\\ 0 & \textcircled{1}\end{pmatrix}$$

and

$$\text{ROB2} = \begin{pmatrix}0 & 1\\ -1 & \textcircled{0}\end{pmatrix}$$

is given by the image 2-vector

$$\text{ROBGRAD}(f) = \left(\text{SCALAR}\left[\frac{\sqrt{2}}{2};\text{FILTER}(f;\text{ROB1})\right],\right.$$

$$\left.\text{SCALAR}\left[\frac{\sqrt{2}}{2};\text{FILTER}(f;\text{ROB2})\right]\right)$$

Pixelwise, we have

$$[\text{ROBGRAD}(f)](i, j) = \left(\frac{\sqrt{2}}{2}[f(i, j) - f(i - 1, j + 1)],\right.$$

$$\left.\frac{\sqrt{2}}{2}[f(i, j + 1) - f(i - 1, j)]\right)$$

The occurrence of the scalar factor $\sqrt{2}/2$ results from the fact that in each case the distance between the two pixels involved in the difference is $\sqrt{2}$. In order to keep the Roberts gradient in scale with the other gradients, scalar multiplication by $\sqrt{2}/2$ is necessary.

Example 2.34

Once again, let f be the image given in Example 2.29. Then ROBGRAD(f) is given by

$$\text{ROBGRAD}(f) = \left(\begin{pmatrix}1.4 & 1.4 & 1.4\\ 1.4 & 0.7 & 1.4\\ 1.4 & 0.7 & 2.1\end{pmatrix}_{0,1}, \begin{pmatrix}-0.7 & 0 & 0.7\\ 0 & -0.7 & 0.7\\ 0.7 & -0.7 & 0\end{pmatrix}_{0,1}\right)$$

Sec. 2.12 Digital Image Gradients

and

$$\text{ROBMAG1}(f) = \begin{pmatrix} 2.1 & 1.4 & 2.1 \\ 1.4 & 1.4 & 2.1 \\ 2.1 & 1.4 & 2.1 \end{pmatrix}_{0,1}$$

Notice how the image tends to get darker as the pixels move downward in a $-45°$ direction. This is reflected by the first component image of ROBGRAD(f). By contrast, the darkness of the image tends to remain stable as the pixels move in a $+45°$ direction. This property is reflected by the second component image of ROBGRAD(f).

Before leaving this section, let us briefly mention that there is a manner in which rate of change is measured in an arbitrary direction θ. This is accomplished by the use of the *directional derivative,* which is defined for the image f in the direction θ and at the pixel (i, j) by

$$[\text{DIRECT}(f;\theta)](i, j) = [\text{DX}(f)](i, j) \cos \theta + [\text{DY}(f)](i, j) \sin \theta$$

where the quantity is considered undefined if either DX(f) or DY(f) is star valued at (i, j). In terms of GRAD(f),

$$[\text{DIRECT}(f;\theta)](i, j) = [\text{GRAD}(f)](i, j) \cdot \begin{pmatrix} \cos \theta \\ \sin \theta \end{pmatrix}$$

where the bold dot refers to the usual Euclidean dot product and the Euclidean vector ($\cos \theta$, $\sin \theta$) is written in column form. DIRECT is a digital variant of the Euclidean directional derivative for a differentiable function $f(x, y)$ of two real variables, which is given by

$$[D_\theta f](x, y) = \nabla f \cdot \begin{pmatrix} \cos \theta \\ \sin \theta \end{pmatrix}$$

where ∇f is the usual Euclidean gradient. However, the meaning of DIRECT is quite different, since a point cannot actually be moved along a directional line in the digital lattice.

By employing the operator ADD, it is easy to implement DIRECT as an image operator. Indeed, the block diagram is given in Figure 2.10.

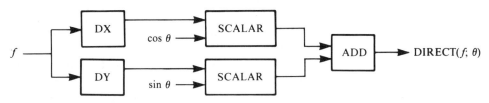

Figure 2.10 Block Diagram of DIRECT

Example 2.35

Referring to Example 2.29, where the image f was given together with $DX(f)$ and $DY(f)$, letting $\theta = 30°$, and realizing that $\cos\theta = \sqrt{3}/2$ and $\sin\theta = 1/2$, we have

$$\text{DIRECT}(f;\theta) = \begin{pmatrix} 0.366 & 0.366 & 1.732 \\ -1 & 0.866 & 0.366 \\ 1.732 & -1.866 & 1.232 \end{pmatrix}_{0,1}$$

Other directional derivatives can be defined using different image gradients, but we shall not pursue the matter any further. The methodology is totally analogous.

2.13 EDGE DETECTION BY IMAGE GRADIENTS

The organization of visual sensory data into patterns is an integral part of human perception. It is upon such patterns, rather than raw sensation, that intelligence operates. Very often these patterns consist of regions that are defined by some form of homogeneity with respect to the data. This homogeneity might result from an essentially uniform level of gray or from some textural properties of certain regions within the overall image. In either case, it might be possible to segment the image relative to some definable homogeneic characteristic. For instance, in thresholding, an effort is made to segment the image into "figure and ground" by a judicious choice of some gray level threshold value. In doing so, one is considering the degree of darkness, in a bivalent sense, as a measure of homogeneity. If an image happens to contain two figures within it which are both substantially darker than the background, then thresholding can detect both of them. However, if one of the figures is darker than the background while the other is lighter, then the single application of THRESH will not serve to detect both images simultaneously. And of course, in "real-life" images the situation is far more complex than merely the occurrence of two gray level, homogeneous figures.

In Figure 2.11, an image f and its pictorial representation are given. Clearly, the letter E is discernible to the eye; yet it would not be discovered in a thresholding operation. This is because the eye is sensitive to the "local" contrast between dark and light, discerning the E in virtue of that contrast, whereas THRESH is a "global" operator relative to gray level variation. In effect, the image of Figure 2.11 has an *illumination gradient*: it is darker on the top than on the bottom. Consequently, it is not receptive to a global operator such as THRESH.

One way to analyze differences in local contrast is through the use of *edge detection*. Intuitively, an edge is a zone of demarcation or transition between two regions that differ according to some measure of homogeneity. If one considers local gray level intensity to be such a measure, then the letter E in Figure 2.11

Sec. 2.13 Edge Detection by Image Gradients

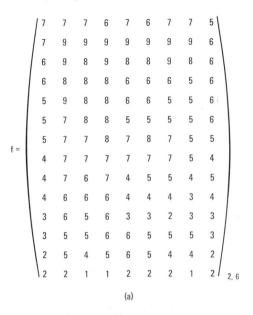

(a)

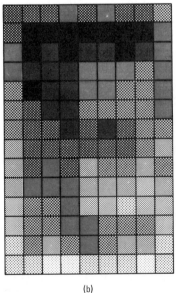

(b)

Figure 2.11 Pictorial Representation of Letter E

differs from its ground accordingly. If we focus our attention on gray level and ignore texture, then an edge may occur in essentially two ways:

1. As the border between two differing regions, each region being homogeneous with regard to some homogeneity criterion.

2. As a thin, dark arc on a light background, or a thin, light arc on a dark background.

The intention in this section is to input an image and output an "edge image," which is construed to be a thin figure that represents the edge of whatever object lies within the input image. Note that in using the term "edge," one should recognize that some latitude in meaning must be allowed. An edge might indeed be a stick figure of single black pixels on a white background; however, it might also be a zone of transition several pixels wide. The latter will often be the case, since what appears thin to the eye might actually be several pixels wide after digitization. The terminology must be sufficiently robust to allow "wide edges."

One of the most fruitful ways to produce an edge image based upon gray level homogeneity is to locate zones of rapid gray level change. This can often be done by applying some image gradient and then thresholding at an appropriate threshold value. The block diagram of Figure 2.12 indicates the methodology.

In applying the detection scheme in the figure, the goodness of the result will usually depend upon the choice of both the gradient and the threshold value t. The latter must be picked so that those pixels at which the magnitude of the gradient is great are clearly separated from those where it is not. In Section 2.16, the gray level histogram will be introduced. This will provide a simple and often satisfactory method for choosing t.

Since the gradient edge detection methodology depends only upon the relative magnitudes of the gradients within an image, the scalar multiplication by factors such as 1/6 in the Prewitt gradient and 1/8 in the Sobel gradient plays no essential role. These factors are necessary merely because of the weighting schemes introduced in the construction of the respective gradients. Consequently, instead of working with the gradient magnitude operators directly, it is common practice to forego the scalar factors. The technique is to apply a magnitude operator to the image vector

$$(\text{FILTER}(f;\text{M1}), \text{FILTER}(f;\text{M2}))$$

where M1 and M2 are the appropriate masks for the gradient operator. The output of this magnitude operation is then thresholded. Thus, instead of the scheme of Figure 2.12, we employ the strategy of Figure 2.13.

Depending upon the choice of masks and magnitude operator applied to the resulting image vector, a host of possible edge detection operators results. Figure 2.14 gives a listing of the most commonly employed operators.

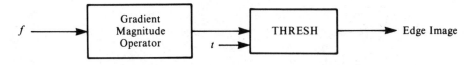

Figure 2.12 Procedure for Producing Edge Image

Sec. 2.13   Edge Detection by Image Gradients   65

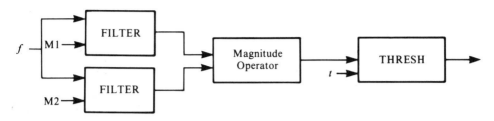

Figure 2.13   Alternate Procedure for Producing Edge Image

Though the detection scheme of Figure 2.13 does not overtly mention the gradient image vectors or the gradient magnitude images, it could have been constructed utilizing these. One would only have to multiply the gradient magnitude by the appropriate scalar factor. For instance, PREWEDGE0 is given by the block diagram

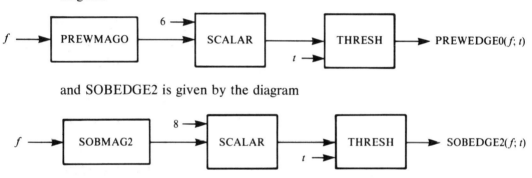

and SOBEDGE2 is given by the diagram

However, these implementations involve redundant scalar multiplication by 1/6 and 6, and 1/8 and 8, respectively. That is why the scheme of Figure 2.13 is employed instead.

Masks	Magnitude Operator	Edge Operator
D1 and D1	MAG0	GRADEDGE0
	MAG1	GRADEDGE1
	MAG2	GRADEDGE2
PREW1 and PREW2	MAG0	PREWEDGE0
	MAG1	PREWEDGE1
	MAG2	PREWEDGE2
SOB1 AND SOB2	MAG0	SOBEDGE0
	MAG1	SOBEDGE1
	MAG2	SOBEDGE2
ROB1 and ROB2	MAG0	ROBEDGE0
	MAG1	ROBEDGE1
	MAG2	ROBEDGE2

Figure 2.14   Listing of Some Edge Detection Operations

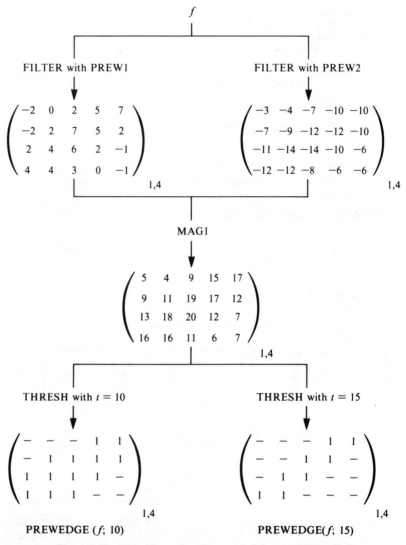

Figure 2.15 Walk-Through of Detection Scheme of Figure 2.13

Example 2.36

Let

$$f = \begin{pmatrix} 2 & 1 & 1 & 1 & 0 & 1 & 4 \\ 2 & 1 & 1 & 2 & 1 & 4 & 4 \\ 2 & 3 & 2 & 2 & 5 & 5 & 5 \\ 4 & 4 & 3 & 6 & 7 & 6 & 6 \\ 5 & 5 & 8 & 8 & 7 & 7 & 7 \\ 6 & 9 & 8 & 8 & 8 & 9 & 8 \end{pmatrix}_{0,5}$$

A walk-through of the detection scheme of Figure 2.13 using the Prewitt masks and the MAG1 operator is given in Figure 2.15. Notice how the choice of threshold value $t = 10$ results in a wider edge than the choice $t = 15$. Notice also that the original image possesses an illumination gradient in that the gray values increase as the image runs from top to bottom. Nevertheless, PREWEDGE(f;15) gives a quite good representation of the diagonal edge.

2.14 EDGE DETECTION BY COMPASS GRADIENTS

The previous section employed image gradients or slight variations thereof to locate edges. The image gradients employed difference operators to detect changes in the horizontal and vertical directions, or, in the case of the Roberts gradient, in the $-45°$ and $45°$ directions. A slightly different approach to the problem is to look for masks that respond to changes in all directions that are multiples of $45°$. In essence, the methodology is to filter by applying masks that respond well to changes in particular compass directions. Since there are eight $45°$ directions, each procedure involves filtering by eight different masks, each mask being a "$45°$ cycling" of the previous one. Once this has been accomplished, an image 8-vector is formed, and the maximum magnitude operator MAG0 is applied to that image vector. Finally, the output image of MAG0 is thresholded to obtain the desired edge image. Figure 2.16 gives the block diagram for the procedure with input masks M1, M2, . . . , M8.

As in the case of image gradient edge detection, there are a number of compass mask collections in common usage. Each such collection defines a particular compass gradient magnitude operator. The Prewitt compass gradient is defined by the masks of Figure 2.17. The Kirsh operator uses the masks of Figure 2.18. In each case, each mask is obtained from the preceding mask by a $45°$ counterclockwise cycling. Other well-known masks are the 3-level masks and the 5-level masks. The 3-level masks are generated by cycling the Prewitt mask

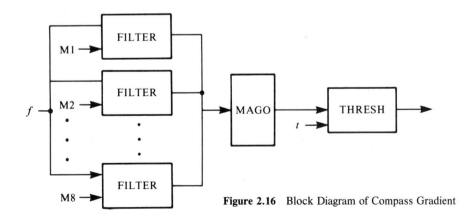

Figure 2.16 Block Diagram of Compass Gradient

$$\begin{pmatrix} -1 & 1 & 1 \\ -1 & \text{\textcircled{-2}} & 1 \\ -1 & 1 & 1 \end{pmatrix} \quad \begin{pmatrix} 1 & 1 & 1 \\ -1 & \text{\textcircled{-2}} & 1 \\ -1 & -1 & 1 \end{pmatrix} \quad \begin{pmatrix} 1 & 1 & 1 \\ 1 & \text{\textcircled{-2}} & 1 \\ -1 & -1 & -1 \end{pmatrix}$$

$$\begin{pmatrix} 1 & 1 & 1 \\ 1 & \text{\textcircled{-2}} & -1 \\ 1 & -1 & -1 \end{pmatrix} \quad \begin{pmatrix} 1 & 1 & -1 \\ 1 & \text{\textcircled{-2}} & -1 \\ 1 & 1 & -1 \end{pmatrix} \quad \begin{pmatrix} 1 & -1 & -1 \\ 1 & \text{\textcircled{-2}} & -1 \\ 1 & 1 & 1 \end{pmatrix}$$

$$\begin{pmatrix} -1 & -1 & -1 \\ 1 & \text{\textcircled{-2}} & 1 \\ 1 & 1 & 1 \end{pmatrix} \quad \begin{pmatrix} -1 & -1 & 1 \\ -1 & \text{\textcircled{-2}} & 1 \\ 1 & 1 & 1 \end{pmatrix}$$

Figure 2.17 Prewitt Masks

$$\begin{pmatrix} -5 & 3 & 3 \\ -5 & \text{\textcircled{0}} & 3 \\ -5 & 3 & 3 \end{pmatrix} \quad \begin{pmatrix} 3 & 3 & 3 \\ -5 & \text{\textcircled{0}} & 3 \\ -5 & -5 & 3 \end{pmatrix} \quad \begin{pmatrix} 3 & 3 & 3 \\ 3 & \text{\textcircled{0}} & 3 \\ -5 & -5 & -5 \end{pmatrix}$$

$$\begin{pmatrix} 3 & 3 & 3 \\ 3 & \text{\textcircled{0}} & -5 \\ 3 & -5 & -5 \end{pmatrix} \quad \begin{pmatrix} 3 & 3 & -5 \\ 3 & \text{\textcircled{0}} & -5 \\ 3 & 3 & -5 \end{pmatrix} \quad \begin{pmatrix} 3 & -5 & -5 \\ 3 & \text{\textcircled{0}} & -5 \\ 3 & 3 & 3 \end{pmatrix}$$

$$\begin{pmatrix} -5 & -5 & -5 \\ 3 & \text{\textcircled{0}} & 3 \\ 3 & 3 & 3 \end{pmatrix} \quad \begin{pmatrix} -5 & -5 & 3 \\ -5 & \text{\textcircled{0}} & 3 \\ 3 & 3 & 3 \end{pmatrix}$$

Figure 2.18 Kirsh Masks

$$\begin{pmatrix} -1 & 0 & 1 \\ -1 & \text{\textcircled{0}} & 1 \\ -1 & 0 & 1 \end{pmatrix} \quad \begin{pmatrix} 0 & 1 & 1 \\ -1 & \text{\textcircled{0}} & 1 \\ -1 & -1 & 0 \end{pmatrix} \quad \begin{pmatrix} 1 & 1 & 1 \\ 0 & \text{\textcircled{0}} & 0 \\ -1 & -1 & -1 \end{pmatrix}$$

$$\begin{pmatrix} 1 & 1 & 0 \\ 1 & \text{\textcircled{0}} & -1 \\ 0 & -1 & -1 \end{pmatrix} \quad \begin{pmatrix} 1 & 0 & -1 \\ 1 & \text{\textcircled{0}} & -1 \\ 1 & 0 & -1 \end{pmatrix} \quad \begin{pmatrix} 0 & -1 & -1 \\ 1 & \text{\textcircled{0}} & -1 \\ 1 & 1 & 0 \end{pmatrix}$$

$$\begin{pmatrix} -1 & -1 & -1 \\ 0 & \text{\textcircled{0}} & 0 \\ 1 & 1 & 1 \end{pmatrix} \quad \begin{pmatrix} -1 & -1 & 0 \\ -1 & \text{\textcircled{0}} & 1 \\ 0 & 1 & 1 \end{pmatrix}$$

Figure 2.19 Eight 3-Level Masks

PREW1, and the 5-level masks are generated by cycling the Sobel mask SOB1. The eight 3-level masks are given in Figure 2.19. Due to symmetry and the use of the absolute value in the computation of MAG0, only the first four masks need be used when applying either the 3-level or the 5-level compass gradient. In each case, filtering by the second four masks and then applying the absolute value operator gives a repetition of the outputs obtained from the first four masks.

Sec. 2.15 Best-Fit Plane

Example 2.37

Let

$$f = \begin{pmatrix} - & 2 & 2 & 2 & 2 & 2 & 2 \\ - & - & 2 & 2 & 2 & 2 & 2 \\ - & - & - & 2 & 2 & 2 & 2 \\ - & - & - & - & 2 & 2 & 2 \\ - & - & - & - & - & 2 & 2 \\ - & - & - & - & - & - & 2 \end{pmatrix}_{0,5}$$

We apply the compass gradient detection scheme of Figure 2.16 to f using the 3-level masks and the threshold value $t = 4$. Recall that only the first four masks are necessary. Denote these by L1, L2, L3, and L4, respectively. We obtain

$$\text{FILTER}(f;L1) = \begin{pmatrix} 4 & 4 & 2 & - & - \\ 2 & 4 & 4 & 2 & - \\ - & 2 & 4 & 4 & 2 \\ - & - & 2 & 4 & 4 \end{pmatrix}_{1,4}$$

$$\text{FILTER}(f;L2) = \begin{pmatrix} 6 & 6 & 2 & - & - \\ 2 & 6 & 6 & 2 & - \\ - & 2 & 6 & 6 & 2 \\ - & - & 2 & 6 & 6 \end{pmatrix}_{1,4}$$

$$\text{FILTER}(f;L3) = \begin{pmatrix} 4 & 4 & 2 & - & - \\ 2 & 4 & 4 & 2 & - \\ - & 2 & 4 & 4 & 2 \\ - & - & 2 & 4 & 4 \end{pmatrix}_{1,4}$$

$$\text{FILTER}(f;L4) = \begin{pmatrix} - & - & - & - & - \\ - & - & - & - & - \\ - & - & - & - & - \\ - & - & - & - & - \end{pmatrix}_{1,4}$$

Applying MAG0 followed by THRESH with $t = 4$ yields the output

$$\begin{pmatrix} 1 & 1 & - & - & - \\ - & 1 & 1 & - & - \\ - & - & 1 & 1 & - \\ - & - & - & 1 & 1 \end{pmatrix}_{1,4}$$

One should use the compass gradient technique with great caution: the technique is highly heuristic and should not be employed imprudently.

2.15 BEST-FIT PLANE

In elementary calculus one may find the tangent plane to the surface (x, y, z), where $z = f(x, y)$ is a differentiable function of two real variables. At the point

(x', y', z'), the tangent plane is given by the equation

$$z = z' + f_x(x', y')(x - x') + f_y(x', y')(y - y')$$
$$= f(x', y') + [\nabla f](x', y') \cdot \begin{pmatrix} x - x' \\ y - y' \end{pmatrix}$$

where the bold dot refers to the usual Euclidean dot product.

The problem in the digital case is somewhat different. Since an image $f(i, j)$ is defined only on the discrete points of the grid, the notion of tangent plane does not apply. Instead, we attempt to find a *best-fit plane*. The problem is to find a plane

$$z = f(i, j) + A(x - i) + B(y - j)$$

that fits the image "best" at the point (i, j) in the lattice. By this, we mean that

1. $z(i, j) = f(i, j)$
2. Relative to some criterion, the plane z "fits the image f the best near (i, j)."

Put another way, the plane z passes through the point $(i, j, f(i, j))$, and near that point it deviates minimally from the image relative to some criterion.

The criterion for fit that we shall employ is the *least-squares criterion*, according to which the sum of the squares of the difference between f and z is minimized over some neighborhood of pixels containing (i, j). The choice of neighborhood will determine the constants A and B. To begin with, we consider the square neighborhood SQUARE(i, j) and we minimize $[f - z]^2$ over that neighborhood. More precisely, we minimize the quantity

$$q = \sum_{(u,v) \in \text{SQUARE}(i,j)} [f(u, v) - z(u, v)]^2$$

in order to determine the best-fit constants A and B. We have

$$q = [f(i + 1, j + 1) - f(i, j) - A - B]^2 + [f(i, j + 1) - f(i, j) - B]^2$$
$$+ [f(i - 1, j + 1) - f(i, j) + A - B]^2$$
$$+ [f(i - 1, j) - f(i, j) + A]^2$$
$$+ [f(i - 1, j - 1) - f(i, j) + A + B]^2$$
$$+ [f(i, j - 1) - f(i, j) + B]^2$$
$$+ [f(i + 1, j - 1) - f(i, j) - A + B]^2$$
$$+ [f(i + 1, j) - f(i, j) - A]^2$$

Sec. 2.15   Best-Fit Plane

To minimize q, we apply max–min theory from calculus to $q(A, B)$. Proceeding,

$$\frac{\partial q}{\partial A} = -2[f(i+1,j+1) - f(i,j) - A - B] + 2[f(i-1,j+1) - f(i,j) + A - B]$$

$$+ 2[f(i-1,j) - f(i,j) + A] + 2[f(i-1,j-1) - f(i,j) + A + B]$$

$$- 2[f(i+1,j-1) - f(i,j) - A + B] - 2[f(i+1,j) - f(i,j) - A]$$

$$= -2[f(i+1,j+1) + f(i+1,j) + f(i+1,j-1)]$$

$$+ 2[f(i-1,j+1) + f(i-1,j) + f(i-1,j-1)] + 12A$$

Setting $\partial q/\partial A = 0$ yields

$$A = \frac{1}{6}[f(i+1,j+1) + f(i+1,j) + f(i+1,j-1)$$

$$- f(i-1,j+1) - f(i-1,j) - f(i-1,j-1)]$$

$$= \frac{1}{6}[\text{FILTER}(f;\text{PREW1})](i,j)$$

$$= [\text{PREWDX}(f)](i,j)$$

where PREW1 is the first Prewitt mask and PREWDX is the Prewitt difference operator in the x-direction. A similar calculation shows that the only solution of the equation $\partial q/\partial B = 0$ is given by

$$B = \frac{1}{6}[f(i+1,j+1) + f(i,j+1) + f(i-1,j+1)$$

$$- f(i-1,j-1) - f(i,j-1) - f(i+1,j-1)]$$

$$= \frac{1}{6}[\text{FILTER}(f;\text{PREW2})](i,j)$$

$$= [\text{PREWDY}(f)](i,j)$$

where PREW2 is the second Prewitt mask and PREWDY is the Prewitt difference operator in the y-direction. Since the second-order partials with respect to A and B satisfy

$$\frac{\partial^2 q}{\partial A^2} = \frac{\partial^2 q}{\partial B^2} = 12$$

and since the mixed second partials are zero, the discriminant

$$q_{AA} q_{BB} - q_{AB}^2 = 144$$

which is greater than zero. Hence, q is minimized at

$$A = [PREWDX(f)](i, j)$$
$$B = [PREWDY(f)](i, j)$$

Consequently, the least-squares best-fit plane over SQUARE(i, j) is given by

$$z = f(i, j) + [PREWDX(f)](i, j) \cdot (x - i) + [PREWDY(f)](i, j) \cdot (y - j)$$
$$= f(i, j) + [PREWGRAD(f)](i, j) \cdot \begin{pmatrix} x - i \\ y - j \end{pmatrix}$$

where the bold dot represents the usual Euclidean dot product. Put into words, for the least-squares best-fit plane over SQUARE(i, j), the Prewitt gradient plays the role that ∇f plays for the tangent plane.

Example 2.38

Let

$$f = \begin{pmatrix} 0 & 2 & 4 & 1 & 3 \\ 2 & -1 & -3 & 2 & 2 \\ -3 & 0 & 2 & 1 & 0 \\ 1 & 0 & 2 & 3 & 7 \\ 2 & 1 & 2 & 3 & 1 \end{pmatrix}_{0,4}$$

For the least-squares best-fit plane at the pixel $(2, 1)$,

$$A = [PREWDX(f)](2, 1)$$
$$= \frac{1}{6} \text{DOT} \left[\begin{pmatrix} -1 & 0 & 1 \\ -1 & 0 & 1 \\ -1 & 0 & 1 \end{pmatrix}_{1,2}, \begin{pmatrix} 0 & 2 & 1 \\ 0 & 2 & 3 \\ 1 & 2 & 3 \end{pmatrix}_{1,2} \right] = 1$$

Similarly, $B = [PREWDY(f)](2, 1) = -1/2$. Therefore,

$$z = f(2, 1) + \begin{pmatrix} 1 \\ -\frac{1}{2} \end{pmatrix} \cdot \begin{pmatrix} x - 2 \\ y - 1 \end{pmatrix}$$
$$= 2 + (x - 2) - \frac{1}{2}(y - 1)$$

The least-squares best-fit plane can be varied by changing the pixel neighborhood over which the fit is measured. Suppose, for instance, that we only want to test for fit at the strong neighbors of (i, j). Then

$$q = [f(i, j + 1) - f(i, j) - B]^2 + [f(i, j - 1) - f(i, j) + B]^2$$
$$+ [f(i - 1, j) - f(i, j) + A]^2 + [f(i + 1, j) - f(i, j) - A]^2$$

Sec. 2.15 Best-Fit Plane

Straightforward calculation yields

$$\frac{\partial q}{\partial A} = 2[f(i-1,j) - f(i+1,j)] + 4A$$

$$\frac{\partial q}{\partial B} = 2[f(i,j-1) - f(i,j+1)] + 4B$$

Setting the partial derivatives equal to zero shows that the best fit is when

$$A = \frac{1}{2}[f(i+1,j) - f(i-1,j)]$$

$$= \frac{1}{2}[\text{FILTER}(f;\text{SYM1})](i,j)$$

$$= [\text{SYMDX}(f)](i,j)$$

and

$$B = \frac{1}{2}[f(i,j+1) - f(i,j-1)]$$

$$= \frac{1}{2}[\text{FILTER}(f;\text{SYM2})](i,j)$$

$$= [\text{SYMDY}(f)](i,j)$$

where SYMDX and SYMDY are the symmetric difference operators in the x- and y-directions, respectively. As a result, the best-fit plane over the strong neighbors of (i, j) is given by

$$z = f(i,j) + [\text{SYMGRAD}(f)](i,j) \cdot \begin{pmatrix} x - i \\ y - j \end{pmatrix}$$

where SYMGRAD is the symmetric gradient. Here, SYMGRAD(f) plays the role that ∇f plays in the continuous case.

One might pose the question as to whether or not the digital gradient GRAD(f) might appear in a best-fit-plane equation in a manner analogous to ∇f. The answer is yes: simply apply the least squares criterion over the neighborhood $\{(i-1, j), (i, j), (i, j-1)\}$. For this neighborhood, one obtains, by max–min theory, the coefficients

$$A = f(i,j) - f(i-1,j) = [\text{DX}(f)](i,j)$$

and

$$B = f(i,j) - f(i,j-1) = [\text{DY}(f)](i,j)$$

Consequently, the best-fit plane is given by

$$z = f(i,j) + [\text{GRAD}(f)](i,j) \cdot \begin{pmatrix} x - i \\ y - j \end{pmatrix}$$

In the definition of the least-squares best fit, the squares of the "errors" between the plane and the image were summed over some neighborhood of the pixel under consideration. This characterization of fit may be altered in a number of ways. One way is to weight the error terms relating to the strong neighbors of the central pixel (i, j). Assuming that the fit is to be taken over $\text{SQUARE}(i, j)$, and assuming that the strong neighbor terms in the least squares sum are to be weighted by 2, then the original evaluation of q given in the beginning of the section for the fit over $\text{SQUARE}(i, j)$ is altered to the extent that the second, fourth, sixth, and eighth summands are each multiplied by a factor of 2. Consequently, the third and sixth summands in the evaluation of $\partial q/\partial A$ have factors of 4 rather than 2. Hence,

$$\frac{\partial q}{\partial A} = -2[f(i + 1, j + 1) + 2f(i + 1, j) + f(i + 1, j - 1)]$$

$$+ 2[f(i - 1, j + 1) + 2f(i - 1, j) + f(i - 1, j - 1)]$$

$$+ 16A$$

Setting $\partial q/\partial A = 0$ yields

$$A = \frac{1}{8}[f(i + 1, j + 1) + 2f(i + 1, j) + f(i + 1, j - 1)$$

$$- f(i - 1, j + 1) - 2f(i - 1, j) - f(i - 1, j - 1)]$$

$$= \frac{1}{8}[\text{FILTER}(f; \text{SOB1})](i, j)$$

$$= [\text{SOBDX}(f)](i, j)$$

Similarly, solving $\partial q/\partial B = 0$ yields $B = [\text{SOBDY}(f)](i, j)$. Hence, the best-fit plane in this case is given by

$$z = f(i, j) + [\text{SOBGRAD}(f)](i, j) \cdot \begin{pmatrix} x - i \\ y - j \end{pmatrix}$$

and the Sobel gradient plays the role of ∇f.

The choice of the weighting factor on the strong neighbors need not be equal to 2. Suppose each strong neighbor term in the least-squares best-fit sum is weighted by some number $\lambda > 0$. Then the preceding method yields

$$A = \frac{1}{(4 + 2\lambda)} \left[\text{FILTER}\left(f; \begin{pmatrix} -1 & 0 & 1 \\ -\lambda & \textcircled{0} & \lambda \\ -1 & 0 & 1 \end{pmatrix}\right) \right](i, j)$$

$$B = \frac{1}{(4 + 2\lambda)} \left[\text{FILTER}\left(f; \begin{pmatrix} 1 & \lambda & 1 \\ 0 & \textcircled{0} & 0 \\ -1 & -\lambda & -1 \end{pmatrix}\right) \right](i, j)$$

But these are precisely the pixelwise components of the gradient that was arrived at in Section 2.12 when the partial difference operators were arrived at by using the masks

$$\begin{pmatrix} -1 & 0 & 1 \\ -\lambda & \textcircled{0} & \lambda \\ -1 & 0 & 1 \end{pmatrix} \quad \text{and} \quad \begin{pmatrix} 1 & \lambda & 1 \\ 0 & \textcircled{0} & 0 \\ -1 & -\lambda & -1 \end{pmatrix}$$

In other words, the case in which the best-fit plane yields the Sobel gradient is just a special case of a weighting scheme that will yield a best-fit plane with a gradient induced by the preceding masks with $\lambda = 2$.

2.16 GRAY LEVEL HISTOGRAM

Very often a great deal of useful information is obtained by examining the distribution of gray values within a digital image. A first-order approach to gray level distribution is through the gray level histogram.

Let us suppose that there is a discrete gray value range $\{0, 1, 2, \ldots, 2^k - 1\}$, where k is a positive integer. Then $[\text{HIST}(f)](y)$ is the number of pixels in the image f which have the gray value y. HIST is a unary operator with an image input and an array output given by

$$[[\text{HIST}(f)](0), [\text{HIST}(f)](1), \ldots, [\text{HIST}(f)](2_k - 1)]$$

It is also possible to view the output of HIST as a 1 by $2^k - 1$ digital image:

$$([\text{HIST}(f)](0) \quad [\text{HIST}(f)](1) \ldots [\text{HIST}(f)](2^k - 1))_{0,0}$$

The latter approach is useful for implementation purposes; indeed, the specification of HIST, using the image-to-image approach, is given in Figure 2.20.

Once $\text{HIST}(f)$ has been computed, it can be graphed in a manner similar to any other discrete histogram. Letting f be the image given in Example 2.17, and assuming a gray level quantization of $\{0, 1, \ldots, 15\}$, the gray level histogram, in graphical form, is given in Figure 2.21. In the figure, there are clearly three

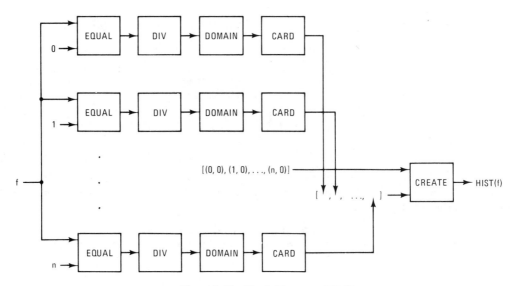

Figure 2.20 Block Diagram of HIST

distinct clusters of gray values; thus, the histogram is trimodal. Descriptively, one might say that HIST(f) consists of three peaks separated by two valleys. Very often, but certainly not always, separated clusters of gray result from distinct features or figures within an image. Therefore, it is often fruitful to apply THRESH with t identical to some value of y within a valley between two gray level peaks. For instance, choices of $t = 4$ and $t = 10$ result in the two thresholded black-and-white images given in Figure 2.22. This technique must be utilized with care since very often separated peaks within a gray level histogram result not from distinct gray level figures but rather from high-frequency gray value fluctuation due to textural properties within the image. In the latter instance, the choice of t based upon a valley within the histogram may simply result in a high-frequency

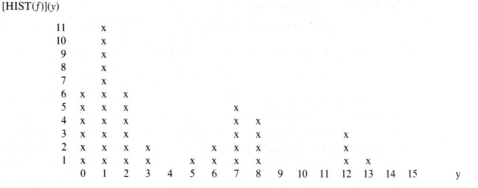

Figure 2.21 Gray Level Histogram

Sec. 2.16   Gray Level Histogram

$$\text{THRESH}(f;4) = \begin{pmatrix} - & - & - & - & - & - & - \\ - & - & 1 & 1 & 1 & 1 & * \\ - & - & 1 & 1 & 1 & 1 & - \\ - & - & 1 & 1 & 1 & 1 & - \\ - & - & 1 & 1 & 1 & 1 & - \\ - & - & - & - & - & - & - \end{pmatrix}_{2,6}$$

$$\text{THRESH}(f;10) = \begin{pmatrix} - & - & - & - & - & - & - \\ - & - & - & - & - & - & * \\ - & - & - & 1 & 1 & - & - \\ - & - & - & 1 & 1 & - & - \\ - & - & - & - & - & - & - \\ - & - & - & - & - & - & - \end{pmatrix}_{2,6}$$

Figure 2.22 Thresholded Images

black-and-white image that may very likely lack the fundamental textural distinctions of the original.

One frequently employed application of the aforementioned technique is in gradient, or compass gradient, edge detection methodology. In the scheme of Figure 2.13, for example, a threshold value t must be chosen in order to produce the edge image from the gradient magnitude image. Application of HIST to the output of the image vector magnitude operator can be very useful in determining the choice of t which should be input into THRESH.

Another common application of the gray level histogram involves the enhancement of contrast within a digital image. Suppose, for instance, f is an image whose gray values are skewed toward the light end of the gray scale. Then it is possible that there are features within the image which could be more readily discernible had the entire gray spectrum been utilized. *Histogram equalization* is a method by which the gray values within the image are rescaled in order to take advantage of the entire quantization scale. The gray values are spread evenly over the gray scale in order to make uniform use of all quantization levels. This revaluation is done in a monotonic fashion. The darkest of the original pixels are reassigned to the darkest quantized levels insofar as those levels have the capacity to accept them. Lighter pixels are reassigned to lighter quantized levels. The following example illustrates the procedure.

Example 2.39

Let

$$f = \begin{pmatrix} 3 & 0 & 1 & 2 & 1 & 0 & 1 \\ 1 & 1 & 2 & 3 & 2 & 2 & 1 \\ 1 & 2 & 3 & 4 & 3 & 2 & 2 \\ 2 & 3 & 3 & 3 & 5 & 4 & 2 \\ 2 & 2 & 2 & 1 & 0 & 1 & 3 \end{pmatrix}_{0,4}$$

Assuming a gray scale running from 0 to 7, absolute equalization would require that each gray value be assigned

$$Q = \frac{\text{CARD}(f)}{8} = \frac{35}{8} = 4.375$$

pixels, where CARD(f) is the number of pixels in the domain of f and 8 is the number of quantization levels. Since in this case Q is not an integer, eight classes of pixels, $C_0, C_1, \ldots, C_7$, are formed in such manner that C_0 contains the four lightest pixels, C_1 contains the next five lightest, C_2 contains the next four lightest, etc., the purpose being to divide up the 35 domain pixels as evenly as possible. The immediate problem is that for C_0 to contain four pixels it must contain one pixel with gray value 1. Though one might wish to utilize prior knowledge in making the selection, it is most straightforward to choose the last pixel in class C_0 by some randomization routine applied to all 1-valued pixels. Randomization tends to minimize, in a statistical sense, the "error" introduced by making pixels that originally possessed the same gray level have different gray levels after equalization. Assuming, in this example, that pixel (1, 3) is chosen by randomization, then

$$C_0 = \{(1, 4), (1, 3), (4, 0), (5, 4)\}$$

Now, since class C_1 is supposed to contain five pixels, and since there remain to be assigned eight pixels having gray value 1, five of the eight must be chosen by randomization. Assuming this to have been done, C_1 might be

$$C_1 = \{(0, 2), (2, 4), (4, 4), (5, 0), (6, 4)\}$$

The process of filling the classes is continued, the randomization routine being employed whenever necessary. When the process has been completed and all eight classes are filled, then a new image g is created such that $g(i, j) = y$ if and only if (i, j) is an element of C_y. A possible output of the histogram equalization process might be

$$g = \begin{pmatrix} 6 & 0 & 1 & 3 & 1 & 0 & 1 \\ 2 & 0 & 4 & 6 & 4 & 3 & 2 \\ 1 & 3 & 6 & 7 & 5 & 3 & 5 \\ 5 & 7 & 6 & 6 & 7 & 7 & 4 \\ 4 & 2 & 4 & 2 & 0 & 1 & 5 \end{pmatrix}_{0,4}$$

Notice that the somewhat darker triangular figure in the middle of the original image possesses a greater contrast with its background in the equalized image g. In array form,

$$\text{HIST}(f) = [3, 9, 12, 8, 2, 1, 0, 0]$$

which is markedly skewed to the left, while

$$\text{HIST}(g) = [4, 5, 4, 4, 5, 4, 5, 4]$$

In applying histogram equalization, a couple of points should be kept in mind. First, if the original histogram happens to be strongly bimodal because it contains a distinct figure and background, then that distinction can be made less marked by histogram equalization, and the figure might tend to "fade out" into its background. Consequently, if a figure is quite intricate with respect to its ground, that intricacy might become muted and its fragile characteristics lost. Second, histogram equilization, as just described, is not deterministic. Due to the employment of a randomization procedure, the output image might be any one of a number of different images. This indeterminism will likely be of no consequence in applications where equalization is clearly advantageous.

2.17 CO-OCCURRENCE MATRIX

While the gray level histogram gives information concerning the overall gray level distribution, it provides no information regarding the manner in which gray levels are related to one another throughout the image. That is, it does not address the question of spatial relationships between gray levels. Yet it is precisely such relationships that result in the perceptual notion of texture. The so-called *co-occurrence matrices* have proven to be a valuable tool in the quantification of textural information. These matrices focus on the spatial relationships between gray levels through the calculation of second-order gray level distributions.

There are many co-occurrence matrices for a given image. Each is determined by some prespecified spatial relationship between pairs of pixels. For instance, pixel (i, j) might be related to pixel (i', j') if (i', j') is the right strong neighbor of (i, j). Once such a relation, say, R, is established, it is used to construct a particular co-occurrence matrix for an image f in the following manner. For any pair of gray levels r and s, $Q_R(r, s)$ is the number of pixel pairs (i, j) and (i', j') for which

1. $(i, j)R(i', j')$ ((i, j) is R-related to (i', j'))
2. $f(i, j) = r$ and $f(i', j') = s$.

Assuming the total gray level range to be $\{0, 1, \ldots, m\}$, the co-occurrence matrix $C[f;R]$ is defined by

$$C[f;R] = \begin{pmatrix} Q_R(0,0) & Q_R(0,1) & \cdots & Q_R(0,m) \\ Q_R(1,0) & Q_R(1,1) & \cdots & Q_R(1,m) \\ \vdots & \vdots & & \vdots \\ Q_R(m,0) & Q_R(m,1) & \cdots & Q_R(m,m) \end{pmatrix}$$

Although there are many possible relations R that might be defined among pixel pairs, the most commonly employed are the eight *adjacency*, or neighbor, relations. These are induced by the eight noncentric pixels of SQUARE(i, j). For $k = 1, 2, \ldots, 8$, the relation R_k is defined by $(i, j)R_k(i', j')$ if (i', j') occupies the kth position as one proceeds counterclockwise around the outside pixels of SQUARE(i, j) starting from the upper right-hand corner pixel $(i + 1, j + 1)$. For example, $(i, j)R_1(i + 1, j + 1)$ and $(i, j)R_4(i - 1, j)$. For notational clarity, we will let Q_k denote the entries in the co-occurrence matrix induced by R_k and $C[f;k]$ denote the matrix itself.

Example 2.40

Assuming the gray level quantization {0, 1, 2, 3, 4}, let

$$f = \begin{pmatrix} 4 & 0 & 4 & 0 & 4 & 1 \\ 0 & 3 & 0 & 4 & 1 & 3 \\ 4 & 1 & 4 & 1 & 3 & 0 \\ 0 & 4 & 0 & 3 & 0 & 3 \end{pmatrix}_{2,5}$$

Then the co-occurrence matrix for the relation R_8, "(i',j') is the right strong neighbor of (i,j)," is given by

$$C[f;8] = \begin{pmatrix} 0 & 0 & 0 & 3 & 4 \\ 0 & 0 & 0 & 2 & 1 \\ 0 & 0 & 0 & 0 & 0 \\ 3 & 0 & 0 & 0 & 0 \\ 3 & 4 & 0 & 0 & 0 \end{pmatrix}$$

Moreover, using the relation R_2, "is the strong neighbor above," we obtain

$$C[f;2] = \begin{pmatrix} 0 & 0 & 0 & 2 & 4 \\ 0 & 0 & 0 & 1 & 2 \\ 0 & 0 & 0 & 0 & 0 \\ 2 & 3 & 0 & 0 & 0 \\ 3 & 1 & 0 & 0 & 0 \end{pmatrix}$$

Notice that in both of the co-occurrence matrices the entries are off the main diagonal. This reflects the fact that there is high-frequency gray level fluctuation in both the horizontal and vertical directions. On the other hand, consider the relation R_1, "(i',j') is to the upper right of (i,j)." The corresponding co-occurrence matrix is

$$C[f;1] = \begin{pmatrix} 3 & 2 & 0 & 0 & 0 \\ 1 & 2 & 0 & 0 & 0 \\ 0 & 0 & 0 & 0 & 0 \\ 0 & 0 & 0 & 2 & 1 \\ 0 & 0 & 0 & 1 & 3 \end{pmatrix}$$

In this instance, the entries are clustered along the main diagonal. This distribution reflects the fact that there is very little gray level fluctuation along 45° lines of pixels within the image.

There is generally a good deal of textural information contained within co-occurrence matrices. Unfortunately, the computation time and storage demands of these matricies are both high. Therefore, one is usually restricted to examining only the eight adjacency relations, or even some subset of these, such as the strong neighbor relations. Even in the latter case, some further compression might be desirable. For instance, one might average the corresponding entries in the four strong neighbor matrices and store only the matrix of averages.

2.18 BEST QUADRIC FIT

In Section 2.15 several least-squares best-fit planes were developed according to the choice of the neighborhood over which the fit was to be measured. Taken from the general perspective of continuous image construction from a digital image, the best-fit plane gives a local Euclidean image—approximation of the given digital image. The reason for the presentation in Section 2.15 was to demonstrate the manner in which different digital image gradients arise naturally from the solution of the best-fit-plane problem.

In this section, a best-fit-quadric-surface problem will be solved. Since a quadric surface is a more general surface than a plane, the solution to be found herein will give a closer local fit to the digital image than the corresponding best-fit plane. Two points should be noted at the outset:

1. Surfaces other than quadric surfaces may give a better Euclidean image construction.
2. We shall not consider the most general quadric surface, but rather, one in which the variable z is isolated from the other variables.

As in the case of the best-fit plane, a least-squares criterion will be used to measure fit. Moreover, the only neighborhood that will be employed to make the fit will be SQUARE(i, j).

At a pixel (i, j) for which SQUARE(i, j) is in the domain of the digital image, the problem is to find the constants A, B, C, D, and E such that

$$q = \sum_{(u,v) \in \text{SQUARE}(i,j)} [f(u, v) - z(u, v)]^2$$

is minimized, where $z(x, y)$ is the quadric surface given by

$$z = z_0 + A(x - i)^2 + B(y - j)^2 + C(x - i) + D(y - j) + E(x - i)(y - j)$$

with $z_0 = f(i, j)$ being the gray value at (i, j). The solution is given by finding the partial derivatives of q with respect to A, B, C, D, and E, and then setting each equal to zero. In order to simplify notation, we will temporarily employ the notation $f(\mu, \nu)$ to denote $f(i + \mu, j + \nu)$. For instance, $f(1,1)$ denotes $f(i + 1, j + 1)$ and $f(0,1)$ denotes $f(i, j + 1)$.

To begin, since (x, y) must vary over SQUARE(i, j),

$$q = [f(1,1) - z_0 - A - B - C - D - E]^2$$
$$+ [f(0,1) - z_0 - B - D]^2 + [f(-1,1) - z_0 - A - B + C - D + E]^2$$
$$+ [f(-1,0) - z_0 - A + C]^2 + [f(-1,-1) - z_0 - A - B + C + D - E]^2$$
$$+ [f(0,-1) - z_0 - B + D]^2 + [f(1,-1) - z_0 - A - B - C + D + E]^2$$
$$+ [f(1,0) - z_0 - A - C]^2$$

Taking the partial of q with respect to E yields

$$\frac{\partial q}{\partial E} = -2[f(1,1) - z_0 - A - B - C - D - E]$$
$$+ 2[f(-1,1) - z_0 - A - B + C - D + E]$$
$$- 2[f(-1,-1) - z_0 - A - B + C + D - E]$$
$$+ 2[f(1,-1) - z_0 - A - B - C + D + E]$$

Setting the expression for $\partial q/\partial E$ equal to zero, we obtain

$$-f(1,1) + f(-1,1) - f(-1,-1) + f(1,-1) + 4E = 0$$

Solving for E, we get

$$E = \frac{1}{4}[f(1,1) + f(-1,-1) - f(-1,1) - f(1,-1)]$$

Now we take the partial derivative of q with respect to C:

$$\frac{\partial q}{\partial C} = -2[f(1,1) - z_0 - A - B - C - D - E]$$
$$+ 2[f(-1,1) - z_0 - A - B + C - D + E]$$
$$+ 2[f(-1,0) - z_0 - A + C] + 2[f(-1,-1) - z_0 - A - B + C + D - E]$$
$$- 2[f(1,-1) - z_0 - A - B - C + D + E] - 2[f(1,0) - z_0 - A - C]$$

Setting the expression for $\partial q/\partial C$ equal to zero, we obtain

$$-f(1,1) + f(-1,1) + f(-1,0) + f(-1,-1) - f(1,-1) - f(1,0) + 6C = 0$$

Therefore,

$$C = \frac{1}{6}[f(1,1) + f(1,0) + f(1,-1) - f(-1,1) - f(-1,0) - f(-1,-1)]$$

Sec. 2.18   Best Quadric Fit   83

As for the partial with respect to D, we have

$$\frac{\partial q}{\partial D} = -2[f(1,1) - z_0 - A - B - C - D - E] - 2[f(0,1) - z_0 - B - D]$$

$$- 2[f(-1,1) - z_0 - A - B + C - D + E]$$

$$+ 2[f(-1,-1) - z_0 - A - B + C + D - E] + 2[f(0,-1) - z_0 - B + D]$$

$$+ 2[f(1,-1) - z_0 - A - B - C + D + E]$$

Setting the expression for $\partial q/\partial D$ equal to zero gives

$$-f(1,1) - f(0,1) - f(-1,1) + f(-1,-1) + f(0,-1) + f(1,-1) + 6D = 0$$

Hence,

$$D = \frac{1}{6}[f(1,1) + f(0,1) + f(-1,1) - f(-1,-1) - f(0,-1) - f(1,-1)]$$

Proceeding with the partial of q with respect to A, we have

$$\frac{\partial q}{\partial A} = -2[f(1,1) - z_0 - A - B - C - D - E]$$

$$- 2[f(-1,1) - z_0 - A - B + C - D + E] - 2[f(-1,0) - z_0 - A + C]$$

$$- 2[f(-1,-1) - z_0 - A - B + C + D - E]$$

$$- 2[f(1,-1) - z_0 - A - B - C + D + E] - 2[f(1,0) - z_0 - A - C]$$

Setting the expression for $\partial q/\partial A$ equal to zero gives

$$-f(1,1) - f(-1,1) - f(-1,0) - f(-1,-1)$$
$$- f(1,-1) - f(1,0) + 6z_0 + 6A + 4B = 0 \quad (1)$$

Since the solution of $\partial q/\partial A = 0$ has resulted in an equation involving two unknowns, A and B, we proceed to the solution of $\partial q/\partial B = 0$. To that end,

$$\frac{\partial q}{\partial B} = -2[f(1,1) - z_0 - A - B - C - D - E] - 2[f(0,1) - z_0 - B - D]$$

$$- 2[f(-1,1) - z_0 - A - B + C - D + E]$$

$$- 2[f(-1,-1) - z_0 - A - B + C + D - E]$$

$$- 2[f(0,-1) - z_0 - B + D] - 2[f(1,-1) - z_0 - A - B - C + D + E]$$

Setting this expression equal to zero gives

$$-f(1,1) - f(0,1) - f(-1,1) - f(-1,-1) - f(0,-1) - f(1,-1)$$
$$+ 6z_0 + 4A + 6B = 0 \quad (2)$$

Equations (1) and (2) are two equations in the two unknowns A and B. Multiplying equation (1) by two and equation (2) by three and then subtracting (2) from (1) gives

$$f(1,1) + f(-1,1) + 3f(0,1) - 2f(-1,0) + f(-1,-1) + f(1,-1)$$
$$+ 3f(0,-1) - 2f(1,0) - 6z_0 - 10B = 0$$

Consequently,

$$B = \frac{1}{10}\{f(1,1) + f(-1,1) + f(-1,-1)$$
$$+ f(1,-1) + 3f(0,-1) + 3f(0,1) - 2f(-1,0) - 2f(1,0) - 6z_0\}$$

Now subtract equation (1) from equation (2) to obtain

$$-f(0,1) - f(0,-1) + f(-1,0) + f(1,0) - 2A + 2B = 0$$

so that

$$A = B + \frac{1}{2}[f(-1,0) + f(1,0) - f(0,1) - f(0,-1)]$$

$$= \frac{1}{10}\{f(1,1) + f(-1,1) + f(-1,-1) + f(1,-1)$$

$$- 2f(0,1) - 2f(0,-1) + 3f(-1,0) + 3f(1,0) - 6z_0\}$$

The equation for the best-fit quadric surface is now obtained by putting the solutions for A, B, C, D, and E back into the original equation for z. However, before doing this, it should be noted that each constant can be written as a scalar times $[\text{FILTER}(f;M)](i, j)$, where M is an appropriate mask. To that end, define the masks

$$\text{QUAD1} = \begin{pmatrix} 1 & -2 & 1 \\ 3 & -6 & 3 \\ 1 & -2 & 1 \end{pmatrix}$$

$$\text{QUAD2} = \begin{pmatrix} 1 & 3 & 1 \\ -2 & -6 & -2 \\ 1 & 3 & 1 \end{pmatrix}$$

$$\text{QUAD3} = \begin{pmatrix} -1 & 0 & 1 \\ 0 & 0 & 0 \\ 1 & 0 & -1 \end{pmatrix}$$

Sec. 2.18   Best Quadric Fit

Then

$$z = \frac{1}{10} [\text{FILTER}(f;\text{QUAD1})](i, j)(x - i)^2$$

$$+ \frac{1}{10} [\text{FILTER}(f;\text{QUAD2})](i, j)(y - j)^2$$

$$+ \frac{1}{6} [\text{FILTER}(f;\text{PREW1})](i, j)(x - i)$$

$$+ \frac{1}{6} [\text{FILTER}(f;\text{PREW2})](i, j)(y - j)$$

$$+ \frac{1}{4} [\text{FILTER}(f;\text{QUAD3})](i, j)(x - i)(y - j)$$

where PREW1 and PREW2 are the Prewitt masks.

It is of interest to note that, just like the Prewitt and Sobel masks, the quadratic masks arise from (weighted) average rates of change over unit digital intervals (Section 2.15). The difference, however, is that the rates involved in the quadratic masks are directional; for instance,

$$\frac{1}{10}[\text{FILTER}(f;\text{QUAD1})](i,j) = \frac{1}{10}([\text{DX}(f)](i + 1, j + 1) - [\text{DX}(f)](i, j + 1)$$

$$+ 3[\text{DX}(f)](i + 1, j) - 3[\text{DX}(f)](i, j)$$

$$+ [\text{DX}(f)](i + 1, j - 1) - [\text{DX}(f)](i, j - 1))$$

Example 2.41

Consider the image

$$f = \begin{pmatrix} 6 & 2 & 6 \\ 2 & ⓪ & 2 \\ 6 & 2 & 6 \end{pmatrix}_{-1,1}$$

The only point at which a quadric surface of the preceding type z can fit is the origin pixel $(0, 0)$. Moreover, $z_0 = f(0, 0) = 0$. Applying the solution just derived,

$$z = 2.8x^2 + 2.8y^2$$

which is the equation of a paraboloid. This solution reflects the symmetry of the image on SQUARE$(0, 0)$.

Example 2.42

Let

$$g = \begin{pmatrix} 2 & 3 & 4 \\ 2 & 3 & 4 \\ 2 & 3 & 4 \end{pmatrix}_{-1,1}$$

In this case,

$$z = 3 + x$$

which reflects the fact that the image is increasing linearly in the x-direction at a constant rate $DX(f) = 1$.

A simple generalization of the best quadric fit just given is the weighted least-squares quadric fit. In this case, the coefficients A, B, C, D, and E of the quadric surface

$$z = z_0 + A(x - i)^2 + B(y - j)^2 + C(x - i) + D(y - j) + E(x - i)(y - j)$$

are found which minimize

$$q = \sum_{(u,v) \in \text{SQUARE}(i,j)} w(u, v)(f(u, v) - z(u, v))^2$$

Again, $z_0 = f(i, j)$ is utilized, and the weights $w(u, v)$ are specified a priori.

Example 2.43

Let the weights of strong neighbors be two and the weights of weak neighbors be one, as illustrated in Figure 2.23. By taking the partial derivatives specified above and setting them to zero, a set of five simultaneous equations arise involving A, B, C, D, and E. Solving these equations yields

$$E = \frac{1}{4}[f(1,1) + f(-1,-1) - f(1,-1) - f(-1,1)]$$

$$C = \frac{1}{6}[f(1,1) + 2f(1,0) + f(1,-1) - f(-1,1) - 2f(-1,0) - f(-1,-1)]$$

$$D = \frac{1}{6}[f(1,1) + 2f(0,1) + f(-1,1) - f(-1,-1) - 2f(0,-1) - f(1,-1)]$$

$$B = \frac{1}{10}[f(1,1) + f(-1,1) + f(-1,-1) + f(1,-1) + 6f(0,-1)$$
$$+ 6f(0,1) - 4f(-1,0) - 4f(1,0) - 6z_0]$$

$$A = \frac{1}{10}[f(1,1) + f(-1,1) + f(-1,-1) + f(1,-1) - 4f(0,1)$$
$$- 4f(0,-1) + 6f(-1,0) + 6f(1,0) - 6z_0]$$

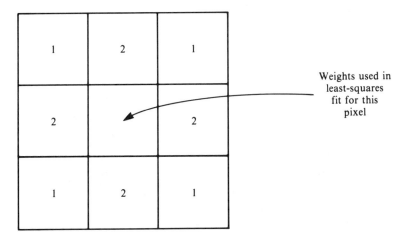

Figure 2.23 Weighting Scheme

EXERCISES

2.1. Let

$$f = \begin{pmatrix} * & * & 0 & -1 \\ \frac{1}{2} & 8 & 1 & 2 \\ -3 & 2 & * & * \end{pmatrix}_{-1,1} \qquad g = \begin{pmatrix} 1 & 0 & 9 \\ 2 & 1 & -4 \\ * & 1 & 0 \\ * & 3 & 1 \end{pmatrix}_{0,1}$$

Find:
- (a) ADD(f, g)
- (b) EXTADD(f, g)
- (c) MULT(f, g)
- (d) EXTMULT(f, g)
- (e) MAX(f, g)
- (f) EXTMAX(f, g)
- (g) MIN(f, g)
- (h) EXTMIN(f, g)
- (i) SCALAR(3; f)
- (j) DIV(g)
- (k) SUB(f)
- (l) TRAN(f; 2, 3)
- (m) SELECT(f; 2, 2, 1, 1)
- (n) NINETY(f)
- (o) FLIP(f)
- (p) NINETY^2(f)
- (q) EXTEND(f, g)
- (r) CREATE([(1, 2), (2, 2)], [3, 4])
- (s) DOMAIN(f)
- (t) RANGE(f)

2.2. Using the images f and g of Exercise 2.1, find:
- (a) TRAN[ADD(f, g); 1, −1]
- (b) EXTADD[SUB(f), DIV(f)]
- (c) MAX[TRAN(f; 1, 1), SCALAR(2; g)]
- (d) ADD[ADD(f, g), SELECT(f; 3, 2, 0, 0)]
- (e) NINETY[FLIP(NINETY(f))]

2.3. Let

$$h = \begin{pmatrix} 0 & 1 & 1 & 0 & 1 & 4 \\ 0 & 2 & 3 & 1 & 5 & 5 \\ * & 3 & 3 & 7 & 6 & 6 \\ 5 & 5 & 8 & 7 & 7 & 8 \\ 6 & 9 & 9 & 7 & 8 & 8 \end{pmatrix}_{0,4}$$

Find:
- (a) THRESH(h; 4)
- (b) THRESH(h; 5)
- (c) THRESH(h; 6)
- (d) EQUAL(h; 6)
- (e) GREATER(h; 5)
- (f) BETWEEN(h; 5, 7)

2.4. Give block diagram specifications for EQUAL, GREATER, and BETWEEN.

2.5. Let f and g be the images given in Exercise 2.1. Find DOT[f, REST(g, f)], NORM[ADD(f, g)] and NORM(f) + NORM(g).

2.6. Consider the masks

$$M = \begin{pmatrix} 0 & \frac{1}{8} & 0 \\ \frac{1}{8} & \textcircled{\frac{1}{2}} & \frac{1}{8} \\ 0 & \frac{1}{8} & 0 \end{pmatrix} \quad N = \begin{pmatrix} 0 & -1 & 0 \\ -1 & \textcircled{6} & -1 \\ 0 & -1 & 0 \end{pmatrix}$$

Letting h be the image given in Exercise 2.3, find FILTER(h; M), FILTER(h; N), and THRESH[FILTER(h; M); 5].

2.7. Using the image h and mask M of Exercise 2.6, do a walk-through of FILTER as it is given in Figure 2.8. Use only the point (2, 2) in the domain of h.

2.8. Using the image f of Exercise 2.1, find MAG0(f), MAG1(f), and MAG2(f).

2.9. Using the image h of Exercise 2.3, find:
- (a) GRAD(h)
- (b) GRADMAG0(h)
- (c) GRADMAG1(h)
- (d) SYMGRAD(h)
- (e) SYMMAG0(h)
- (f) SYMMAG2(h)
- (g) PREWGRAD(h)
- (h) PREWMAG1(h)
- (i) SOBGRAD(h)
- (j) SOBMAG2(h)
- (k) ROBGRAD(h)
- (l) ROBMAG1(h)
- (m) DIRECT(h; 30°)
- (n) GRADEDGE0(h; t) for t = 2, 4, 6
- (o) PREWEDGE0(h; t) for t = 10, 12, 14
- (p) PREWEDGE1(h; t) for t = 10, 14, 18
- (q) SOBEDGE1(h; t), where t is chosen by using the gray scale histogram method
- (r) ROBEDGE2(h; t), where t is chosen by using the gray scale histogram method

2.10. Using the image h of Exercise 2.3, apply the compass gradient edge detection methodology of Figure 2.16 using the:
- (a) Kirsh masks
- (b) Prewitt compass masks
- (c) 3-level masks
- (d) 5-level masks

2.11. Using the image h of Exercise 2.3, find the least-squares best-fit plane at the pixel (2, 2) using the minimization neighborhood
  (a) SQUARE(2, 2)
  (b) consisting of strong neighbors

2.12. Find the equation of the least-squares best-fit plane if the minimization neighborhood for the pixel (i, j) is given by $\{(i - 1, j), (i + 1, j)\}$.

2.13. Find the gray level histogram of the image h given in Exercise 2.3. Use that histogram to pick an appropriate threshold value. Compare your choice to the threshold parameters used in Exercise 2.3.

2.14. Let

$$f = \begin{pmatrix} 0 & 0 & 0 & 0 \\ 0 & 2 & 1 & 0 \\ 0 & 1 & 1 & 0 \\ 0 & 1 & 2 & 0 \\ 1 & 1 & 1 & 1 \\ 2 & 1 & 1 & 5 \end{pmatrix}_{3,4}$$

Assuming a gray level quantization scale running from 0 through 7, apply histogram equalization to f.

2.15. Let

$$g = \begin{pmatrix} 0 & 0 & 1 & 2 & 1 & 1 \\ 4 & 4 & 3 & 4 & 4 & 5 \\ 0 & 1 & 0 & 0 & 1 & 1 \\ 5 & 4 & 5 & 4 & 4 & 5 \\ 0 & 0 & 0 & 1 & 1 & 2 \\ 5 & 4 & 3 & 5 & 5 & 4 \end{pmatrix}_{3,-5}$$

Assuming a quantization scale of 0 through 5, find the co-occurrence matrices $C[g; 1]$, $C[g; 2]$, $C[g; 3]$, and $C[g; 4]$. Interpret the results.

2.16. Using the image h of Exercise 2.3, find the best-fit quadric surface, as discussed in Section 2.18, at the pixel (2, 2).

2.17. Let Q be the 5 by 5 image

$$Q = \begin{pmatrix} * & * & 1 & * & * \\ * & * & 1 & * & * \\ 1 & 1 & ① & 1 & 1 \\ * & * & 1 & * & * \\ * & * & 1 & * & * \end{pmatrix}$$

Find the equation of the least-squares best-fit plane when the minimization neighborhood is the domain of Q.

2.18. In Section 2.11, we considered the l_∞, l_1, and l_2 norms. More generally, we can consider the l_P norm, $P > 0$, given by

$$\|V\|_P = \left[\sum_{k=1}^{n} |x_k|^P \right]^{1/P}$$

where $V = (x_1, x_2, \ldots, x_n)$. For a given P, this P-norm leads to a corresponding magnitude operator, MAGP, as well as to edge operators. Give block diagram specifications of PREWMAGP and PREWEDGEP. For $P = 3$ and image h of Exercise 2.3, find PREWMAGP(h) and PREWEDGEP($h; t$), where t is found by using the gray scale histogram method.

2.19. Consider the 4 by 4 mask

$$N = \begin{pmatrix} \frac{1}{32} & \frac{1}{32} & \frac{1}{32} & \frac{1}{32} \\ \frac{1}{16} & \frac{1}{16} & \frac{1}{16} & \frac{1}{32} \\ \frac{1}{16} & \textcircled{\frac{1}{4}} & \frac{1}{16} & \frac{1}{32} \\ \frac{3}{32} & \frac{1}{16} & \frac{1}{16} & \frac{1}{32} \end{pmatrix}$$

Since PIXSUM(N) = 1, N is called a *4 by 4 averaging mask*. For the image h of Exercise 2.3, find FILTER($h; N$).

2.20. [SYMDX(f)](i, j) gives the gray level average of [DX(f)]($i + 1, j$) and [DX(f)](i, j), which in turn gives the arithmetic mean of the left and right gray level differences at the pixel (i, j). Define the operator FOURDX by

$$[\text{FOURDX}(f)](i, j) = \frac{1}{4} \sum_{k=-1}^{2} [\text{DX}(f)](i + k, j)$$

FOURDX gives the arithmetic mean of four gray level differences, these being symmetric about (i, j) in the x-direction.

(a) Define the y-directional operator FOURDY.
(b) Define the corresponding gradient FOURGRAD.
(c) Give a block diagram specification for the magnitude operator FOURMAG0.
(d) Given a block diagram specification for the edge operator FOUREDGE.
(e) Using the image

$$h = \begin{pmatrix} * & 2 & 1 & 6 & 7 & 1 & 0 & 1 \\ 2 & 1 & 3 & 6 & 8 & 2 & 1 & 7 \\ 1 & 1 & 2 & 8 & 7 & 1 & 1 & 0 \\ 9 & 7 & 7 & 8 & 7 & 7 & 8 & 9 \\ 9 & 7 & 8 & 6 & 6 & 6 & 8 & 7 \\ 1 & 0 & 0 & 6 & 7 & 0 & 1 & 0 \\ 8 & 1 & 0 & 8 & 8 & 0 & 0 & 0 \\ 1 & 1 & 8 & 8 & 8 & 1 & 1 & 1 \end{pmatrix}_{2,2}$$

find FOURGRAD(h), FOURMAG1(h), and FOUREDGE1($h; t$), where t is found by using the gray scale histogram method.

(f) In a manner exactly analogous to the definition of FOURDX, define an operator EIGHTDX.

3

Morphological Image Processing

3.1 MINKOWSKI ALGEBRA

Whereas the previous chapter was entirely digital in nature, the present one begins with a treatment of morphology in the Euclidean model. The fundamental operations employed in morphological image analysis are Minkowski addition and Minkowski subtraction.[1] These operations, together with their basic properties, will be presented in the Euclidean framework. Their digital counterparts, necessary for machine implementation, will be presented afterward.

This chapter differs significantly from the previous one in style as well as substance. Chapter 2 dealt with digital image operations and digital image processing algorithms. The current chapter concerns constant Euclidean images, which are subsets of R^2. (A precise logical formulation of the subset–image relation is provided in Section 3.7.) The chapter contains fundamental set-theoretic results important to image processing and hence it will be more mathematical in character. However, it is hoped that the supporting explanations and examples will provide sufficient augmentation to the essentially propositional presentation.

As a common scientific term, *morphology* denotes the study of form and structure. In image processing specifically, it refers to a particular approach to the analysis of geometric structure, or texture, within an image. The underlying strategy is to understand the textural or geometric properties of an image by probing the microstructure of the image with various forms. These probes, known as *structuring elements*, are examined in terms of the manner in which they "fit" into a set or into the complement of a set. The analysis is geometric in character, approaching image processing from the vantage point of human perception. The intent is to derive quantitative measures of natural perceptual categories and thereby exploit whatever inherent congruences exist between image structure and

[1] H. Minkowski, "Volumen and Oberflache," *Math. Ann. Vol 57*, (1903):447–95.

ordinary human recognition. Such an approach is well suited to eventual integration into an artificial intelligence schema. For a computer vision system to yield image-based decisions resembling those which result from direct human understanding, the categories upon which that system operates must correspond well to native human perceptual categories.

In searching for a given pattern within an image, a person perceives the image through the filter of his or her own motivation. Sensory data are not passively received and acted upon by intelligence; rather, they are organized by the brain into "percepts," and it is these percepts which are the raw material for intelligence. To employ engineering terminology, one might loosely refer to the act of perception (i.e., of rendering data into percepts) as a form of data compression. This act of compression involves a choice, prior to sensory reception, as to what manner and what end the compression is to take place. Moreover, not only are the sensory data organized, but even higher level filtering must take place in order to search for desired patterns. The elaboration of structure, which is after all the intent, involves an analysis of the relationships between the component parts of whatever object is under investigation. But these relationships are imposed upon the image by intelligence; they do not exist independently of that intelligence. The founder of mathematical morphology, G. Matheron, puts the matter succinctly when he states that the choice of what structural relationships are to be used "plays an a priori constitutive role."[2] Indeed, only those aspects of the image which conform to some predetermined set of relational categories are relevant. In that sense, the image engineer's choice of those categories *constitutes*, or frames, the image. For practical morphological image processing, this means that the type of filtering or probing of an image depends upon the particular knowledge desired. Once the image is constituted in terms of the relational base of this desired knowledge, other characteristics of the image are no longer accessible.

Morphological analysis begins with a set A in the plane R^2. For x an element of R^2, the *translation* of A by x is defined to be

$$A + x = \{a + x: a \in A\}$$

where "$+$" within the set notation denotes vector addition. If x is a vector in the plane, then $A + x$ is A translated along the vector x. (See Figure 3.1.) We also write $x + A$ to denote the translation of A by x. Notice that a point z is an element of $A + x$ if and only if there exists a point $a \in A$ such that $z = a + x$.

Given two images A and B in R^2, the *Minkowski addition of A by B* is defined to be

$$A \oplus B = \bigcup_{x \in B} A + x$$

---

[2] G. Matheron. *Random Sets and Integral Geometry* (New York: Wiley, 1975).

Sec. 3.1   Minkowski Algebra

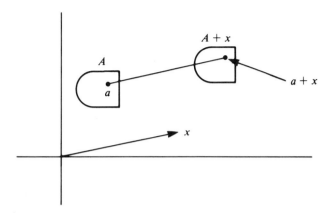

Figure 3.1   Translation of Euclidean Image

In other words, $A \oplus B$ is constructed by translating A by each element of B and then taking the union of all the resulting translations. (See Figure 3.2.) Some immediate properties of Minkowski addition are:

(i) $A \oplus \{\overline{0}\} = A$, where $\overline{0}$ denotes the origin $(0, 0)$.
(ii) $A \oplus \{x\} = A + x$ for any element $x \in R^2$.

Proposition 3.1.

$$A \oplus B = \bigcup_{\substack{x \in A \\ y \in B}} \{x + y\}$$

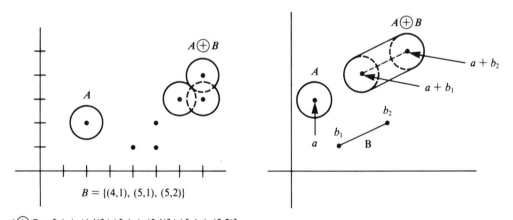

$A \oplus B = [A + (4,1)] \cup [A + (5,1)] \cup [A + (5,2)]$

(a)   (b)

Figure 3.2   Minkowski Addition

Proof.

$$A \oplus B = \bigcup_{y \in B} (A + y)$$
$$= \bigcup_{y \in B} \left(\left[\bigcup_{x \in A} \{x\}\right] + y\right)$$
$$= \bigcup_{y \in B} \left(\bigcup_{x \in A} \{x + y\}\right)$$
$$= \bigcup_{\substack{x \in A \\ y \in B}} \{x + y\}$$

Proposition 3.1 states that Minkowski addition can be accomplished by adding (as 2-vectors) all the elements of A and B and then taking the union of all the resultants. Immediate corollaries are that Minkowski addition is both commutative and associative.

Proposition 3.2.

(i) $A \oplus B = B \oplus A$ (Commutativity)
(ii) $(A \oplus B) \oplus C = A \oplus (B \oplus C)$ (Associativity)

Proof. By Proposition 3.1,

$$A \oplus B = \bigcup_{\substack{x \in A \\ y \in B}} \{x + y\} = \bigcup_{\substack{y \in B \\ x \in A}} \{y + x\} = B \oplus A$$

As for associativity,

$$(A \oplus B) \oplus C = \left(\bigcup_{\substack{x \in A \\ y \in B}} \{x + y\}\right) \oplus C$$
$$= \bigcup_{\substack{x \in A \\ y \in B \\ z \in C}} \{(x + y) + z\}$$
$$= \bigcup_{\substack{x \in A \\ y \in B \\ z \in C}} \{x + (y + z)\}$$
$$= A \oplus \left(\bigcup_{\substack{y \in B \\ z \in C}} \{y + z\}\right)$$
$$= A \oplus (B \oplus C)$$

Example 3.1

Let $A = \{(1, 1), (1, 2), (2, 1)\}$, $B = \{(0, 0), (1, 3)\}$, and $C = \{(2, 1), (2, 2)\}$. Then

$$A \oplus B = \{(1, 1), (1, 2), (2, 1), (2, 4), (2, 5), (3, 4)\}$$

and

$$(A \oplus B) \oplus C = \{(3, 2), (3, 3), (4, 2), (4, 5), (4, 6), (5, 5), (3, 4), (4, 3), (4, 7), (5, 6)\}$$

On the other hand,

$$B \oplus C = \{(2, 1), (2, 2), (3, 4), (3, 5)\}$$

and

$$A \oplus (B \oplus C) = \{(3, 2), (3, 3), (4, 5), (4, 6), (3, 4), (4, 7), (4, 2), (4, 3), (5, 5), (5, 6)\}$$

Note that $(A \oplus B) \oplus C = A \oplus (B \oplus C)$.

The next proposition states the translational invariance of Minkowski addition. Put another way, one can either translate and then apply Minkowski addition, or apply Minkowski addition first and then translate. (See Figure 3.3.)

Proposition 3.3.
$$A \oplus [B + x] = [A \oplus B] + x$$

Proof. This is a special case of the associative law, Proposition 3.2(ii). Specifically,

$$A \oplus [B + x] = A \oplus [B \oplus \{x\}]$$
$$= [A \oplus B] \oplus \{x\}$$
$$= [A \oplus B] + x$$

The second fundamental morphological operation is *Minkowski subtraction*,

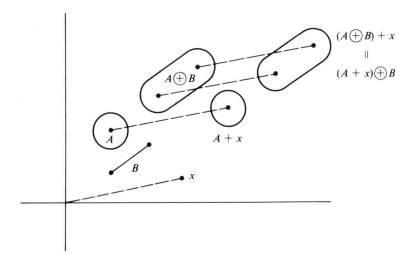

Figure 3.3 Translational Invariance of Minkowski Addition

defined to be

$$A \ominus B = \bigcap_{x \in B} A + x$$

$A \ominus B$ is the intersection over all elements of B of the translates $A + x$. (See Figure 3.4.)

Before proceeding to the next result, we define the *scalar multiplication* of a real number times an image A in R^2. For any real number t,

$$tA = \{tx: x \in A\}$$

Put simply, tA is obtained by multiplying each vector x in A by the scalar t. A particularly important instance is scalar multiplication by -1, in which case we write $-A$. By definition, $-A = \{-x: x \in A\}$. (See Figure 3.5.) $-A$ is a 180° rotation of A about the origin and is called the *reflection of A*.

The next two propositions give the so-called *duality* relations between Minkowski addition and Minkowski subtraction. Each is given in terms of complements involving the other. The complement of a set A is defined to be $A^c = \{x: x \notin A\}$. (See Figure 3.6 for an illustration of Proposition 3.4.)

Proposition 3.4.

$$A \oplus B = [A^c \ominus B]^c \quad \text{(Duality)}$$

Proof. Let x be any point in R^2. Then $x \in [A^c \ominus B]^c$ if and only if $x \notin [A^c \ominus B]$

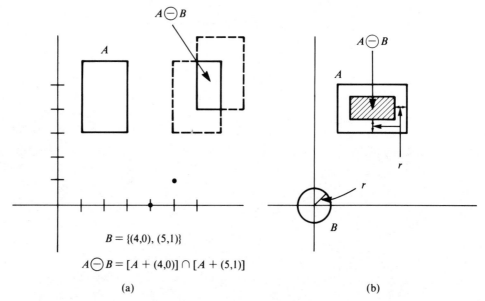

Figure 3.4 Minkowski Subtraction

Sec. 3.1   Minkowski Algebra   97

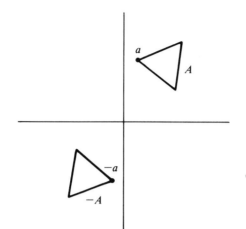

Figure 3.5 Reflection Operation

if and only if there exists $b \in B$ such that $x \notin A^c + b$ if and only if there exists $b \in B$ such that $x \in A + b$ if and only if $x \in A \oplus B$.

Proposition 3.5.
$$A \ominus B = [A^c \oplus B]^c \quad \text{(Duality)}$$

Proof. Applying Proposition 3.4 and the fact that successive complementations produce the original set, we have

$$A^c \oplus B = [(A^c)^c \ominus B]^c = [A \ominus B]^c$$

Complementing both sides yields Proposition 3.5.

The next result states that, like Minkowski addition, Minkowski subtraction possesses invariance with respect to translation.

Proposition 3.6.

(i) $A \ominus [B + x] = [A \ominus B] + x$
(ii) $[A + x] \ominus B = [A \ominus B] + x$

Proof. We apply duality together with translational invariance for Minkowski addition:

$$A^c \oplus [B + x] = [A^c \oplus B] + x$$

Taking complements, we obtain (i); that is,

$$A \ominus [B + x] = ([A^c \oplus B] + x)^c$$
$$= [A^c \oplus B]^c + x = [A \ominus B] + x$$

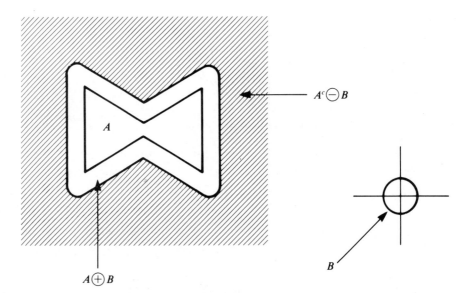

Figure 3.6 Illustration of Proposition 3.4

where we have used the fact that complementation can be interchanged with translation. As for (ii), apply complementation to

$$[A^c + x] \oplus B = [A^c \oplus B] + x$$

The result follows in a similar manner to the proof of part (i).

An alternative characterization of Minkowski subtraction will now be presented. It is this characterization which is most useful in image processing. It says that the Minkowski subtraction of B from A is composed of all elements x in R^2 such that $-B + x$ is a subset of A.

Proposition 3.7.
$$A \ominus B = \{x: -B + x \subset A\}$$

Proof.
$$\{x: -B + x \subset A\} = \bigcap_{y \in B} \{x: -y + x \in A\}$$
$$= \bigcap_{y \in B} \{x: x \in A + y\}$$
$$= \bigcap_{y \in B} A + y = A \ominus B$$

Since $-(-B) = B$, the preceding proposition could also be written as

$$A \ominus (-B) = \{x: B + x \subset A\}$$

Sec. 3.1   Minkowski Algebra

It is in fact this latter set which is used in the digital implementation of Minkowski algebra. Indeed, we define the operation of *erosion* of A by B (see Figure 3.7) as

$$\mathcal{E}(A, B) = \{x: B + x \subset A\}$$

Thus, Proposition 3.7 may be reformulated as Proposition 3.8.

Proposition 3.8.
$$\mathcal{E}(A, B) = A \ominus (-B)$$

Corresponding to the terminology of erosion, there is a term commonly employed with regard to Minkowski addition. Instead of $A \oplus B$, it is common in image processing to say the *dilation* of A by B, written $\mathcal{D}(A, B)$. Then $\mathcal{D}(A, B) = A \oplus B$, while $\mathcal{E}(A, B) = A \ominus (-B)$. The next proposition gives a characterization of dilation which is the dual of Proposition 3.7 for $A \oplus B$. (See Figure 3.8.)

Proposition 3.9.
$$A \oplus B = \{x: [(-B) + x] \cap A \neq \emptyset\}$$

Proof.

$\{x: [(-B) + x] \cap A \neq \emptyset\}$

$= \{x: \text{there exists an } a \in A \text{ such that } a \in [(-B) + x]\}$

$= \{x: \text{there exists an } a \in A \text{ such that } a - x \in (-B)\}$

$= \{x: \text{there exists an } a \in A \text{ such that } x \in B + a\}$

$= \{x: x \in \bigcup_{a \in A} B + a\} = B \oplus A = A \oplus B$

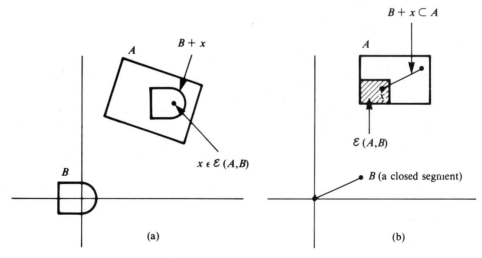

Figure 3.7   Erosion Operation

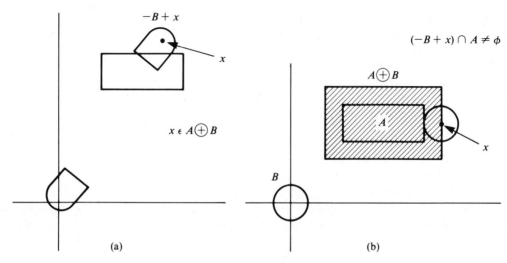

Figure 3.8 Dilation Operation

Theorem 3.1 collects some of the earlier propositions and puts them in terms of dilation and erosion.

Theorem 3.1.

(i) $\mathcal{D}(A, B) = \mathcal{D}(B, A)$

(ii) $\mathcal{D}(A, B) = [\mathcal{E}(A^c, -B)]^c$

(iii) $\mathcal{E}(A, B) = [\mathcal{D}(A^c, -B)]^c$

(iv) $\mathcal{D}(A, B + x) = \mathcal{D}(A, B) + x$

(v) $\mathcal{E}(A, B + x) = \mathcal{E}(A, B) - x$

Before proceeding, it should be mentioned that if B contains the origin $\bar{0}$, then $\mathcal{E}(A, B)$ is a subimage (subset) of A. To see this, simply note that $z \in \mathcal{E}(A, B)$ implies that $B + z \subset A$, and hence $z = \bar{0} + z \in A$. This property of erosion is significant in applications.

An operation $P(A)$ on sets is said to be *increasing* if $A \subset B$ implies $P(A) \subset P(B)$. $P(A)$ is said to be *decreasing* if $A \subset B$ implies $P(A) \supset P(B)$. Increasing set operations play a significant role in morphological image processing. This is reflected in Proposition 3.10, which states that, for a fixed set B, both dilation $\mathcal{D}(\cdot, B)$ and erosion $\mathcal{E}(\cdot, B)$ are increasing set functions in the first variable. Proposition 3.11 states that erosion is also decreasing in the second variable.

Proposition 3.10. Suppose B is fixed and $A_1 \subset A_2$. Then

(i) $\mathcal{D}(A_1, B) \subset \mathcal{D}(A_2, B)$

(ii) $\mathcal{E}(A_1, B) \subset \mathcal{E}(A_2, B)$

Proof. (i) follows at once from the definition of Minkowski addition since each of $\mathcal{D}(A_1, B)$ and $\mathcal{D}(A_2, B)$ is the union over the elements of B and in each case $A_1 + x \subset A_2 + x$. The second statement follows in like manner since each of $\mathcal{E}(A_1, B)$ and $\mathcal{E}(A_2, B)$ is an intersection over the elements of B.

Proposition 3.11. Suppose A is fixed and $B_1 \subset B_2$. Then $\mathcal{E}(A, B_1) \supset \mathcal{E}(A, B_2)$.

Proof. Using the definition of erosion directly, $B_2 + x \subset A$ implies $B_1 + x \subset A$. Hence, $x \in \mathcal{E}(A, B_2)$ implies $x \in \mathcal{E}(A, B_1)$, and the proposition follows.

The next results show that both Minkowski addition and Minkowski subtraction satisfy a type of homogeneity. (See Figure 3.9.)

Proposition 3.12. Let t be a real number, $t \neq 0$. Then

$$t\left[\left(\frac{1}{t}\right) A \oplus B\right] = A \oplus (tB)$$

Proof.

$$z \in t\left[\left(\frac{1}{t}\right) A \oplus B\right]$$

if and only if there exists $a \in A$ and $b \in B$ such that

$$z = t\left[\left(\frac{1}{t}\right) a + b\right] = a + tb$$

which means precisely that $z \in A \oplus (tB)$.

Proposition 3.13. Let t be a real number, $t \neq 0$. Then

$$t\left[\left(\frac{1}{t}\right) A \ominus B\right] = A \ominus (tB)$$

Proof.

$$z \in t\left[\left(\frac{1}{t}\right) A \ominus B\right]$$

if and only if

$$\left(\frac{1}{t}\right) z \in \left(\frac{1}{t}\right) A \ominus B.$$

But the latter is true if and only if

$$-B + \left(\frac{1}{t}\right) z \subset \left(\frac{1}{t}\right) A$$

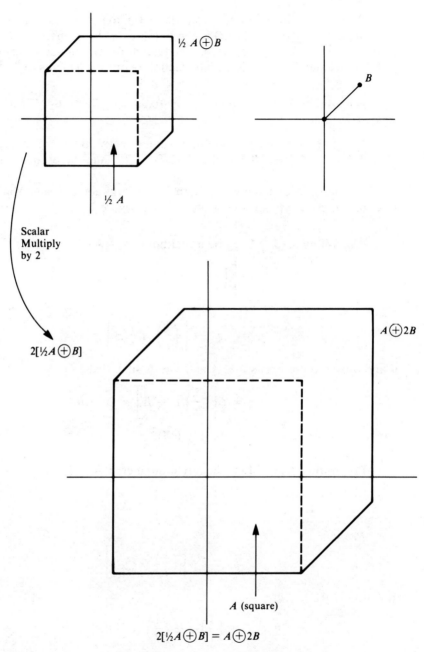

Figure 3.9 Special Case of Proposition 3.12

Sec. 3.1 Minkowski Algebra

which is equivalent to asserting that for any point b in B, there is a corresponding point a_b in A such that

$$-b + \left(\frac{1}{t}\right) z = \left(\frac{1}{t}\right) a_b$$

or, in other words,

$$-tb + z = a_b$$

But this means precisely that $-tB + z \subset A$, which in turn means that $z \in A \ominus tB$.

We now present some further algebraic properties of Minkowski algebra. Though some of these play a lesser role in the current application of morphological methods, they are a fundamental part of the overall structure of Minkowski algebra. Moreover, it is becoming ever more apparent that the algebraic nature of image processing operations will likely occupy a central position in both theory and application. As a result, it is vital that these properties be part of the presentation. However, since they are included only for the purpose of completeness, and since they will not play a role in what is to follow, they, together with their proofs, will be presented without extensive comment. For those interested mainly in the algorithmic side of morphology, the rest of this section may be skimmed.

Proposition 3.14.

(i) $A \oplus (B \cup C) = (A \oplus B) \cup (A \oplus C)$

(ii) $(B \cup C) \oplus A = (B \oplus A) \cup (C \oplus A)$

(Distributivity of Minkowski addition over union)

Proof.

$$A \oplus (B \cup C) = \bigcup_{z \in A} (B \cup C) + z = \bigcup_{z \in A} (B + z) \cup (C + z)$$

$$= [\bigcup_{z \in A} B + z] \cup [\bigcup_{z \in A} C + z]$$

$$= (A \oplus B) \cup (A \oplus C)$$

Hence, (i) is proven. Part (ii) follows at once from the commutativity of Minkowski addition.

It should be noted that there is no general distributivity of dilation over intersection. One can only get a partial, inclusionary, relation. The next proposition gives this relation from both the left and the right.

Proposition 3.15.

(i) $A \oplus (B \cap C) \subset (A \oplus B) \cap (A \oplus C)$

(ii) $(B \cap C) \oplus A \subset (B \oplus A) \cap (C \oplus A)$

Proof.

$$\begin{aligned}
A \oplus (B \cap C) &= \bigcup_{z \in A} (B \cap C) + z \\
&= \bigcup_{z \in A} (B + z) \cap (C + z) \\
&\subset [\bigcup_{z \in A} B + z] \cap [\bigcup_{z \in A} C + z] \\
&= (A \oplus B) \cap (A \oplus C)
\end{aligned}$$

Thus, (i) is proven. Part (ii) follows by the commutativity of Minkowski addition.

Proposition 3.16.

$$A \ominus (B \cup C) = (A \ominus B) \cap (A \ominus C)$$

Proof. Applying duality, the previous result, and De Morgan's Law, we have

$$\begin{aligned}
A \ominus (B \cup C) &= [A^c \oplus (B \cup C)]^c \\
&= [(A^c \oplus B) \cup (A^c \oplus C)]^c \\
&= (A^c \oplus B)^c \cap (A^c \oplus C)^c \\
&= (A \ominus B) \cap (A \ominus C)
\end{aligned}$$

Proposition 3.16 asserts a sort of left antidistributivity of Minkowski subtraction over union in the sense that the union becomes an intersection after distribution. If Minkowski subtraction is applied on the right to a union, antidistributivity does not hold; indeed, in general, the best that can be said is that

$$(B \cup C) \ominus A \supset (B \ominus A) \cap (C \ominus A)$$

Proposition 3.17.

$$(B \cap C) \ominus A = (B \ominus A) \cap (C \ominus A)$$

Proof.

$$\begin{aligned}
(B \cap C) \ominus A &= [(B \cap C)^c \oplus A]^c = [(B^c \cup C^c) \oplus A]^c \\
&= [(B^c \oplus A) \cup (C^c \oplus A)]^c \\
&= (B^c \oplus A)^c \cap (C^c \oplus A)^c \\
&= (B \ominus A) \cap (C \ominus A)
\end{aligned}$$

Proposition 3.17 asserts a right distributivity of Minkowski subtraction over intersection. In this case, left distributivity does not hold. In general, the best that can be said is that the following partial left antidistributivity holds:

$$A \ominus (B \cap C) \supset (A \ominus B) \cup (A \ominus C)$$

Proposition 3.18.

$$(A \ominus B) \ominus C = A \ominus (B \oplus C)$$

Proof.

$$(A \ominus B) \ominus C = \bigcap_{z \in C} (A \ominus B) + z$$

$$= \bigcap_{z \in C} [(\bigcap_{y \in B} A + y) + z]$$

$$= \bigcap_{z \in C} [\bigcap_{y \in B} A + (z + y)]$$

$$= \bigcap_{x \in B \oplus C} A + x = A \ominus (B \oplus C)$$

3.2 OPENING AND CLOSING

Minkowski addition and subtraction are the primitive morphological operations. Two other operations, each of which is defined in terms of the primitive operations, also play a central role in image analysis. These two secondary operations are called the *opening* and *closing* of one image by another and are respectively defined as

$$O(A, B) = [A \ominus (-B)] \oplus B$$

and

$$C(A, B) = [A \oplus (-B)] \ominus B$$

It follows immediately from the definitions of dilation and erosion that the preceding definitions can be written as

$$O(A, B) = \mathcal{D}[\mathcal{E}(A, B), B]$$

and

$$C(A, B) = \mathcal{E}[\mathcal{D}(A, -B), -B]$$

(See Figure 3.10.)

The next two propositions give the duality relations between the opening and the closing. They can be most useful, since they result in dual properties for the closing when properties of the opening are known. They also allow the com-

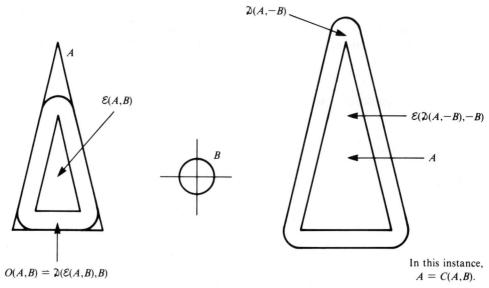

Figure 3.10 Opening and Closing Operations

putation of the closing to be done in terms of the opening, and conversely. Proposition 3.19 says that the complement of the closing is equal to the opening of the complement. Proposition 3.20 says that the complement of the opening is equal to the closing of the complement.

Proposition 3.19.

$$C(A, B)^c = O(A^c, B) \quad \text{(Duality)}$$

Proof. Applying the duality criteria for Minkowski subtraction and Minkowski addition, we obtain

$$C(A, B)^c = [(A \oplus (-B)) \ominus B]^c$$
$$= [A \oplus (-B)]^c \oplus B$$
$$= [A^c \ominus (-B)] \oplus B = O(A^c, B)$$

Proposition 3.20.

$$O(A, B)^c = C(A^c, B) \quad \text{(Duality)}$$

Proof.
$$O(A, B)^c = O[(A^c)^c, B]^c$$
$$= [C(A^c, B)^c]^c \quad \text{(by Proposition 3.19)}$$
$$= C(A^c, B)$$

Theorem 3.2 is fundamental to the digital implementation of the opening. It is preceded by a preliminary proposition and is followed by an explanatory figure. In so many words, it says that the opening of A by B is the union of all translates of B which are subsets of A.

Proposition 3.21, to follow, states that z is an element of the opening if and only if the translate of $-B$ by z intersects the erosion of A by B.

Proposition 3.21. $z \in O(A, B)$ if and only if $[(-B) + z] \cap [A \ominus (-B)] \neq \emptyset$.

Proof. $z \in O(A, B)$ if and only if $z \in \cup_{y \in B} [A \ominus (-B)] + y$ if and only if there exists $b \in B$ with $z \in [A \ominus (-B)] + b$ if and only if there exists $b \in B$ and $w \in [A \ominus (-B)]$ with $z = w + b$, or, equivalently, with $w = -b + z$. But this means precisely that $w \in [(-B) + z] \cap [A \ominus (-B)]$, which means that the intersection in the proposition is nonempty.

Theorem 3.2.

$$O(A, B) = \cup \{B + y : B + y \subset A\}$$

Proof. By Proposition 3.21, $z \in O(A, B)$ if and only if there exists a point y such that $y \in [(-B) + z] \cap [A \ominus (-B)]$, that is, if and only if there is a point y such that $z \in B + y$ and $B + y \subset A$. But this last statement means precisely that z is in the union specified in the statement of the theorem.

Figure 3.11 shows the opening in terms of the union of translates.

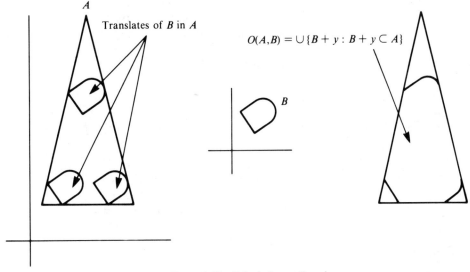

Figure 3.11 Calculation at Opening

Propositions 3.22 and 3.23 give pointset characterizations of the closing. Proposition 3.22 is the dual of Theorem 3.2. Figure 3.12 illustrates Proposition 3.23.

Proposition 3.22.

$$C(A, B) = \cap\{(B + y)^c : B + y \subset A^c\}$$

Proof. We apply duality to Theorem 3.2, yielding

$$C(A, B)^c = O(A^c, B) = \cup\{B + y : B + y \subset A^c\}$$

Application of De Morgan's Law gives Proposition 3.22.

Proposition 3.23. $z \in C(A, B)$ if and only if $(B + y) \cap A \neq \emptyset$ for any translate $B + y$ containing z.

Proof. By Proposition 3.22, $z \in C(A, B)$ if and only if $B + y \subset A^c$ implies $z \in (B + y)^c$, which is itself true, by contraposition, if and only if $z \in B + y$ implies $(B + y) \cap A \neq \emptyset$.

The next two theorems give the fundamental algebraic properties of the opening and the closing, respectively. These properties, simple but powerful, are of paramount importance for both theoretical and practical reasons. A full appreciation of the genesis of the morphological methodology that will shortly be discussed cannot be had without a thorough understanding of them. Theorem 3.3 states (i) that the opening produces a smaller set, (ii) that it is increasing in the first variable, and (iii) that successive openings have no effect after the first one. Theorem 3.4 states (i) that the closing produces a larger set, (ii) that it is increasing in the first variable, and (iii) that it has no effect after the first application.

Theorem 3.3. The opening satisfies

(i) $O(A, B) \subset A$ (Antiextensive)

Translates of B containing z. Each intersects A, and therefore, $z \in C(A,B)$.

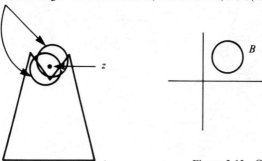

Figure 3.12 Closing Using Proposition 3.23

Sec. 3.2 Opening and Closing

(ii) $A_1 \subset A_2$ implies $O(A_1, B) \subset O(A_2, B)$ (Increasing)

(iii) $O[O(A, B), B] = O(A, B)$ (Idempotent)

Proof.

(i) $z \in O(A, B)$ implies that there exists a translate $B + y$ such that $z \in B + y \subset A$. Hence, $O(A, B) \subset A$.

(ii) Once again applying Theorem 3.2, $B + y \subset A_1$ implies $B + y \subset A_2$.

(iii) By (i), $O[O(A, B), B] \subset O(A, B)$. For the converse,

$$O[O(A, B), B] = \{[((A \ominus (-B)) \oplus B] \ominus (-B)\} \oplus B$$
$$= C[A \ominus (-B), (-B)] \oplus B$$
$$\supset [A \ominus (-B)] \oplus B = O(A, B)$$

where the containment follows from the fact that the closing of a given set by another contains the original set. (This will be proven independently in Theorem 3.4, part i.)

Theorem 3.4.   The closing satisfies

(i) $C(A, B) \supset A$ (Extensive)

(ii) $A_1 \subset A_2$ implies $C(A_1, B) \subset C(A_2, B)$ (Increasing)

(iii) $C[C(A, B), B] = C(A, B)$ (Idempotent)

Proof.

(i) $C(A, B)^c = O(A^c, B) \subset A^c$ by duality and the antiextensivity of the opening. Taking complements reverses the inclusion sign, and (i) follows.

(ii) $C(A_1, B)^c = O(A_1^c, B) \supset O(A_2^c, B) = C(A_2, B)^c$ by duality. Taking complements gives (ii).

(iii) $C[C(A, B), B] = O[C(A, B)^c, B]^c$ (by duality)
$ = O[O(A^c, B), B]^c$
$ = O(A^c, B)^c$ (Theorem 3.3, part (iii))
$ = C(A, B)$

The idempotence properties in Theorems 3.3 and 3.4 motivate the following definitions. Set A is said to be *open with respect to set B* if $O(A, B) = A$. In other words, A is open with respect to B if opening A by B has no effect. Similarly, set A is said to be *closed with respect to set B* if $C(A, B) = A$. The idempotence properties can be restated with respect to this new terminology by saying that $O(A, B)$ is open with respect to B (see Figure 3.13) and $C(A, B)$ is closed with respect to B. The following proposition follows from duality.

Proposition 3.24. A is open with respect to B if and only if A^c is closed with respect to B.

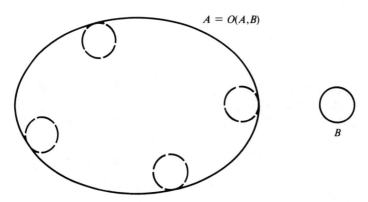

Figure 3.13 Set A Open with Respect to Set B

Proof. A is open with respect to B if and only if $O(A, B) = A$. Taking complements and applying duality gives logical equivalence with $A^c = C(A^c, B)$, which is precisely the criterion for A^c being closed with respect to B.

Henceforth we shall say that A is B-*open* if A is open with respect to B and that A is B-*closed* if A is closed with respect to B. The next two propositions give characterizations of these concepts in terms of the primitive morphological operations.

Proposition 3.25. A is B-open if and only if there exists a set E such that $A = E \oplus B$.

Proof. If A is B-open, then

$$A = O(A, B) = [A \ominus (-B)] \oplus B$$

Simply letting $E = A \ominus (-B)$ yields the forward implication. As for the converse, suppose that $A = E \oplus B$. Then

$$O(A, B) = O(E \oplus B, B) = [(E \oplus B) \ominus (-B)] \oplus B$$
$$= C(E, -B) \oplus B \supset E \oplus B = A$$

But by antiextensivity,

$$O(A, B) = O(E \oplus B, B) \subset E \oplus B = A$$

Consequently, $O(A, B) = A$, and A is B-open.

Proposition 3.26. A is B-closed if and only if there exists a set E such that $A = E \ominus B$.

Proof. If A is B-closed, then

$$A = C(A, B) = [A \oplus (-B)] \ominus B$$

Simply take $E = A \oplus (-B)$. Conversely, suppose $A = E \ominus B$. Then, by duality, $A = [E^c \oplus B]^c$, and taking complements gives $A^c = E^c \oplus B$. By Proposition 3.25, it follows that A^c is B-open, and hence, by Proposition 3.24, A is B-closed.

The next proposition states that if A is B-open, then opening by A produces a smaller image than opening by B, and conversely for closing by A and B. The proof is rather technical; however, it is included for the sake of completeness.

Proposition 3.27. Suppose A is B-open. Then, for any image F,

$$O(F, A) \subset O(F, B) \subset F \subset C(F, B) \subset C(F, A)$$

Proof. The principal part of the proof is to show the first inclusion. To that end, let $z \in O(F, A)$. Then, by the characterization of opening given in Theorem 3.2, there exists some point y such that

$$z \in A + y \subset F$$

However, since A is B-open, there exists an E such that $A = E \oplus B$. Consequently,

$$z \in (E \oplus B) + y \subset F$$

The first inclusion means that there exist points b' in B and e' in E such that $z = e' + b' + y$. The second inclusion means that $[B + (e + y)] \subset F$ for any e in E. Since $e' \in E$, this applies to e'. Hence, taken together, the two inclusions yield the corresponding relations

$$z = (e' + y) + b'$$

and

$$B + (e' + y) \subset F$$

Together, these imply that

$$z \in B + (e' + y) \subset F$$

which in turn implies that $z \in O(F, B)$. Hence, $O(F, A) \subset O(F, B)$.

The next two inclusions follow respectively from the antiextensivity and extensivity of the opening and the closing. We obtain the last inclusion by duality; in particular,

$$C(F, B) = O(F^c, B)^c \subset O(F^c, A)^c = C(F, A)$$

The inclusion follows from the first part of the proof applied to F^c where one must keep in mind that complementing both sides of a containment relation yields an inclusion relation.

The next proposition gives a generalization of the idempotent laws for the opening.

Proposition 3.28. If A is B-open, then for any image F,

(i) $O[O(F, B), A] = O(F, A)$

(ii) $O[O(F, A), B] = O(F, A)$

Proof.

(i) By antiextensivity of the opening, $O(F, B) \subset F$. But the opening is increasing; hence,

$$O[O(F, B), A] \subset O(F, A)$$

Moreover, by Proposition 3.27, since A is B-open, $O(F, A) \subset O(F, B)$. Consequently, by idempotence,

$$O(F, A) = O[O(F, A), A] \subset O[O(F, B), A]$$

and (i) is proven.

(ii) Suppose $z \in O(F, A)$. Then there exists y such that $z \in A + y \subset F$. But A is B-open. Hence,

$$z \in O(A, B) + y \subset F$$

Consequently, there exists some point w such that

$$z \in (B + w) + y \subset A + y \subset F$$

Since $O(F, A)$ consists of the union of all translates of A which are in F, this last relation implies that

$$z \in B + (w + y) \subset O(F, A)$$

Therefore, $z \in O[O(F, A), B]$, and it follows that

$$O(F, A) \subset O[O(F, A), B]$$

The reverse inclusion is immediate by the antiextensivity of the opening, and (ii) follows thereupon.

Proposition 3.29. If A^c is B-closed, then for any image F,

(i) $C[C(F, B), A] = C(F, A)$

(ii) $C[C(F, A), B] = C(F, A)$

Proof. From part (i) of Proposition 3.28, with F^c in place of F, we obtain, by duality,

$$C[O(F^c, B)^c, A] = C(F, A)$$

But, again by duality, $O(F^c, B)^c = C(F, B)$, and (i) follows. The second part follows similarly; in particular, by duality and part (ii) of Proposition 3.28, we have

$$C[O(F^c, A)^c, B] = C(F, A)$$

The desired result follows immediately by another application of duality.

3.3 CONVEX SETS

Convex sets play a significant role in Euclidean morphology, from both a granulometric and a measurement point of view. For that reason, some basic definitions and properties need to be presented. Except for those propositions which are particular to morphology, proofs will be omitted. They can be found in any standard treatise on convexity.

If x and y are points in the plane R^2, then the line segment between x and y is denoted by $\overline{xy}$ and is given by

$$\overline{xy} = \{ax + by: a \geq 0, b \geq 0, \text{ and } a + b = 1\},$$

where the '+' in '$ax + by$' actually denotes vector addition. A set A in R^2 is said to be *convex* if, for any points x and y in A, $\overline{xy} \subset A$. (See Figure 3.14.)

Given any set A, it is often useful to consider the "smallest" convex set containing A. This minimal convex enclosing set is called the *convex hull* of A and is denoted by $H(A)$. In general, any intersection of convex sets is convex. In particular, $H(A)$ is given by the intersection of all convex sets containing A. (See Figure 3.15.) A is convex if and only if $H(A) = A$. An analytic formulation of the convex hull can be provided by employing the notion of a *convex combination*, which is a vector sum of the form

$$z = t_1 x_1 + t_2 x_2 + \cdots + t_m x_m$$

where t_k is a nonnegative real number, $x_k \in R^2$, and $t_1 + t_2 + \cdots + t_m = 1$.

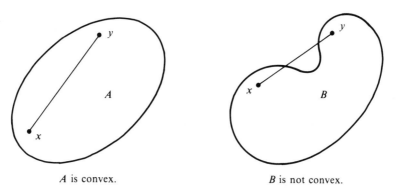

A is convex. B is not convex.

Figure 3.14 Convex and Nonconvex Sets

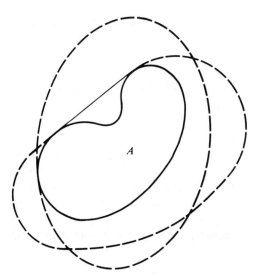

$H(A) = \cap \{B\colon B \supset A \text{ and } B \text{ is convex}\}$ **Figure 3.15**   Convex Hull

Proposition 3.30.[3] $H(A)$ is given by the set of all convex combinations of points in A.

An immediate consequence of Proposition 3.30 is that set A is convex if and only if every convex combination of points in A lies in A, for this means precisely that $H(A) \subset A$ and, consequently, that $A = H(A)$.

Proposition 3.30 is well known; a less known result, due to Caratheodory, states that every point $x \in H(A)$ in R^2 can be represented as a convex combination of at most three points in A.[4]

Example 3.2

Consider $S = \{(1, 1), (5, 1), (1, 5), (5, 5)\}$, the four corners of a square in the first quadrant. Then $H(S)$ is precisely the closed square. To illustrate the theorem of Caratheodory, consider the point $(2, 2)$, which is within the triangular convex hull of $\{(1, 1), (5, 1), (1, 5)\}$. We need to solve the vector equation

$$\begin{pmatrix} 2 \\ 2 \end{pmatrix} = t_1 \begin{pmatrix} 1 \\ 1 \end{pmatrix} + t_2 \begin{pmatrix} 5 \\ 1 \end{pmatrix} + t_3 \begin{pmatrix} 1 \\ 5 \end{pmatrix}$$

together with the constraint equation

$$1 = t_1 + t_2 + t_3$$

---

[3] Steven Lay, *Convex Sets and Their Applications* (New York: Wiley, 1982), 17.
[4] Ibid., 17.

Sec. 3.3 Convex Sets

But the vector equation yields the two equations

$$2 = t_1 + 5t_2 + t_3$$
$$2 = t_1 + t_2 + 5t_3$$

Simultaneous solution gives

$$t_1 = \frac{1}{2}, t_2 = \frac{1}{4}, t_3 = \frac{1}{4}$$

The preceding example illustrates two especially important types of convex sets. The convex hull of a finite collection of points is known as a *polytope*. The hull of three noncollinear points is called a *2-simplex*, and the points are called the *vertices* of the simplex. Notice in Example 3.2 that the solution for t_1, t_2, and t_3 was uniquely determined. In general, if $S = \{x_1, x_2, x_3\}$, then each point in the 2-simplex $H(S)$ has a unique convex combination representation in terms of the vertices. The coefficients in that representation are referred to as the *barycentric coordinates* of the point. In the preceding example, the barycentric coordinates of (2, 2) were 1/2, 1/4, and 1/4.

Of particular importance in morphology are convex sets that are *compact*. A compact set is one that is both *closed* and *bounded*. A closed set is one that contains its boundary, while a bounded set is one that is contained in a finite disk centered at the origin. (A point x is in the boundary of A if and only if each disk centered at x intersects both A and its complement A^c.) Two other topological definitions are the following. The *closure* of set A, denoted $\overline{A}$, is equal to the union of A with its boundary. A well-known result in topology states that A is closed if and only if $A = \overline{A}$. Finally, the interior of set A is the set-theoretic subtraction of the boundary of A from A.

Example 3.3

Let

$$A = \{(x, y): 0 \leq x \leq 1, 0 \leq y < 1\}$$
$$B = \{(x, y): x^2 + y^2 \leq 1\}$$
$$C = \{(x, y): y = 3x + 2\}$$

Then A is not closed since $\{(x, y): 0 \leq x \leq 1, y = 1\}$ is part of the boundary of A and is not contained in A. Moreover, the closure of A and the interior of A are respectively given by

$$\{(x, y): 0 \leq x \leq 1, 0 \leq y \leq 1\}$$

and

$$\{(x, y): 0 < x < 1, 0 < y < 1\}$$

The set B is closed and bounded, and hence compact. Set C, which is a line, is closed but not bounded. (See Figure 3.16.)

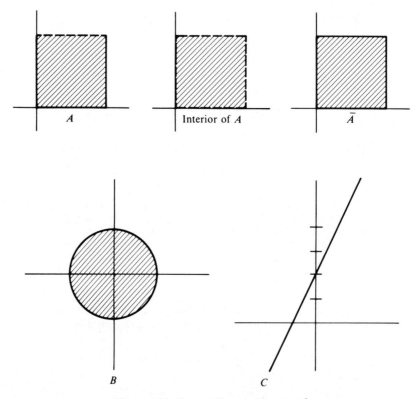

Figure 3.16 Several Types of Sets in R^2

Consider a set A and a line L in the Euclidean plane. For any point $x \in A$, let $\mathbf{n}(x)$ be the vector determined by the perpendicular line segment from x to L, where the initial point is the point on L and the terminal point is x. (See Figure 3.17.) In the event that $x \in L$, then $\mathbf{n}(x)$ is the zero vector. If it happens that A lies on one side of L, then all the nonzero vectors $\mathbf{n}(x)$ point in the same direction. Formally, consider the dot product $\mathbf{n}(x) \cdot \mathbf{n}(y)$. There are three possibilities.

(i) $\mathbf{n}(x) \cdot \mathbf{n}(y) = \|x\| \|y\| > 0$, which means that $\mathbf{n}(x)$ and $\mathbf{n}(y)$ point in the same direction.

(ii) $\mathbf{n}(x) \cdot \mathbf{n}(y) = -\|x\| \|y\| < 0$, which means that $\mathbf{n}(x)$ and $\mathbf{n}(y)$ point in opposite directions.

(iii) $\mathbf{n}(x) \cdot \mathbf{n}(y) = 0$, which means that either x or y is the zero vector.

L *bounds* A whenever either (i) or (iii) holds for every $x, y \in A$.

A line L is said to *support* a set A at $x \in A$ if $x \in L$ and L bounds A. (See Figure 3.18.) The following proposition is fundamental.

Sec. 3.3 Convex Sets

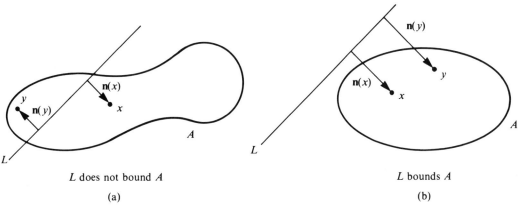

L does not bound A   L bounds A
(a)   (b)

Figure 3.17 Lines that Do and Do Not Bound the Set A

Proposition 3.31.[5] Suppose x is a boundary point of the closed convex set A. Then there exists at least one line supporting A at x.

To get an intuitive understanding of Proposition 3.31, suppose that the boundary of A is given by the parametric representation $x = (x_1, x_2)$, where $x_1 = x_1(s)$ and $x_2 = x_2(s)$ have arc length as parameter, and where the curve is traversed in a counterclockwise direction as s goes from zero to $\Lambda(A)$, the length of the boundary of A. (See Figure 3.19.) Because A is convex, the curvature κ, if it exists, must be nonnegative. If we suppose that κ is actually positive at some point $x(s_0)$ on the boundary, then there exists a tangent vector $\mathbf{T} = (x_1'(s_0), (x_2'(s_0))$ at $x(s_0)$, and the tangent line at that point is given by the equation

$$x_2'(s_0)[z_1 - x_1(s_0)] = x_1'(s_0)[z_2 - x_2(s_0)]$$

where (z_1, z_2) denotes an arbitrary point on the line and the prime denotes differentiation with respect to s. It is precisely this line which supports A at $x(s_0)$.

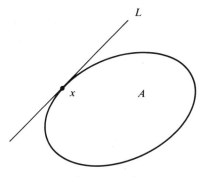

L supports A at x   **Figure 3.18** Line that Supports Set A

[5] Ibid., 41.

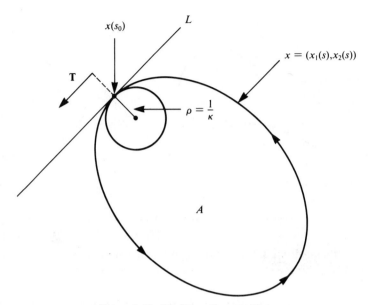

Figure 3.19 Finding a Support Line

Also, since $\kappa > 0$, there is an osculating circle supported by the same line at x and also lying on the same side of the line.

If we consider sets with nonempty interiors, then the supporting-line criterion can be used to characterize closed convex sets. Indeed, it can be shown that if A is closed and has a nonempty interior, then A is convex if and only if, for any boundary point $x \in A$, there exists at least one line supporting A at x.[6]

We next proceed to some properties of convex sets which are of specific interest in morphological image analysis.

Proposition 3.32. If A and B are convex, then so are $A \oplus B$, $A \ominus B$, $O(A, B)$, and $C(A, B)$.

Proof. Let $z, w \in A \oplus B$, $r + s = 1$, and $r, s \geq 0$. Since z and w are elements of the Minkowski addition, $z = a + b$ and $w = a' + b'$, where $a, a' \in A$ and $b, b' \in B$. Hence, due to the convexity of A and B,

$$rz + sw = r(a + b) + s(a' + b')$$
$$= (ra + sa') + (rb + sb') \in A \oplus B$$

and $A \oplus B$ is convex. Since the intersection of convex sets is again a convex set, and since translation is a special case of Minkowski addition,

$$A \ominus B = \bigcap_{b \in B} A + b$$

[6] Ibid., 41.

Sec. 3.3   Convex Sets   119

is convex. The convexity of the opening and closing is immediate since each is an iteration of a Minkowski sum and a Minkowski subtraction.

It should be noted that the proof of Proposition 3.32 actually shows that $A \ominus B$ is convex whenever A alone is convex. Before proceeding, we mention the fact that for any $t > 0$, tA is convex whenever A is convex.

In general, $(r + s)A \subset rA \oplus sA$ since $x \in (r + s)A$ means that there exists an $a \in A$ such that

$$x = (r + s)a = ra + sa \in rA \oplus sA$$

What is of interest is that actual equality is achieved if A is convex.

Proposition 3.33.   If A is convex, then $(r + s)A = rA \oplus sA$ for $r, s \geq 0$.

Proof. Let $x = ra + sa' \in rA \oplus sA$. Since A is convex,

$$\left(\frac{r}{r+s}\right)a + \left(\frac{s}{r+s}\right)a' \in A$$

Scalar multiplication by $r + s$ then shows that $x \in (r + s)A$, so that $rA \oplus sA \subset (r + s)A$. It follows immediately that $(r + s)A = rA \oplus sA$.

Now suppose $r \geq s > 0$; then, according to the preceding result, if A is convex,

$$rA = [s + (r - s)]A = sA \oplus (r - s)A$$

Consequently, according to Proposition 3.25, rA is sA-open. Hence, we have the following proposition, which is fundamental to the method of granulometries.

Proposition 3.34.   If $r \geq s > 0$ and B is convex, then for any set A, $O(A, rB) \subset O(A, sB)$.

Proof. By the remarks preceding Proposition 3.34, rB is sB-open. The result follows at once from Proposition 3.27.

Figure 3.20 gives an illustration of Proposition 3.34 for the case of a unit disk B.

The remarks preceding Proposition 3.34 can be restated by saying that if A is convex, then tA is A-open for any $t \geq 1$. Matheron has shown that if we assume that A is compact then the converse holds. Though the proof is beyond the level of the present text, the theorem can be stated. Its fundamental importance will be seen later in this chapter.

Theorem 3.5[7] **(Matheron).**   Let A be a compact set. Then tA is A-open for any $t \geq 1$ if and only if A is convex.

[7] Matheron, *Random Sets and Integral Geometry,* 196.

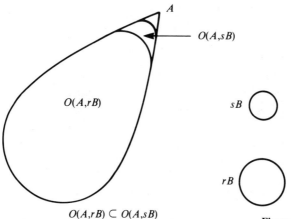

Figure 3.20 Illustration of Proposition 3.34

3.4 GRANULOMETRIES

Suppose A is a compact set (image) in R^2. The fundamental approach of morphological image analysis is to operate on A by some image-to-image transformation Ψ and then to analyze the transformed image. The most basic manner in which to study the effect of Ψ on A is to evaluate the area of $\Psi(A)$. The whole procedure then takes the form

$$A \xrightarrow{\Psi} \Psi(A) \xrightarrow{m} m[\Psi(A)]$$

where m denotes area (for those with measure theory, the Lebesgue measure). In fact, one does not merely make a single transformation, but instead, one studies a whole family of parameterized transformations. The measurement of the outputs of the transformations results in a so-called size distribution, which contains textural or other image-analytic information about the original image.

In this section, we shall study the class of transformations generated by the successive openings of some image A by a family of sets known as *structuring elements*. The procedure is quite straightforward. Let E, the *generating* structuring element, be a convex set. Then, according to Proposition 3.34, the mapping $t \to O(A, tE)$ is decreasing, in the sense that $t \geq t'$ implies that $O(A, t'E) \supset O(A, tE)$. As a consequence, the real-valued function $t \to u_A(t) = m[O(A, tE)]$ is also decreasing. Moreover, as long as E consists of more than a single point, the opening of A by tE will be empty for sufficiently large t. Hence, $u_A(t) = 0$ for sufficiently large t. The mapping $t \to O(A, tE)$ is known as a *granulometry* and the function $A \to u_A(t)$ is known as the *size distribution* generated by the granulometry. Note that when no possible confusion can arise we simply write $u(t)$.

Sec. 3.4   Granulometries   121

Example 3.4

Consider the image A which is a rectangle of length 2 and width 3. Suppose E is a square of edge length 1. Then

$$O(A, tE) = \begin{cases} A & \text{if } t \leq 2 \\ \emptyset & \text{if } t > 2. \end{cases}$$

Hence,

$$u(t) = \begin{cases} 6 & \text{if } t \leq 2 \\ 0 & \text{if } t > 0 \end{cases}$$

The graph of $u(t)$ is given in Figure 3.21(b).

Example 3.5

Consider the same image as in the previous example, only this time let E be the closed unit disk. A typical opening of A by E is given in Figure 3.21(a). In general,

$$u(t) = \begin{cases} \pi t^2 + 2(3 - 2t) + 2t(2 - 2t) & \text{if } t \leq 1 \\ 0 & \text{if } t > 1 \end{cases}$$

The graph of $u(t)$ for this example is given in Figure 3.21(c).

The preceding examples demonstrate the manner in which the shape of an object can lead to different size distributions depending upon the structuring element employed. What is of interest is that different shapes lead to different size distributions when opened by the same one-parameter family of structuring elements. Should an image contain intricate textural properties, these can be revealed by granulometric analysis. Because of the complexity of the calculations involved, digital morphological processing needs to be utilized. This will be discussed in Section 3.7.

From an intuitive standpoint, the granulometric process represents a type of filtering in which those sections of the image which are not sufficiently large to hold the structuring element tE are removed from the image. Indeed, Figure 3.22 gives an illustration of just such a filtering when the structuring elements E, $2E$, and $3E$ are applied to the image in the figure. Put in classical engineering terminology, a granulometry is a one-parameter family of low-pass filters. The filters are low pass because it is high-frequency fluctuation between a set and its complement which is attenuated in the output images when the structuring element is sufficiently smooth.

Returning to the granulometries themselves, let $\Psi_t(A)$ denote the opening $O(A, tE)$. Then the family $\{\Psi_t\}$ has numerous algebraic properties. These will be discussed in detail from an axiomatic point of view in Section 3.12; however, the following will be listed here without proof for those who are not concerned with the theoretical underpinnings:

(i) $r \geq s$ implies that $\Psi_r(A) \subset \Psi_s(A)$.

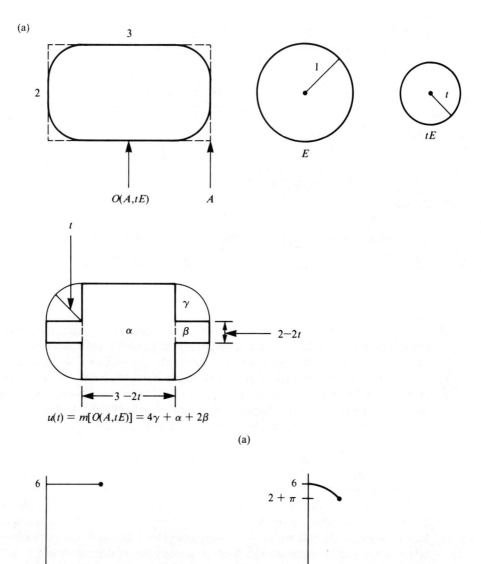

Figure 3.21 Size Distributions

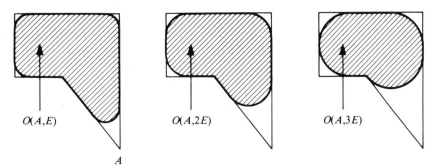

Figure 3.22 Granulometry as a Form of Filtering

(ii) $A \subset B$ implies that $\Psi_t(A) \subset \Psi_t(B)$.
(iii) For any $r, s \geq 0$, $\Psi_r[\Psi_s(A)] = \Psi_s[\Psi_r(A)] = \Psi_{\max(r,s)}(A)$.
(iv) $\Psi_t(A + x) = \Psi_t(A) + x$.

In brief, (i) has already been mentioned, (ii) states the fact that opening by tE is an increasing transformation, and (iii) states that opening by an iteration of rE and sE is equivalent to opening solely by the scalar multiple of the maximum of r and s times E. Property (iv) says that opening a translate is the same as translating the corresponding opening.

Returning to Examples 3.4 and 3.5, in both, the graph of $u(t)$ is continuous from the left. Although it will not be proved here, as long as A is compact and the structuring element is compact and convex, as is the case in these examples, the resulting size distribution $u(t)$ will be continuous from the left.

3.5 IMAGE FUNCTIONALS

Letting $\mathcal{K}$ denote the class of nonempty compact sets in R^2, we can consider the area (Lebesgue measure) m as a real-valued mapping of compact images, $m: \mathcal{K} \to R$. Any such real-valued mapping on images will be called an *image functional*. The measure m is only one of many such image functionals. When applying intelligence techniques to images, it is necessary to work on the output of image functionals since an image itself contains far too much information. However, not all image functionals are suitable for morphological analysis. In the first place, certain requirements must be imposed. And secondly, a given functional may behave differently on some classes of images than it does on others. Consequently, we seek to characterize those functional properties that are image relevant. In order to give structure to the presentation, we shall enumerate the various properties that image functionals might satisfy by the letter Q followed by a number.

The most obvious requirement that might be imposed upon an image functional q is invariance under translation. Indeed, imagine an image recognition algorithm operating on some input photograph, and suppose it is trying to recognize the existence of some object such as a bridge. Whatever feature parameters might be involved in the recognition procedure, it is unlikely, without a priori information, that we wish the position of the bridge and its background within the actual photograph to be of consequence. Similarly, if the image is to be used to characterize regions of uniform textural variation, the positions of the regions within the overall grid structure should not influence textural parameters. Accordingly, we shall often require *translational invariance* of image functionals, that is,

Q1: $q(A + x) = q(A)$ for any $x \in R^2$

Some immediate examples of functionals satisfying Q1 are the area $m(A)$, the perimeter $\Lambda(A)$ (if it exists) and the length of any projection of A onto a line. The last functional turns out to be important in morphological analysis. If **u** is a unit vector, then Proj(A, **u**) denotes the length of the projection of A onto any line perpendicular to **u**. (See Figure 3.23.) Since each **u** is determined by an angle θ with the x-axis, where $0 \leq \theta < 2\pi$, we sometimes write Proj(A, θ).

A second property an image functional might satisfy is *rotational invariance*. Thus, we have

Q2: $q[R(A)] = q(A)$, where $R(A)$ is a rotation of A

The desirability of rotational invariance is not as clear cut as that of translational invariance. For instance, suppose that recognition depends upon the vertical and

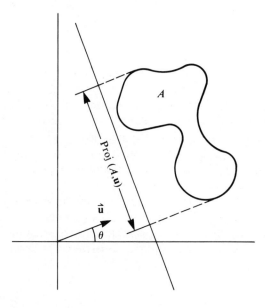

Figure 3.23 Length of Projection

horizontal relation between component parts of an image. Then rotation of the image would change these relationships. On the other hand, if a missile is using image information to home in on a target, the polar orientation of the object will sometimes be of no consequence. It should be obvious that m and Λ are rotationally invariant but that Proj is not. However, if the projections are averaged over all unit vectors, then the resulting average projection, given by

$$\frac{1}{2\pi} \int_0^{2\pi} \text{Proj}(A, \theta) \, d\theta = \frac{1}{\pi} \int_0^{\pi} \text{Proj}(A, \theta) \, d\theta$$

is rotationally invariant.

The third property we wish to consider is called *homogeneity*. Suppose that a toxicological slide is viewed under different magnifications. Certainly we do not wish intelligence-based decisions regarding the microscopic structure to be dependent upon the degree of magnification. Hence, consider a scalar multiple of the set A, say tA, for $t > 0$. It is well known that the area satisfies the relationship $m(tA) = t^2 m(A)$. This equation can be used to relate the operation of the functional under different magnifications. This notion leads to the image functional property of *homogeneity of degree k*:

Q3: There exists some positive constant k such that $q(tA) = t^k q(A)$.

Area m is homogeneous of degree 2.

Now consider the perimeter Λ. To avoid undue analytical difficulties, suppose that the boundary of the set A is piecewise smooth and is given by the parametric representation $x = x(v)$, $y = y(v)$, for $v_1 \leq v \leq v_2$. Then the boundary consists of a finite number of arcs on which both x and y have continuous derivatives. Moreover, tA has boundary given by $tx(v)$ and $ty(v)$. Hence,

$$\Lambda(tA) = \int_{v_1}^{v_2} [(tx'(v))^2 + (ty'(v))^2]^{1/2} \, dv = t\Lambda(A)$$

so that Λ is homogeneous of degree 1.

In order to examine the homogeneity of the projection onto a line, consider the projection of A vertically onto the x-axis. Assume that A consists of a single connected component and is compact. Then there exist points $z_1 = (x_1, y_1)$ and $z_2 = (x_2, y_2)$ in A such that x_1 is the minimum x-coordinate for all z in A and x_2 is the maximum x-coordinate for all z in A, and it follows that

$$\text{Proj}(A, \mathbf{u}) = x_2 - x_1$$

and

$$\text{Proj}(tA, \mathbf{u}) = tx_2 - tx_1 = t \, \text{Proj}(A, \mathbf{u})$$

Since the length of the projection in any direction can be found by rotating and then projecting vertically, we conclude that Proj is homogeneous of degree 1 whenever it is applied to compact sets consisting of a single component. A similar argument applies to compact sets consisting of a finite number of components.

To this point we have not emphasized the fact that the validity of any image-functional property depends upon the class over which it is to be applied. The matter is of some concern, however, in regard to the remaining properties to be discussed. Those familiar with measure theory are aware that m satisfies the *additivity* property

$$\text{Q4: } q(A \cup B) = q(A) + q(B) - q(A \cap B)$$

Implicit in the additivity of m is the measure-theoretic proposition that if A and B are measurable then so are $A \cup B$ and $A \cap B$. (If A and B have areas, so do $A \cup B$ and $A \cap B$.) A similar remark in fact applies to Q2 in that if A is measurable then tA is measurable. In other words, the area m satisfies Q2 and Q4 provided the sets under consideration are measurable. (See Figure 3.24.)

The difficulty inherent in the underlying collection of sets (images) comes to the fore in the morphological study of convex sets. Simply because A and B are convex does not guarantee that $A \cup B$ is convex. To apply Q4 to the collection of convex compact sets, we must state additivity under the assumption that $A \cup B$ is convex. The result is the property of *C-additivity*, which states that q satisfies Q4 as long as A, B, and $A \cup B$ are convex. (If A and B are convex, then $A \cap B$ is convex.) In effect, C-additivity is a weaker condition than measure additivity since a functional may be C-additive but not additive over arbitrary unions of measurable sets. It is of consequence that Λ is C-additive. The notion of the underlying collection of sets upon which an image functional is to be considered is not purely academic; in fact, it determines the domain over which certain artificial intelligence (AI) decision techniques may be applied.

The next property of functionals we wish to present is the *increasing* property:

$$\text{Q5: } A \subset B \text{ implies that } q(A) \leq q(B)$$

Because of their favorable behavior, increasing image functionals are employed extensively in morphological analysis. Both m and Proj are increasing. Though Λ is not increasing in general, it is increasing on convex compact sets.

The last two image functional properties to be presented are topological in

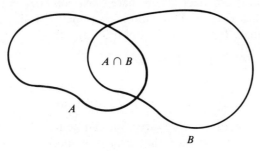

$m(A \cup B) = m(A) + m(B) - m(A \cap B)$
$(\text{Area}(A \cup B) = \text{Area}(A) + \text{Area}(B) - \text{Area}(A \cap B))$ **Figure 3.24** Additivity of Area

Sec. 3.5 Image Functionals

that they involve limits. A notion of "distance" between two nonempty compact sets shall be introduced. For any nonempy compact set $A \in \mathcal{H}$ and for any point $x \in R^2$, we define the *distance* from x to A as

$$d(x, A) = \inf\{\,|x - a| : a \in A\}$$

Using d, we define the *Hausdorff metric* on the class $\mathcal{H}$ of nonempty compact sets. For any two sets A and B in $\mathcal{H}$,

$$h(A, B) = \max(\sup_{b \in B} d(b, A),\ \sup_{a \in A} d(a, B))$$

Intuitively, the Hausdorff metric measures the greatest distance by which A differs from B. (See Figure 3.25.)

In terms of the behavior of image functionals, the Hausdorff metric plays a key role; indeed, to say that $\lim_{n \to \infty} h(A_n, A) = 0$, where $\{A_n\}$ is a sequence of compact sets and A is a compact set, is to say that for large n, A_n is nearly identical to A. Hence, as an image, A_n appears very much like A. This is why the Hausdorff metric is used to define image convergence for compact sets; i.e., we define $\lim_{n \to \infty} A_n = A$ if $h(A_n, A) \to 0$ as $n \to \infty$. (See Figure 3.26.)

The concept of continuity in calculus says that a real-valued function f is continuous at point a if $\lim_{x \to a} f(x) = f(a)$. This limit notion is generalized topologically into many areas of mathematics. Insofar as image functionals are concerned, *continuity* is given by

$$Q6: \lim_{n \to \infty} q(A_n) = q(A) \text{ whenever } \lim_{n \to \infty} A_n = A$$

Intuitively, the image functional q is continuous if, whenever the compact images A_n approach the compact image A, the functional values of A_n approach the functional value of A. But care must be taken: once again, the class of images under

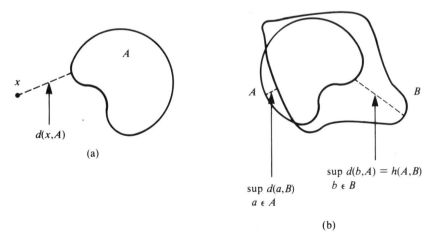

Figure 3.25 Hausdorff Metric

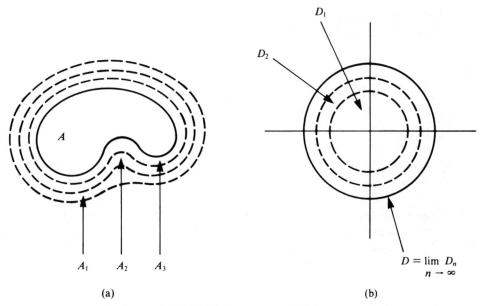

Figure 3.26 Convergence of Sets

consideration is vitally important. For instance, if we restrict our attention to convex compact sets, then Λ is continuous; that is, $\lim_{n\to\infty} A_n = A$ implies that $\lim_{n\to\infty} \Lambda(A_n) = \Lambda(A)$. However, this is not true for compact sets in general.[8]

Example 3.6

Consider the sequence of unit disks of radius $1 - 1/2n$ and centered at the origin. (See Figure 3.26(b).) Call these disks D_n, and let D denote the unit disk centered at the origin. Then $D_n \to D$, and also,

$$\Lambda(D_n) = 2\pi(1 - 1/2n) \to 2\pi = \Lambda(D)$$

as it must since Λ is continuous on the class $\mathcal{H}_c$ of nonempty convex compact sets.

Example 3.7

Consider again the closed unit disk D. Suppose a grid with mesh $1/2^k$ is placed over D in such a way that the origin of the grid is situated on the center of D. Let A_k be the closed polygon that is obtained by taking the union of all grid squares whose interiors intersect D. (See Figure 3.27.) Then, as grids of finer and finer mesh size are taken (i.e., as $k \to \infty$), $\lim_{k\to\infty} A_k = D$ (using the Hausdorff metric). But $\Lambda(A_k) \geq 8$ for all k, while $\Lambda(D) = 2\pi$. Hence, Λ is not continuous over the collection $\mathcal{H}$ of nonempty compact sets.

---

[8] For an advanced calculus-level description of the Hausdorff metric and some of its basic properties with regard to morphology, see E. R. Dougherty and C. R. Giardina, "Binary Euclidean Images and Convergence," *Intelligent Systems and Machines* (1986).

Sec. 3.5   Image Functionals   129

It turns out that a weaker condition than image-functional continuity is suitable for morphological analysis. Consider a collection of nonempty compact sets A_n which are *nested;* i.e., for which $A_1 \supset A_2 \supset A_3 \supset \ldots$. The Cantor Intersection Theorem, a well-known theorem from advanced calculus, states that there is a nonempty intersection A which is also compact. An example of such a sequence is given in Example 3.26(a). Another well-known theorem, this time from measure theory, states that for such a nested collection, $\lim_{n\to\infty} m(A_n) = m(A)$. In accordance with these theorems, an image functional q is said to be *continuous from above* if

Q7: q is increasing and, for any nested sequence of compact nonempty sets,
$$\lim_{n\to\infty} q(A_n) = q(A), \text{ where } A = \bigcap_{n=1}^{\infty} A_n$$

As with the other criteria presented, one might choose to reduce the class of sets over which Q7 is to be employed. For instance, one might speak of an image functional's being continuous from above over convex compact sets. Note that

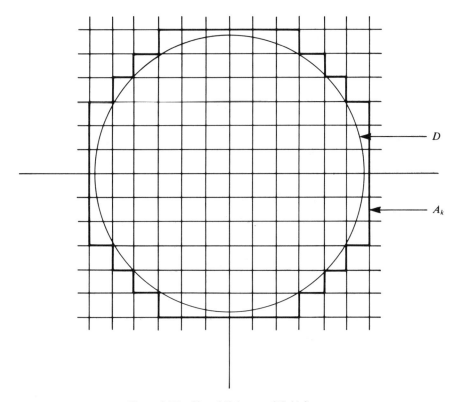

Figure 3.27 Closed Polygon of Grid Squares

Q7 assumes that q is increasing. Note also that if q is increasing and continuous, then it must be continuous from above.

Morphological feature generation involves the application of some image functional to the output of some collection of image-to-image transformations. For instance, the granulometries were given by the operator sequence

$$A \xrightarrow{\Psi_t} O(A, tE) \xrightarrow{m} u_A(t) = m[O(A, tE)]$$

In this instance, the composition $m \circ \Psi_t$ can be considered an image functional since its input is an image A and its output is a real parameter $u_A(t)$. If, as a case in point, we consider only the compact sets of $\mathcal{H}$, then for fixed t, $u(t): \mathcal{H} \to R$. An important question is how the properties Q1 through Q7 apply to such composition-type image functionals. For example, in the case of the area of the opening, assuming the structuring element to be compact and convex, the opening is increasing on $\mathcal{H}$. Hence, since m is increasing, then so is the composite $u(t) = m \circ \Psi_t$. Indeed, if $A \subset B$, then $O(A, tE) \subset O(B, tE)$, and this in turn implies that

$$u_A(t) = m[O(A, tE)] \leq m[O(B, tE)] = u_B(t)$$

Fortunately, we do not have to always consider such specific cases as the area of the opening. The next two propositions are very helpful. In each case, the function compositions are well defined since it can be shown that whenever A and B are compact, $A \oplus B$, $A \ominus B$, $O(A, B)$, and $C(A, B)$ are also compact.

Proposition 3.35. Suppose q is an increasing image functional defined on $\mathcal{H}$. Then the composition of q with dilation by a compact set, erosion by a compact set, opening by a compact set, or closing by a compact set results in an increasing image functional on $\mathcal{H}$.

Proof. Consider the mapping $q \circ \mathcal{D}_B$ on $\mathcal{H}$, where $\mathcal{D}_B$ represents dilation by B. By Proposition 3.10, together with the fact that q is increasing, if $A_1 \subset A_2$,

$$[q \circ \mathcal{D}_B](A_1) = q[\mathcal{D}(A_1, B)] \leq q[\mathcal{D}(A_2, B)] = [q \circ \mathcal{D}_B](A_2)$$

The other cases are proved similarly.

Proposition 3.36. Suppose q is a translationally invariant image functional on $\mathcal{H}$. Then the composition of q with dilation by a compact set, erosion by a compact set, opening by a compact set, or closing by a compact set results in a translationally invariant image functional on $\mathcal{H}$.

Proof. Proposition 3.3, together with the translational invariance of q, implies that, for dilation $\mathcal{D}_B$,

$$[q \circ \mathcal{D}_B](A + x) = q[(A + x) \oplus B] = q[(A \oplus B) + x]$$
$$= q(A \oplus B) = [q \circ \mathcal{D}_B](A)$$

A similar argument applies to the other cases.

Example 3.8

Since the length of a projection in a given direction, Proj(A, **u**), is translationally invariant, then so is an opening followed by a projection; indeed, according to the previous result,

$$\text{Proj}[O(A + x, B), \mathbf{u}] = \text{Proj}[O(A, B), \mathbf{u}]$$

We state the following proposition regarding the property of continuity from above without proof. The proof requires techniques from advanced calculus.

Proposition 3.37. If q is continuous from above on $\mathcal{H}$, then so are the compositions of q with dilation by a compact set, erosion by a compact set, opening by a compact set, and closing by a compact set.

3.6 INTEGRAL GEOMETRY AND IMAGE FUNCTIONALS

Image functionals that have been adopted from the area of integral geometry have proven to be instrumental in the morphological analysis of images. Five salient reasons for the success of these functionals are the following:

1. They represent quantifications of geometrically intuitive, and hence perceptually relevant, image characteristics.
2. They are closely related to each other and their properties have been extensively investigated, especially by the German mathematician H. Hadwiger.
3. Fundamental theorems of integral geometry relate parameters as they apply to different-dimensional spaces. Consequently, integral geometric image functionals are *stereological* in that parameters computed for two-dimensional images taken by sectioning three-dimensional bodies may be utilized to estimate related three-dimensional parameters of the original body, a problem central to biology and the material sciences. Since the exposition at hand is concerned solely with image processing, stereological considerations will not be specifically pursued.
4. Image functionals from integral geometry often have representations in terms of the morphological operations of Minkowski algebra.
5. Upon digitization of an image, the digital versions of the Minkowski algebra operations can be implemented in a highly parallel fashion, and hence, digital approximations of many integral geometric functionals are obtainable under acceptable time constraints.

In this section, we shall continue discussing the image functionals m, Λ, and Proj, which are parameters central to integral geometry. We shall also present some basic results concerning integral geometry and morphological analysis.

In the preceding section, the projection length Proj(A, θ) of A onto a line perpendicular to the direction θ was mentioned frequently. We also mentioned the rotational invariance of the average of these projections. In fact, for a convex compact set $A \in \mathcal{K}_c$,

$$\Lambda(A) = \pi \frac{1}{2\pi} \int_0^{2\pi} \text{Proj}(A, \theta)\, d\theta$$

In words, the length of the boundary equals π times the average length of the projection. This is the well-known Cauchy projection theorem for convex compact sets in R^2. Note that because of symmetry, the average need be taken only over 0 to π. In such a case, the denominator of the coefficient of the integral is π.

Example 3.9

Let A be a square of edge length 1. Then $\Lambda(A) = 4$. Since the orientation of A is inconsequential when averaging the projections, suppose it is situated with one side horizontal as in Figure 3.28. Then, from the figure, it can be seen that whatever the value of θ between 0 and π, the length of the projection is given by $|\cos \theta|$ + $|\sin \theta|$. Therefore, the average projection length is given by

$$\frac{1}{\pi} \int_0^{\pi} [\, |\cos \theta| + |\sin \theta|\,]\, d\theta = \frac{4}{\pi} = \frac{1}{\pi} \Lambda(A)$$

as the Cauchy theorem states it must.

Numerous image-functional properties relating to m and Λ have been mentioned thus far. If the underlying collection of sets is $\mathcal{K}_c$, so that m and Λ are

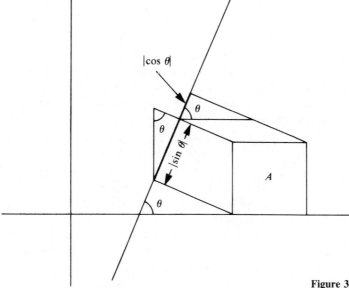

Figure 3.28 Square with Edge 1

restricted to convex compact sets, then both image functionals satisfy Q1 through Q6, where homogeneity for m is of degree 2 and for Λ is of degree 1. (Note that Q7 is also satisfied since the stronger continuity condition Q6 is satisfied.) A fundamental characterization theorem has been given by H. Hadwiger which in essence states a converse in terms of Q1, Q2, Q4, and Q6.

Theorem 3.6[9] **(Hadwiger).** Suppose q is an image functional on $\mathcal{H}_c$ which is invariant with respect to translation and rotation, is C-additive, and is continuous. Then q can be represented as a linear combination of m, Λ, and 1. That is, there exist constants a, b, and c such that for any convex compact set A,

$$q(A) = am(A) + b\Lambda(A) + c$$

Moreover, if the continuity assumption is replaced by an assumption that q is increasing, then not only is q also continuous on $\mathcal{H}_c$, but the constants a, b, and c are nonnegative.

Several comments concerning Hadwiger's Theorem are in order. We shall concentrate on the form where q is assumed to be increasing, since this is a more natural requirement than is continuity.

The Hadwiger Theorem limits the form of those image functionals for which we require monotonicity (that the functional be increasing), additivity, and translational and rotational invariance. Given any image functional q on some class of compact sets, say $\tilde{\mathcal{H}}$, which contains $\mathcal{H}_c$, the restriction of q to $\mathcal{H}_c$ (not q itself!) must be of the form given in the theorem. While this is quite a strong conclusion, one must be careful not to overplay the consequences of the Hadwiger Theorem. First, note that it requires rotational invariance, which, as has been previously mentioned, is certainly not desirable in all circumstances. Moreover, the assumption of rotational invariance cannot be dropped from the hypothesis. Second, the characterization of an image functional on $\mathcal{H}_c$ does not uniquely determine it on some larger domain such as $\mathcal{H}$ itself. For consider the class $\mathcal{R}_c$ consisting of finite unions of convex compact sets. This class is known as the *convex ring* since if A, $B \in \mathcal{R}_c$ then also $A \cap B$, $A \cup B \in \mathcal{R}_c$ (as long as we allow the null set to be an element of $\mathcal{R}_c$). For any functional of the form $am + b\Lambda + c$ on $\mathcal{H}_c$, there exist numerous extensions to $\mathcal{R}_c$.[10] Finally, the theorem demands the additivity requirement. Nonadditive functionals are not restricted by Hadwiger's Theorem. Indeed, continuity from above, such as that mentioned in Proposition 3.37 for compositions including erosion, opening, and closing, appears to be a much more natural requirement than additivity. This coincides with a current trend in AI where the notion of a "fuzzy measure" is nothing more than a generalization of continuity from above.

---

[9] H. Hadwiger, *Vorslesungen Uber Inhalt, Oberflache Und Isoperimetric* (Berlin: Springer, 1957), 221.

[10] Matheron, *Random Sets and Integral Geometry*, 119.

Before we leave Hadwiger's theorem, its relation to the Cauchy theorem should be noted. The two functionals in the Hadwiger characterization are related in that $\Lambda(A)$ can be obtained by averaging lengths of projections of A, each of which is actually a one-dimensional measure parameter, whereas $\Lambda(A)$ is a two-dimensional parameter. This is one of the stereological relations mentioned earlier. If we were to consider three-dimensional parameters, Hadwiger's Theorem would generalize and each of the relevant parameters would be an average of two-dimensional parameters, in one case of areas and in another of perimeters. In fact, the theorem is generalizable to n-dimensional Euclidean space, and the parameters involved are known as the *Minkowski functionals*.[11]

It was previously mentioned that the Minkowski algebra operators can be employed to find geometric parameters. As an illustration, let us suppose that A consists of a single simply connected component. (That is, A is connected and has no holes, but is not necessarily convex.) Let B be the unit disk centered at the origin. Then $A \oplus tB$ equals the set of all points within t of the set A; graphically, it is A together with a band C of width t about it. Letting m denote the area (measure), we have

$$m(A \oplus tB) = m(A) + m(C)$$

By means of the unit normal **N**, the unit tangent **T**, and the curvature κ, we see (Figure 3.29) that

$$m(C) = \int_0^{\Lambda(A)} \left[t\, ds + \frac{1}{2} \text{sign}(\kappa)\, t^2 \sin \theta \right]$$

But

$$\sin \theta = |\mathbf{T}(s) \times \mathbf{T}(s + ds)| = |\mathbf{T}(s) \times [\mathbf{T}(s) + \kappa\, ds\, \mathbf{N}(s)]|$$
$$= |\kappa\, ds|$$

since **T** is perpendicular to **N**. Consequently,

$$m(C) = \int_0^{\Lambda(A)} \left[t\, ds + \frac{1}{2} t^2 \kappa(s)\, ds \right]$$
$$= t \int_0^{\Lambda(A)} ds + \frac{1}{2} t^2 \int_0^{\Lambda(A)} \kappa(s)\, ds$$
$$= t\Lambda(A) + \pi t^2$$

(Note that the above demonstration holds with only slight modification as long

---

[11] Hadwiger, *Vorslesungen Uber Inhalt, Oberflache Und Isoperimetric*, 221.

Sec. 3.6   Integral Geometry and Image Functionals

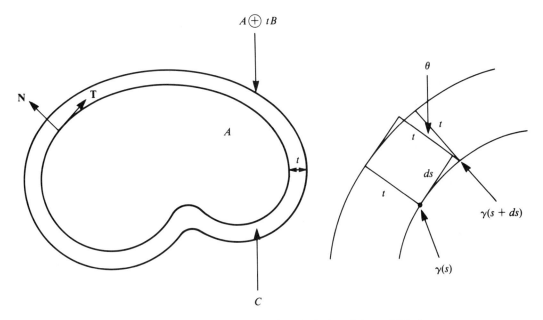

Figure 3.29   Parameters in Calculation of Area of Dilation

as there exists a tangent at all but a finite number of points of the boundary of A.) As a result,

$$\lim_{t \to 0} \frac{m(A \oplus tB) - m(A)}{t} = \lim_{t \to 0} \frac{t\Lambda(A) + \pi t^2}{t} = \Lambda(A)$$

In other words, using the notion of a one-sided derivative, we have

$$\Lambda(A) = \frac{d}{dt} [m(A \oplus tB)] \bigg|_{t=0}$$

Though the derivation has been carried out for a set having a single simply connected component, it can be carried out for reasonably well-behaved sets that consist of more than one component or are multiply connected (have holes). In sum, the total perimeter of a set can be found by using dilation. In practice, this methodology must be accomplished digitally.

Example 3.10

Let A be a 2 by 3 rectangle. (See Figure 3.30.) Then $m(A \oplus tB) = 6 + 10t + \pi t^2$. Therefore,

$$\lim_{t \to 0} \frac{1}{t} [m(A \oplus tB) - m(A)] = \lim_{t \to 0} [10 + \pi t] = 10$$

which is the perimeter of the rectangle.

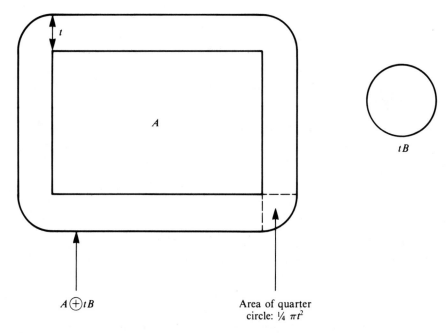

Figure 3.30 Area of Dilation of Rectangle

It is also possible to find a projection length of a simply connected component by use of the dilation.[12] Consider the Minkowski sum $A \oplus tE$, where E is a unit interval starting at the origin and in the same direction as a unit vector **u**. Although we shall not go through the theoretical details, as long as A is sufficiently regular, the length Proj(A, **u**) of the projection can be obtained as a one-sided derivative of $m(A \oplus tE)$, viz.,

$$\text{Proj}(A, \mathbf{u}) = \frac{d}{dt}[m(A \oplus tE)]\Big|_{t=0}$$

$$= \lim_{t \to 0} \frac{m(A \oplus tE) - m(A)}{t}$$

Example 3.11

Let A be a square of edge length 1 sitting with its edges parallel to the coordinate axes. Let **u** be the unit vector pointing in the 45° direction. (See Figure 3.31.) Then

$$m(A \oplus tE) = 1 + t\sqrt{2}$$

Hence,

$$\lim_{t \to 0} \frac{m(A \oplus tE) - m(A)}{t} = \sqrt{2}$$

[12] G. Watson, "Mathematical Morphology," *Tech. Report No. 21*, Dept. of Stat., (1973):11.

which is exactly equal to the length of the diagonal, which is in turn equal to Proj(A, **u**).

Many other morphological formulations of geometric parameters are possible. Whereas the preceding methodologies have involved dilation by a parameterized family of balls and a dilation by a parameterized family of line segments, the next one involves erosion by a family of two point structuring elements generated by some set $\{0, w\}$. Indeed, we consider the eroded sets $A \ominus (-tE) = A \cap (A - tw)$, where $E = \{0, w\}$ and $|w| = 1$. We define the *covariance* of A to be

$$[\text{Cov}(A, w)](t) = m[A \ominus (-tE)]$$

Like a granulometric size distribution, the covariance, as a function of t, gives a type of size distribution. Hence, when no confusion can arise, we write $C(t)$. Though the covariance function can be employed in a manner similar to the granulometric size distributions, we will not pursue this aspect of it; however, a digital example will be presented at the appropriate time.

With regard to geometric parameters, though no proof will be given, it is known that for sufficiently regular compact sets, the perimeter can be found by using averages of the derivative of the covariance at 0.[13] Indeed,

$$\frac{1}{2\pi} \int_0^{2\pi} C'(0) \, d\theta = -\Lambda(A)/\pi$$

where the average is taken over all Cov(A, w) such that $|w| = 1$.

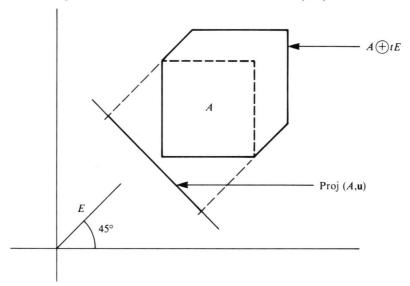

Figure 3.31 Projection of Square

[13] Ibid., 13.

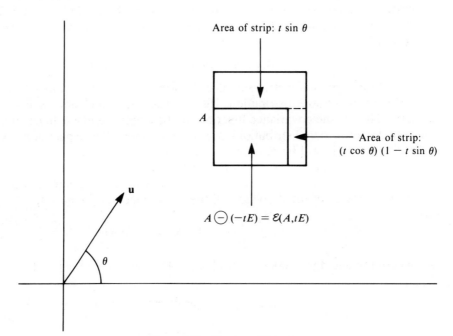

Figure 3.32 Erosion of Square

Example 3.12

Consider a unit square with sides parallel to the coordinate axes. Then for a given angle θ and sufficiently small t (Figure 3.32),

$$[\text{Cov}(A, \theta)](t) = 1 - [t\,|\sin\theta| + t\,|\cos\theta| - t^2\,|\sin\theta|]$$

Hence,

$$[\text{Cov}(A, \theta)]'(0) = -[\,|\sin\theta| + |\cos\theta|\,]$$

Averaging over all θ between 0 and 2π yields

$$\frac{1}{2\pi}\int_0^{2\pi} C'(0)\,d\theta = \frac{-1}{2\pi}\int_0^{2\pi}[\,|\sin\theta| + |\cos\theta|\,]\,d\theta$$

$$= \frac{-4}{\pi} = \frac{-\Lambda(A)}{\pi}$$

3.7 DIGITAL MINKOWSKI ALGEBRA

To this point, the development of morphology has been strictly Euclidean. Theory has been presented together with numerous graphical examples. In practice, it is the digital version of Minkowski algebra which is applied to digitizations of continuous images. In this section we shall assume that such a digitization has taken place, and the digital counterparts of the Euclidean morphological operators will be applied. As is always the case with digital algorithms that represent continuous procedures, there is a problem of sampling: what "error" is introduced by going from the continuous model to the digital model? The sampling problem for morphological operators is addressed in Section 3.13.

Insofar as Euclidean images are concerned, we have to this point considered an image to be a subset of R^2. In terms of gray values, such images are "constant" images in the sense that they take on only a single gray value, say 1. To be precise, a subset A in R^2 defines a constant gray value image by identifying A with the image A' defined by

$$A'(x, y) = \begin{cases} 1 & \text{if } (x, y) \in A \\ \text{undefined} & \text{if } (x, y) \notin A \end{cases}$$

Employing this identification, the set-theoretic operations of union and intersection become the pointwise operations $\vee$ and $\wedge$ respectively given by

$$(A' \vee B')(x, y) = \begin{cases} 1 & \text{if } A'(x, y) = 1 \text{ or } B'(x, y) = 1 \\ \text{undefined} & \text{if } A'(x, y) \text{ and } B'(x, y) \text{ are undefined} \end{cases}$$

$$(A' \wedge B')(x, y) = \begin{cases} 1 & \text{if } A'(x, y) = B'(x, y) = 1 \\ \text{undefined} & \text{otherwise} \end{cases}$$

It follows that the collection of subsets of R^2 under the operations of $\cup$ and $\cap$ is algebraically isomorphic to the collection of 1-valued constant images under $\vee$ and $\wedge$.

Example 3.13

Let A be the unit circle centered at the origin, and let B be a square of edge length 1 centered at $(\frac{1}{2}, \frac{1}{2})$. Then $A \cap B$ is the quarter circle in the first quadrant. From the point of view of A' and B',

$$(A' \wedge B')(x, y) = \begin{cases} 1 & \text{if } x \geq 0, y \geq 0 \text{ and } x^2 + y^2 \leq 1 \\ \text{undefined} & \text{elsewhere} \end{cases}$$

Note that under the identification, $A' \wedge B'$ is identical to $A \cap B$. Similar comments apply to the union. But this is precisely what the notion of isomorphism means in this instance: $(A \cap B)' = A' \wedge B'$ and $(A \cup B)' = A' \vee B'$.

Now consider a constant digital image with single gray value 1. (In this chapter, all constant images will have gray value 1.) When employing the bound matrix representation it is assumed that the image is defined everywhere, although it has a gray value only at a finite number of pixels; at the rest it is star valued. If we were to apply the set-theoretic morphology directly to the grid in R^2, then a subset S of the grid would correspond to an image S' defined as 1 on S and undefined outside of S. However, by utilizing the bound matrix identification, we will associate with S the image S' defined by

$$S'(i,j) = \begin{cases} 1 & \text{if } (i,j) \in S \\ * & \text{if } (i,j) \notin S \end{cases}$$

Under this identification, the union and intersection operations respectively become

$$(S' \vee T')(i,j) = \begin{cases} 1 & \text{if } S'(i,j) = 1 \text{ or } T'(i,j) = 1 \\ * & \text{otherwise} \end{cases}$$

and

$$(S' \wedge T')(i,j) = \begin{cases} 1 & \text{if } S'(i,j) = T'(i,j) = 1 \\ * & \text{otherwise} \end{cases}$$

In terms of operations on bound matrices, $S' \vee T' = \text{EXTMAX}(S', T')$ and $S' \wedge T' = \text{MIN}(S', T')$. Because of these identifications, we will henceforth assume that a constant image with single gray value 1 is a bound matrix S and that for two such images the operations $\vee$ and $\wedge$ are defined as above with output images $S \vee T$ and $S \wedge T$. We need merely keep in mind that $\vee$ is EXTMAX and $\wedge$ is MIN.

Example 3.14

Let

$$S = \begin{pmatrix} 1 & * & 1 \\ * & ① & 1 \end{pmatrix} \quad \text{and} \quad T = \begin{pmatrix} ① & * \\ 1 & 1 \\ * & 1 \end{pmatrix}$$

Then

$$S \vee T = \begin{pmatrix} 1 & * & 1 \\ * & ① & 1 \\ * & 1 & 1 \\ * & * & 1 \end{pmatrix}$$

and

$$S \wedge T = (①)$$

The digital version of Minkowski addition is defined by

$$S \boxplus E = \bigvee_{(i,j) \in D_E} \text{TRAN}(S; i, j)$$

where D_E is the domain of E and $\vee$ denotes the extended maximum over all

Sec. 3.7 Digital Minkowski Algebra

translates of S by pairs in D_E. As previously noted, in this instance the extended maximum acts exactly like a union. Moreover, if the origin is in the domain of E, then S is a subimage of $S \boxplus E$. In practice, E plays the role of a template, or structuring element, and S is "expanded" by E. S is translated to the activated pixels and the resulting translates are unioned.

Example 3.15

Consider the two images

$$S = \begin{pmatrix} 1 & 1 & 1 & * \\ 1 & 1 & 1 & * \\ * & 1 & * & 1 \\ \circledast & 1 & 1 & 1 \end{pmatrix} \text{ and } E = \begin{pmatrix} * & 1 \\ \textcircled{1} & 1 \end{pmatrix}$$

$S \boxplus E$ is found by successively translating S to the 1-valued pixels of E. There are three such translates: TRAN$(S; 0, 0) = S$,

$$\text{TRAN}(S; 1, 0) = \begin{pmatrix} * & 1 & 1 & 1 & * \\ * & 1 & 1 & 1 & * \\ * & * & 1 & * & 1 \\ \circledast & * & 1 & 1 & 1 \end{pmatrix}$$

and

$$\text{TRAN}(S; 1, 1) = \begin{pmatrix} * & 1 & 1 & 1 & * \\ * & 1 & 1 & 1 & * \\ * & * & 1 & * & 1 \\ * & * & 1 & 1 & 1 \\ \circledast & * & * & * & * \end{pmatrix}$$

Application of $\vee$ to the three translates yields

$$S \boxplus E = \begin{pmatrix} * & 1 & 1 & 1 & * \\ 1 & 1 & 1 & 1 & * \\ 1 & 1 & 1 & 1 & 1 \\ * & 1 & 1 & 1 & 1 \\ \circledast & 1 & 1 & 1 & 1 \end{pmatrix}$$

Since E has value 1 at the origin, S is a subimage of $S \boxplus E$.

In terminology similar to that of Euclidean morphology, $S \boxplus E$ is called the *dilation* of S by E. It has a block diagram given by

In line with our customary notation for digital image operators, we often write DILATE(S, E) instead of $S \boxplus E$. Figure 3.33 gives an implementation of dilation in terms of the operators DOMAIN, TRAN, and EXTMAX ($\vee$). The high degree

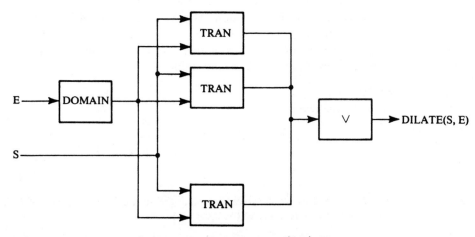

Figure 3.33 Block Diagram of DILATE

of parallelism within the diagram indicates the potential for specifically designed massively parallel image processing architecture.

Digital Minkowski addition was defined by taking a digital version of the Euclidean definition. One can proceed similarly for digital Minkowski subtraction to get

$$S \boxminus E = \bigwedge_{(i,j) \in D_E} \text{TRAN}(S; i, j)$$

where $\bigwedge$ = MIN. In contrast to Minkowski addition, the subtraction operation yields a "smaller" image; indeed, if the origin is an activated pixel of E, then $S \boxminus E$ is a subimage of S.

Example 3.16

Take S and E as in Example 3.15. Applying $\bigwedge$ to the three translates found in that example yields

$$S \boxminus E = \begin{pmatrix} * & 1 & 1 \\ * & * & 1 \\ * & * & * \\ \circledast & * & * \end{pmatrix}$$

According to Proposition 3.7, as applied to the digital case,

$$[S \boxminus E](i, j) = \begin{cases} 1 & \text{if TRAN[NINETY}^2(E); i, j] \vee S = S \\ * & \text{otherwise} \end{cases}$$

where $T \vee T' = T'$ means that T is a subimage of T'. This correspondence with Proposition 3.7 follows from the fact that the domain of $\text{NINETY}^2(E)$ equals, in set theoretic terms, minus the domain of E. (As will be the case with other digital properties that are related to Euclidean properties, no proof will be pro-

Sec. 3.7 Digital Minkowski Algebra

vided.) In an analogous manner with the Euclidean case, $S \ominus \text{NINETY}^2(E)$ will be called the *erosion* of S by E and will be denoted by $\text{ERODE}(S, E)$. Thus, we have

$$[\text{ERODE}(S, E)](i, j) = \begin{cases} 1 & \text{if } \text{TRAN}(E; i, j) \vee S = S \\ * & \text{otherwise} \end{cases}$$

The simplicity of this representation motivates the use of the erosion in morphological applications. The erosion of S by a structuring element E is simply the set of centers of all those translates of E which "fit" into S, which are subimages of S. By going back to the original definition of Minkowski subtraction, we have

$$\text{ERODE}(S, E) = \bigwedge_{(i,j) \in \text{DOMAIN}(E)} \text{TRAN}(S; -i, -j)$$

$$= \bigwedge_{(i,j) \in \text{DOMAIN}[\text{NINETY}^2(E)]} \text{TRAN}(S; i, j)$$

In Figure 3.34, this identity is employed to give a specification of the operator ERODE.

Intuitively, an erosion can be obtained by template translation, where the template E is moved about the image in such a manner as to see where it fits. It is precisely this geometric interpretation which has led to the successful application of Minkowski algebra to the textural analysis of images. Those points at which the image is "narrow" with respect to the structuring element are "eroded" (removed) from the image.

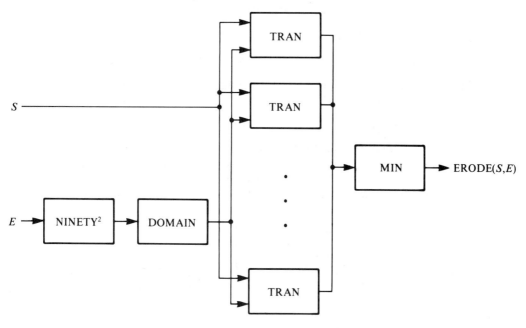

Figure 3.34 Block Diagram of ERODE

For the purposes of application and intuition, one should recognize that a similar analysis applies to dilation. That is, since Minkowski addition is commutative,

$$\text{DILATE}(S, E) = \bigvee_{(i,j) \in D_S} \text{TRAN}(E; i, j)$$

Hence the dilation is formed by "adjoining" small template pieces together, and, although dilation is commutative, one of the input images, usually the second, is called a structuring element. In this instance, certain pieces of the grid which lie outside the image and which are small in comparison to the structuring element are adjoined, so that the output image is "larger" than the original image. Notice that in the case of both erosion and dilation, the alteration of the input image is highly dependent upon the size and shape of the structuring element.

3.8 DIGITAL OPENINGS AND CLOSINGS

As in the case of the Euclidean opening, the digital opening acts like a filter in that it eliminates from a constant image any portion of that image into which a particular structuring element does not fit. It outputs the "union" of all those translates of the structuring element which do fit into the input image. Opening differs from erosion in that the actual center of the structuring element plays no role in the formation of the output.

Given two images S and E, the former to be opened and the latter to do the opening, we define, in accordance with the Euclidean definition,

$$\text{OPEN}(S, E) = [S \ominus \text{NINETY}^2(E)] \oplus E$$
$$= \text{DILATE}[\text{ERODE}(S, E), E]$$

Insofar as the fundamental notion of template fitting is concerned, the digital version of Theorem 3.2 states that

$$\text{OPEN}(S, E) = \bigvee \{\text{TRAN}(E; i, j): \text{TRAN}(E; i, j) \vee S = S\}$$

which means precisely that a pixel (u, v) is activated in $\text{OPEN}(S, E)$ if and only if there exists some translate $\text{TRAN}(E; i, j)$ of E such that (u, v) is in the domain of $\text{TRAN}(E; i, j)$ and $\text{TRAN}(E; i, j)$ is a subimage of S.

For implementation purposes (see Figure 3.35), the definition of the opening

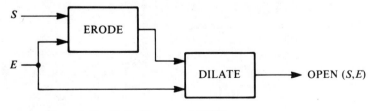

Figure 3.35 Block Diagram of OPEN

Sec. 3.8   Digital Openings and Closings   145

serves quite well; however, for an appreciation of the geometric content of the opening operation, it is its characterization in terms of fitted translates which is of consequence. The next example emphasizes this approach.

Example 3.17

Consider S and E as given in Example 3.15. The translates of E by the input pairs (2, 0), (0, 2), and (1, 2) "fit" into S. Consequently,

$$\text{OPEN}(S, E) = \text{TRAN}(E; 2, 0) \vee \text{TRAN}(E; 0, 2) \vee \text{TRAN}(E; 1, 2)$$

$$= \begin{pmatrix} * & 1 & 1 & * \\ 1 & 1 & 1 & * \\ * & * & * & 1 \\ \circledast & * & 1 & 1 \end{pmatrix}$$

Again employing the algebraic formulation utilized in the Euclidean case to motivate the digital algorithm, we define the digital closing by

$$\text{CLOSE}(S, E) = [S \boxplus \text{NINETY}^2(E)] \boxminus E$$

$$= \text{ERODE}[\text{DILATE}(S, \text{NINETY}^2(E)), \text{NINETY}^2(E)]$$

Like the digital opening, the digital closing also has a template version. In accordance with Proposition 3.23,

$$[\text{CLOSE}(S, E)](u, v) = \begin{cases} 1 & \text{if for all } (i, j) \text{ such that} \\ & [\text{TRAN}(E; i, j)](u, v) = 1, \\ & \text{TRAN}(E; i, j) \wedge S \neq \varnothing \\ * & \text{otherwise} \end{cases}$$

In other words, pixel (u, v) is activated in CLOSE(S, E) if and only if it is activated in any translate of E which "intersects" S. As in the Euclidean case, whereas the opening eliminates portions of the image which are inconsequential with respect to the structuring element, the closing adjoins pixels which are "close" to the image with respect to the size and shape of the structuring element. Like that of the opening, the definition of the closing serves well for implementation purposes (see Figure 3.36), while the preceding template characterization yields the geometric basis of application techniques.

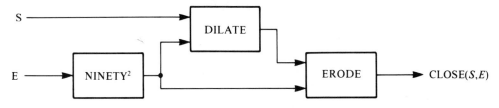

Figure 3.36 Block Diagram of CLOSE

Example 3.18

Once again employing image S of Example 3.15,

$$\text{CLOSE}(S, E) = \begin{pmatrix} 1 & 1 & 1 & * \\ 1 & 1 & 1 & * \\ * & 1 & 1 & 1 \\ \circledast & 1 & 1 & 1 \end{pmatrix}$$

Note that pixel (2, 1) has been activated in the closing since there are three translates of E, namely, TRAN(E; 1, 0), TRAN(E; 1, 1) and TRAN(E; 2, 1), which have (2, 1) in their domains, and for each of these, TRAN(E; i, j) $\wedge$ $S \neq \varnothing$. Notice, on the other hand, that (0, 1) has not been adjoined to the image since [TRAN(E; $-1, 0$)] (0, 1) = 1 but TRAN(E; $-1, 0$) $\wedge$ $S = \varnothing$.

Observe that fundamental algebraic properties of the basic Minkowski operators—dilation, erosion, opening, and closing—which have been demonstrated for the Euclidean case have counterparts in the digital case; indeed, some of these have already been exploited. Though we will not provide a specific list of these in terms of the digital terminology, it would be most useful for the reader to do so on his or her own. It is often the formulation of the algebraic properties in the digital domain which provides advantageous structuring in the areas of algorithm development, programming, and architectural specification.

3.9 DIGITAL SIZE AND SHAPE DISTRIBUTIONS

Size distributions resulting from granulometries were discussed in Section 3.4. In this section, three different morphological approaches to size and shape quantification will be considered.

If A is a compact set and B is both compact and convex, then the mappings $t \to O(A, tB)$, $t > 0$, are the granulometries of Section 3.4. In moving to the digital setting, one must take care in defining a counterpart to the scalar multiple of a Euclidean set. If Q is a collection of pixels in the grid, $tQ = \{(ti, tj): (i, j) \in Q\}$ is not necessarily a subset of the set of all lattice points. Moreover, even if it is a subset, the original image may have no "holes," whereas the latter might. Though we will not go into such topological questions at this time, Figure 3.37 should provide sufficient exposure to the difficulty.

When attempting to construct a digital analog of a particular granulometry, not only must A and B be digitized, but there must also be given some method of choosing a digital analogue of tB. Being specific, if S is the digitization of A and E is the digitization of B, we must define a corresponding granulometry by $k \to \text{OPEN}(S, E_k)$, where, for any integer $k > 0$, E_k is an appropriate digitization of some scalar multiple of B. The problem goes to the very heart of sampling theory and is considered in detail in Section 3.13. In essence, one must find a

Sec. 3.9   Digital Size and Shape Distributions   147

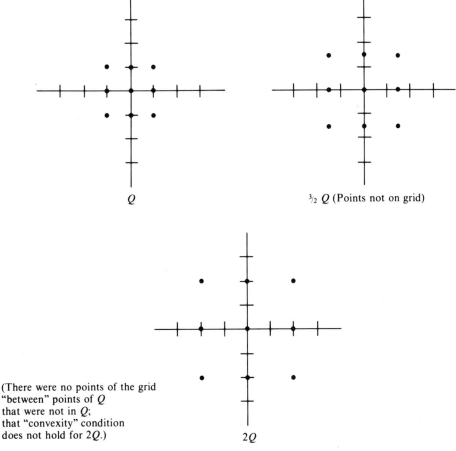

Figure 3.37 Difficulty with Digital Image Scalar Multiplication

satisfactory digital model to approximate in some manner the Euclidean granulometry. An example may help to clarify matters.

Example 3.19

Suppose A and B are the Euclidean images given in Figure 3.38, and let S represent the digitization of A formed by taking all grid points for which the grid square centered at the point has interior intersecting A. Suppose E is formed in a similar manner. Notice that tB is simply a stretched version of B. What is of consequence is the choice of some discrete sequence of digital structuring elements E_k. In this instance the choice appears quite clear: for any integer $k > 0$, let E_k be a vertical string of 1-valued pixels starting at the origin pixel $(0, 0)$ and ending at (and including) the pixel $(0, k - 1)$. E_k thus consists of k activated pixels. Instead of studying the Euclidean granulometry $t \to O(A, tB)$, we would proceed to compute the digital granulometry $k \to \text{OPEN}(S, E_k)$. Be forewarned, however: the relation between the two granulometries is not simple! (See Section 3.13.)

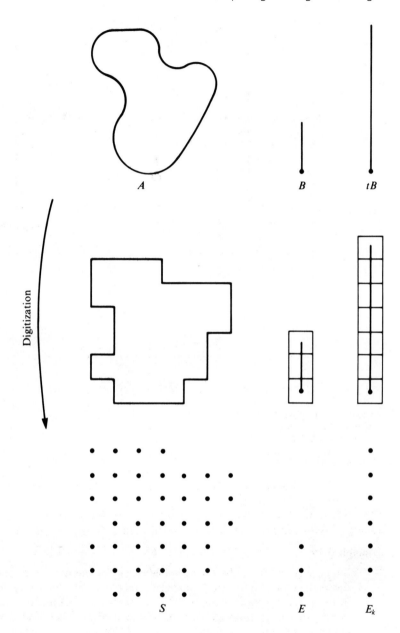

Note: E_k consists of k activated pixels

Figure 3.38 Digitization Process

Using the preceding example as motivation, we define the *linear digital granulometries*

$$k \to \text{OPEN}(S, V(k))$$

and

$$k \to \text{OPEN}(S, H(k))$$

where $k = 1, 2, 3, \ldots,$

$$V(k) = \begin{pmatrix} 1 \\ \vdots \\ 1 \\ 1 \\ ① \end{pmatrix} \quad \text{and} \quad H(k) = (①\, 1 \ldots 1)$$

each with k activated pixels. Note that

$$S \supset \text{OPEN}(S, V(1)) \supset \text{OPEN}(S, V(2)) \supset \text{OPEN}(S, V(3)) \supset \ldots$$

and that a similar relation holds for the openings by $H(k)$.

In the Euclidean case, size distributions were created from granulometries by measuring the succeeding outputs with m, the Lebesgue measure (the area). In the digital case, the size distributions result from "counting" the activated pixels in the succeeding outputs. Put precisely, we define the *digital linear granulometric size distributions*

$$\omega_1(k) = \text{CARD}[\text{OPEN}(S, V(k))]$$

and

$$\omega_2(k) = \text{CARD}[\text{OPEN}(S, H(k))]$$

For the sake of symbological clarity, we have omitted mention of the original set in the size distribution notation. This should present no great difficulty since the set S under consideration should be clearly known from the context of the discussion. Recall that CARD outputs the cardinality of the domain of a digital image.

Example 3.20

Let S be the image given in Figure 3.39. Then the size distributions ω_1 and ω_2 are given by

$k =$	1	2	3	4	5	6	7	8	9	10	11	12	13
$\omega_1(k) =$	80	76	70	70	66	66	54	54	54	45	35	24	0
$\omega_2(k) =$	80	66	52	22	10	0	0	0	0	0	0	0	0

For $k > 13$, $\omega_1(k) = \omega_2(k) = 0$.

Note in the preceding example that the vertical size distribution $\omega_1(k)$ is less affected by small k than is the horizontal size distribution $\omega_2(k)$. This means that

$$S = \begin{pmatrix} * & * & * & 1 & 1 & 1 & * & * & 1 & 1 & 1 & * \\ 1 & 1 & * & 1 & 1 & 1 & 1 & * & 1 & 1 & 1 & 1 \\ 1 & * & 1 & 1 & 1 & * & * & * & 1 & 1 & 1 & * \\ 1 & * & 1 & 1 & 1 & * & * & * & 1 & 1 & 1 & * \\ 1 & * & 1 & 1 & 1 & * & * & * & 1 & 1 & 1 & * \\ 1 & * & 1 & 1 & 1 & * & * & * & 1 & 1 & 1 & * \\ 1 & * & 1 & 1 & * & 1 & 1 & 1 & 1 & 1 & * & * \\ 1 & 1 & 1 & 1 & * & 1 & 1 & 1 & 1 & 1 & * & * \\ 1 & * & 1 & 1 & * & * & 1 & * & 1 & 1 & * & * \\ 1 & * & 1 & 1 & * & * & 1 & * & 1 & 1 & * & * \\ 1 & * & 1 & 1 & * & * & * & * & 1 & * & * & * \\ 1 & * & * & 1 & * & * & * & * & 1 & * & * & * \end{pmatrix}_{3,4}$$

Figure 3.39 Image S

$H(k)$ has a greater filtering effect than $V(k)$. The difference in the behavior of the size distributions results from the fact that the image contains a greater distribution of linear size in the vertical direction than it does in the horizontal direction. The size distributions provide some sort of quantification of textural information. In general, the measure of a granulometry yields a quantification of the manner in which activated pixels are clustered relative to the shape of the structuring element.

One can also construct size distributions by using linear erosion. Instead of employing the granulometric mappings, one could employ the erosion mappings

$$k \rightarrow \text{ERODE}(S, V(k))$$

and

$$k \rightarrow \text{ERODE}(S, H(k))$$

Again utilizing CARD, we have, for the size distributions,

$$\mu_1(k) = \text{CARD}[\text{ERODE}(S, V(k))]$$

and

$$\mu_2(k) = \text{CARD}[\text{ERODE}(S, H(k))]$$

Example 3.21

Let S be the image given in Figure 3.39. Then the size distributions μ_1 and μ_2 are given by

$k =$	1	2	3	4	5	6	7	8	9	10	11	12	13
$\mu_1(k) =$	80	65	54	46	38	31	24	19	14	9	5	2	0
$\mu_2(k) =$	80	44	22	7	2	0	0	0	0	0	0	0	0

For $k > 13$, $\mu_1(k) = \mu_2(k) = 0$.

One can also utilize a digital version of the morphological covariance to create size distributions. One essential difference here is that the resulting dis-

Sec. 3.9   Digital Size and Shape Distributions   151

tributions are not decreasing functions of k. To proceed, define the digital structuring elements

$$v(k) = \begin{pmatrix} 1 \\ * \\ \vdots \\ * \\ \textcircled{1} \end{pmatrix} \quad \text{and} \quad h(k) = (\textcircled{1} * * \ldots * 1)$$

each of which, except for $k = 1$, consists of two activated pixels at either end of a string of k pixels. For the case $k = 1$, $v(k) = h(k) = (\textcircled{1})$. The horizontal digital covariance function and the vertical digital covariance function are then respectively defined by

$$v_1(k) = \text{CARD}[\text{ERODE}(S, v(k))]$$

and

$$v_2(k) = \text{CARD}[\text{ERODE}(S, h(k))]$$

Some of the important properties of the digital covariance, which of course correspond to certain Euclidean properties, are:

(i) $v_1(1) = \text{CARD}(S)$
(ii) $v_1(k) = \text{CARD}[\text{ERODE}(S, -v(k))]$
(iii) $v_1(k) = \text{CARD}[S \wedge \text{TRAN}(S; 0, -k + 1)]$

The last equality follows directly from the fact that

$$S \wedge \text{TRAN}(S; 0, -k + 1) = \text{ERODE}(S, v(k))$$

Analogous properties hold for $v_2(k)$.

Example 3.22

Let S be the image given in Figure 3.40. Then the covariance functions $v_1(k)$ and $v_2(k)$ are given by

$k =$	1	2	3	4	5	6	7	8	9
$v_1(k) =$	43	35	27	21	15	9	6	3	0
$v_2(k) =$	43	29	15	8	14	18	13	5	0

For $k > 9$, $v_1(k) = v_2(k) = 0$. Notice how there is a fairly steady decline in the vertical covariance. This reflects the lack of deactivated pixels separating strings of activated pixels in the vertical direction. On the other hand, there are six different strings of deactivated pixels separating strings of activated pixels in the horizontal direction. This gives rise to two local maxima for the function $v_2(k)$, one at $k = 1$ and another at $k = 6$.

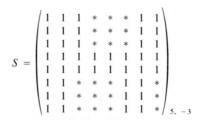

Figure 3.40 Image S for which Covariance Function Is to Be Found

Insofar as interpretation of any of the aforementioned size distributions is concerned, one should always be aware that such interpretation depends heavily on the experience of the investigator relative to the class of images under discussion. While it has been used to identify different images on the basis of texture, the behavior of the size distributions is not describable in elementary geometric terms. It is true that size distributions provide quantitative "feature" information, but it is also true that their utility depends upon a deep understanding of the scientific structures to which they are to be applied.

3.10 INCREASING τ-MAPPINGS

This section is devoted to demonstrating the power of Minkowski addition and Minkowski subtraction by presenting a theorem of G. Matheron which characterizes increasing translation-invariant mappings in terms of these operations. The subject matter is of a theoretical nature and can be read lightly by those lacking experience in abstract algebraic methodology. One should, however, pay attention to the definition of τ-mappings, the definition of the kernel, and the characterization theorem of Matheron (Theorem 3.7).

Let X denote the class of all two-dimensional constant Euclidean images (subsets of R^2). An image-to-image mapping $\Psi: X \to X$ is said to be *compatible with translation* if the application of Ψ can be interchanged with translation, i.e., if

$$\Psi(A + x) = \Psi(A) + x$$

for all images A and points x. According to Propositions 3.3 and 3.6, Minkowski addition and subtraction are both compatible with translation. Any such mapping will be called a τ-*mapping*.

Given a τ-mapping Ψ, we define the *kernel* of Ψ by

$$\text{Ker}[\Psi] = \{A : \overline{0} \in \Psi(A)\}$$

The kernel of Ψ is thus the collection of images that contain the origin $\overline{0} = (0, 0)$ after they have been operated upon by Ψ.

Sec. 3.10   Increasing τ-Mappings   153

Proposition 3.38. $z \in \Psi(A)$ if and only if $A \in \text{Ker}[\Psi] + z$.

Proof. $z \in \Psi(A)$ if and only if $\bar{0} \in \Psi(A) - z$ if and only if $\bar{0} \in \Psi(A - z)$ if and only if $A - z \in \text{Ker}[\Psi]$ if and only if $A \in \text{Ker}[\Psi] + z$.

Proposition 3.39.   Suppose image B is fixed and Ψ is defined in terms of Minkowski addition by $\Psi(A) = A \oplus B$. Then

$$\text{Ker}[\Psi] = \{A: A \cap (-B) \neq \emptyset\}$$

Proof. As mentioned before, Ψ is a τ-mapping. Now, $A \in \text{Ker}[\Psi]$ if and only if $\bar{0} \in \Psi(A)$, which means precisely that $\bar{0} \in A \oplus B$. But the latter holds if and only if there exists an $a \in A$ and a $b \in B$ such that $\bar{0} = a + b$, which means that $a = -b$.

Proposition 3.40.   Suppose image B is fixed and Ψ is defined in terms of Minkowski subtraction by $\Psi(A) = A \ominus B$. Then Ψ is a τ-mapping, and

$$\text{Ker}[\Psi] = \{A: A \supset -B\}$$

Proof. The fact that Ψ is τ-mapping has already been noted. Moreover, $A \in \text{Ker}[\Psi]$ if and only if $\bar{0} \in \Psi(A)$, which means, by Proposition 3.7, that

$$-B = -B + \bar{0} \subset A$$

Proposition 3.41.   Suppose B is fixed and Ψ is defined in terms of the opening by $\Psi(A) = O(A, B)$. Then Ψ is a τ-mapping, and the kernel of Ψ is given by

$$\text{Ker}[\Psi] = \bigcup_{z \in B} \{A: B - z \subset A\}$$

Proof. That the opening $O(\cdot, B)$ is a τ-mapping follows from its definition in terms of the two primitive operators and the fact that they are both τ-mappings; indeed,

$$O(A, B) + x = [(A \ominus (-B)) \oplus B] + x$$
$$= [(A \ominus (-B)) + x] \oplus B$$
$$= [(A + x) \ominus (-B)] \oplus B$$
$$= O(A + x, B)$$

Now, $\bar{0} \in \Psi(A)$ if and only if (Theorem 3.2) there exists a w such that

$$\bar{0} \in B + w \subset A$$

But the latter means precisely that $w \in (-B)$ and $B + w \subset A$, i.e., that there exists a $z \in B$ with $B - z \subset A$.

Proposition 3.42. Suppose B is fixed and Ψ is defined in terms of the closing by $\Psi(A) = C(A, B)$. Then Ψ is a τ-mapping, and

$$\text{Ker}[\Psi] = \bigcap_{z \in B} \{A : A \cap (B - z) \neq \varnothing\}$$

Proof. That the closing $C(\cdot, B)$ is a τ-mapping can be shown in a manner similar to that employed in Proposition 3.41 to show that $O(\cdot, B)$ is a τ-mapping. By Proposition 3.23, $\overline{0} \in C(A,B)$ if and only if $(B + y) \cap A \neq \varnothing$ for any translate $B + y$ containing $\overline{0}$. However, the latter condition means that $-y \in B$. Hence, $\overline{0} \in C(A, B)$ is equivalent to $(B - z) \cap A \neq \varnothing$ for any $z \in B$. But this is precisely what is to be proved.

Proposition 3.43. Let Q_0 denote the collection of all images containing the origin, and let Ψ be a τ-mapping. Then

(i) Ψ is extensive if and only if $Q_0 \subset \text{Ker}[\Psi]$
(ii) Ψ is antiextensive if and only if $Q_0 \supset \text{Ker}[\Psi]$

Proof. Suppose there exists some image A such that A is not properly contained in $\Psi(A)$. Then there exists a point z in $A - \Psi(A)$, where "$-$" denotes set difference. Since Ψ is a τ-mapping,

$$\overline{0} \in [A - \Psi(A)] - z = (A - z) - (\Psi(A) - z)$$
$$= (A - z) - \Psi(A - z)$$

Hence, $A - z$ contains $\overline{0}$, but $\Psi(A - z)$ does not. In other words, $A - z \in Q_0$ but $A - z \notin \text{Ker}[\Psi]$; i.e., the kernel of Ψ does not contain Q_0. The reverse implication in (i) follows by contradiction. Suppose $Q_0 \not\subset \text{Ker}[\Psi]$. Then there exists an image A such that $\overline{0} \in A$ but $\overline{0} \notin \Psi(A)$. It is immediate that A is not contained in $\Psi(A)$; hence, Ψ is not extensive. The second part of the proposition may be demonstrated by a similar argument.

Proposition 3.44. Let Ψ_1 and Ψ_2 be τ-mappings. Then $\text{Ker}[\Psi_1] \subset \text{Ker}[\Psi_2]$ if and only if $\Psi_1(A) \subset \Psi_2(A)$ for any image A.

Proof. Suppose there exists some image A in the kernel of Ψ_1 which is not in the kernel of Ψ_2. Then $\overline{O}$ is in $\Psi_1(A)$ but not $\Psi_2(A)$, and hence, $\Psi_1(A) \not\subset \Psi_2(A)$. Conversely, suppose there exists an element z such that

$$z \in \Psi_1(A) - \Psi_2(A).$$

Then, since Ψ is a τ-mapping,

$$\overline{0} \in [\Psi_1(A) - \Psi_2(A)] - z = \Psi_1(A - z) - \Psi_2(A - z)$$

In other words, $A - z \notin \text{Ker}[\Psi_2]$, but $A - z \in \text{Ker}[\Psi_1]$.

Sec. 3.10   Increasing τ-Mappings   155

An immediate consequence of the preceding proposition is that two τ-mappings with the same kernel must be identical. The next proposition states, in terms of Minkowski subtraction, the first part of the Matheron Representation Theorem for increasing τ-mappings.[14]

Proposition 3.45.   Suppose Ψ is an increasing τ-mapping. Then for any image A,

$$\Psi(A) = \bigcup_{B \in \text{Ker}[\Psi]} A \ominus (-B)$$

Proof. Suppose $B \in \text{Ker}[\Psi]$. If A contains B, then $A \in \text{Ker}[\Psi]$ since $\overline{0} \in \Psi(B) \subset \Psi(A)$. In other words, if B is in the kernel of Ψ, then $\{A: A \supset B\}$ is a subclass of $\text{Ker}[\Psi]$. Hence,

$$\bigcup_{B \in \text{Ker}[\Psi]} \{A: A \supset B\} \subset \text{Ker}[\Psi]$$

The reverse inclusion is trivial since $B \in \{A: A \supset B\}$. Consequently,

$$\bigcup_{B \in \text{Ker}[\Psi]} \{A: A \supset B\} = \text{Ker}[\Psi]$$

But, by Proposition 3.40, the collection to the left of the equal sign is the kernel of the mapping Φ, where $\Phi(A)$ is defined as

$$\bigcup_{B \in \text{Ker}[\Psi]} A \ominus (-B)$$

Therefore, by the remark following Proposition 3.44, Proposition 3.45 is proved.

In order to obtain a dual result for the preceding theorem, we introduce the notion of a *dual mapping* Ψ^*. For any τ-mapping Ψ, Ψ^* is defined by

$$\Psi^*(A) = [\Psi(A^c)]^c$$

Proposition 3.46.   If Ψ is an increasing τ-mapping, then so is its dual Ψ^*. Moreover, the kernel of the dual is given by

$$\text{Ker}[\Psi^*] = \{A: A^c \notin \text{Ker}[\Psi]\}$$

Proof. To see that the dual is increasing, suppose $A \subset B$. Then $B^c \subset A^c$. Since Ψ is increasing, $\Psi(B^c) \subset \Psi(A^c)$. Taking complements yields

$$\Psi^*(A) = [\Psi(A^c)]^c \subset [\Psi(B^c)]^c = \Psi^*(B)$$

Moreover, the dual is a τ-mapping since, if Ψ is a τ-mapping, then for any point x,

$$\Psi(A^c) + x = \Psi(A^c + x)$$

---

[14] Matheron, *Random Sets and Integral Geometry*, 221.

which implies, by taking complements, that

$$\Psi^*(A + x) = [\Psi((A + x)^c)]^c = [\Psi(A^c) + x]^c$$
$$= [\Psi(A^c)]^c + x = \Psi^*(A) + x$$

Finally, the relation involving the kernel of the dual follows from

$$\text{Ker}[\Psi^*] = \{A: \overline{0} \in \Psi^*(A)\}$$
$$= \{A: \overline{0} \notin \Psi(A^c)\}$$
$$= \{A: A^c \notin \text{Ker}[\Psi]\}$$

The dual of Proposition 3.45 can now be obtained. It gives a characterization of an increasing τ-mapping in terms of Minkowski addition.

Proposition 3.47. Suppose Ψ is an increasing τ-mapping. Then for any image A,

$$\Psi(A) = \bigcap_{B \in \text{Ker}[\Psi^*]} A \oplus (-B)$$

Proof. By Proposition 3.45 applied to Ψ^* and A^c,

$$\Psi^*(A^c) = \bigcup_{B \in \text{Ker}[\Psi^*]} A^c \ominus (-B)$$

Taking complements, applying duality, and recognizing that the dual of the dual is the original mapping, we have

$$\Psi(A) = \bigcap_{B \in \text{Ker}[\Psi^*]} [A^c \ominus (-B)]^c$$
$$= \bigcap_{B \in \text{Ker}[\Psi^*]} A \oplus (-B)$$

Propositions 3.45 and 3.47 can be put into the more common morphological terminology of dilations and erosions. They state that an increasing τ-mapping can be represented as a union of erosions and an intersection of dilations, respectively. This is the Matheron Representation Theorem.

Theorem 3.7. (Matheron Representation Theorem). Suppose Ψ is an increasing τ-mapping. Then for any image A,

$$\Psi(A) = \bigcup_{B \in \text{Ker}[\Psi]} \mathscr{E}(A, B) = \bigcap_{B \in \text{Ker}[\Psi^*]} \mathscr{D}(A, -B)$$

Example 3.23

In order to illustrate some of the concepts of this section, define the mapping Γ as follows: the point $(x, y) \in \Gamma(A)$ if and only if there exist points $(a, b), (c, d) \in A$, not necessarily distinct, such that $x = a$, $x \geq c$, $y = d$, and $y \geq b$. (See Figure 3.41.) Graphically, $\Gamma(A)$ is constructed by taking the union of all horizontal rays in the

Sec. 3.10   Increasing τ-Mappings   **157**

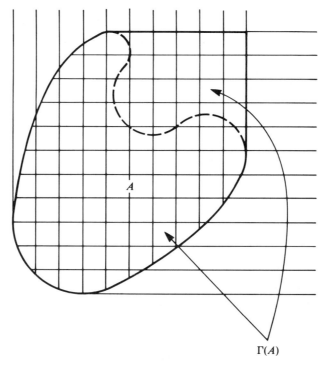

Figure 3.41   Graphical Construction of $\Gamma(A)$

positive direction emanating from points of A, taking the union of all vertical rays in the positive direction emanating from points of A, and then taking the intersection of the two unions. Γ is an increasing τ-mapping. Moreover, set $A \in \text{Ker}[\Gamma]$ if and only if A contains the origin or A contains points on each of the negative coordinate axes.

Suppose we restrict our attention to constant digital images, which, according to the identification discussed in Section 3.7, are finite subsets of R^2 that are restricted to the grid points. In other words, a constant bound matrix S can be viewed as a subset of R^2 in the sense that its activated pixels form such a subset. Γ can now be applied to constant bound matrices. For instance, if

$$S = \begin{pmatrix} * & 1 & 1 & * & * \\ * & * & 1 & 1 & 1 \\ * & 1 & * & 1 & * \\ * & 1 & 1 & 1 & * \\ \circledast & * & * & * & * \end{pmatrix}$$

then

$$\Gamma(S) = \begin{pmatrix} * & 1 & 1 & 1 & 1 \\ * & * & 1 & 1 & 1 \\ * & 1 & 1 & 1 & * \\ * & 1 & 1 & 1 & * \\ \circledast & * & * & * & * \end{pmatrix}$$

When Γ is restricted to constant bound matrices, its kernel becomes the set of all bound matrices that contain a 1 at the origin pixel or contain a 1 to the left and a 1 below the origin pixel. Thus, kernel elements are of the following form:

$$\begin{pmatrix} & & & \vdots & & & & & \\ . & . & 1 & . & . & \fbox{a_{ij}} & . & . & . \\ & & & \vdots & & & & & \\ & & & 1 & & & & & \\ & & & \vdots & & & & & \end{pmatrix}$$

Applying the digital version of Theorem 3.7 to Γ_0, the restriction of Γ to digital images, we have

$$\Gamma_0(S) = \bigvee_{T \in \text{Ker}[\Gamma_0]} \text{ERODE}(S, T)$$

In fact, the EXTMAX operation need not be taken over all of the kernel. Pixel (i, j) is activated in the erosion if and only if $\text{TRAN}(T; i, j)$ is a subimage of S. Hence, if T_1 and T_2 are both in the kernel and T_1 is a subimage of T_2, then $\text{TRAN}(T_2; i, j)$ being a subimage of S implies that $\text{TRAN}(T_1; i, j)$ is also a subimage of S. Thus, translates of T_2 need not be checked, and consequently, one need only find a subclass $\mathcal{M}$ of $\text{Ker}[\Gamma_0]$ such that

(i) For any $T \in \text{Ker}[\Gamma_0]$, there exists a $T' \in \mathcal{M}$ such that T' is a subimage of T.
(ii) No element of $\mathcal{M}$ is a subimage of any other element of $\mathcal{M}$.

$\mathcal{M}$ is called a *basis* for the kernel.

For the operator Γ_0 under consideration in this example, images of the following form provide a basis for the kernel:

$$B_{00} = (\fbox{1}) \qquad B_{11} = \begin{pmatrix} 1 & \fbox{$*$} \\ * & 1 \end{pmatrix} \qquad B_{12} = \begin{pmatrix} 1 & * & \fbox{$*$} \\ * & * & 1 \end{pmatrix}$$

$$B_{21} = \begin{pmatrix} 1 & \fbox{$*$} \\ * & * \\ * & 1 \end{pmatrix} \qquad B_{22} = \begin{pmatrix} 1 & * & \fbox{$*$} \\ * & * & * \\ * & * & 1 \end{pmatrix}$$

In general, for Γ_0, the basis elements are of the form

$$B_{rs} = \begin{pmatrix} 1 & * & * & . & . & \fbox{$*$} \\ * & * & * & . & . & * \\ * & * & * & . & . & * \\ \vdots & \vdots & \vdots & & & \vdots \\ * & * & * & . & . & 1 \end{pmatrix}$$

where the bound matrix B_{rs} has $r + 1$ rows and $s + 1$ columns. Using this basis for

Γ_0, Theorem 3.7 becomes

$$\Gamma_0(S) = \text{ERODE}(S, B_{00}) \vee \left[\bigvee_{i,j=1}^{\infty} \text{ERODE}(S, B_{ij}) \right]$$

where EXTMAX need only be taken over finitely many inputs. For the particular S given,

$$\Gamma_0(S) = \text{ERODE}(S, B_{00}) \vee \text{ERODE}(S, B_{11}) \vee \text{ERODE}(S, B_{12})$$

Though the notions introduced in this rather long example can be formalized and incorporated into a general theory, we shall not do so here.[15]

3.11 ALGEBRAIC OPENINGS AND CLOSINGS OF IMAGES

This section continues the work of G. Matheron regarding the characterization of important classes of morphological image operations. As in Section 3.10, the material is algebraic in nature and the proofs need only be skimmed by someone whose interest is not theoretical. The main result is Theorem 3.8, the τ-opening Representation Theorem.

In general, if Q and Q' are collections of sets, then a mapping $\Psi: Q \to Q'$ is called an *algebraic opening* if the following conditions are satisfied:

(i) Ψ is antiextensive, i.e., $\Psi(A) \subset A$.
(ii) Ψ is increasing, i.e., $A \subset B$ implies $\Psi(A) \subset \Psi(B)$.
(iii) Ψ is idempotent, i.e., $\Psi[\Psi(A)] = \Psi(A)$.

On the other hand, a mapping Ψ is called an *algebraic closing* if Ψ is extensive ($\Psi(A) \supset A$), increasing and idempotent. According to Theorems 3.3 and 3.4, the opening $O(\cdot, B)$ and the closing $C(\cdot, B)$, are an algebraic opening and an algebraic closing, respectively. In each case the domain and range of the mappings is X, the collection of all sets in R^2. In the sequel, we shall assume that $Q = Q' = X$.

Proposition 3.48. Ψ^*, the dual of Ψ, is an algebraic opening if and only if Ψ is an algebraic closing.

Proof. The three aforementioned conditions need to be checked. (i) Ψ^* is antiextensive if and only if $\Psi^*(A) \subset A$ for any A if and only if $\Psi^*(A^c) \subset A^c$ for any A if and only if $\Psi(A) = [\Psi^*(A^c)]^c \supset A$ (by taking complements). But the latter means that Ψ is extensive. (ii) Let $A \subset B$. Then $A^c \supset B^c$. Now, $\Psi^*(A) \subset \Psi^*(B)$ if and only if $[\Psi^*(A)]^c \supset [\Psi^*(B)]^c$ if and only if $\Psi(A^c) \supset \Psi(B^c)$ (by the definition of the dual). (iii) is simply an exercise in complementation and will be left to the reader.

---

[15] E. R. Dougherty and C. R. Giardina, "A Digital Version of the Matheron Representation Theorem for Increasing τ-Mappings in Terms of a Basis for the Kernel," *IEEE Computer Vision and Pattern Recognition* (1986).

Note that since the dual of the dual is the original mapping, Proposition 3.48 holds with the roles of Ψ and Ψ^* reversed.

If Ψ is an algebraic opening (closing) on X, then it is called a τ-*opening* (τ-*closing*) if it is compatible with translation. Thus, $O(\cdot, B)$ is a τ-opening and $C(\cdot, B)$ is a τ-closing. Matheron's theorem states that all τ-openings and τ-closings can be represented in terms of these elementary openings and closings, respectively.

Now suppose Ψ is either an algebraic opening or an algebraic closing. Then the class of *invariant* sets under Ψ is the class Inv[Ψ] of all images A such that $\Psi(A) = A$. That is, Inv[Ψ] consists precisely of those images which are unaffected by Ψ. For the opening $O(\cdot, B)$, the invariant images are those which are B-open; for the closing $C(\cdot, B)$, the invariant images are those which are B-closed.

Proposition 3.49. An algebraic opening is a τ-opening if and only if Inv[Ψ] is closed under translation. (Inv[Ψ] is closed under translation means that $A \in$ Inv[Ψ] if and only if $A + x \in$ Inv[Ψ] for any $x \in R^2$.)

Proof. Suppose Ψ is a τ-opening. Then $\Psi(A) = A$ if and only if $\Psi(A) + x = A + x$ if and only if $\Psi(A + x) = A + x$ if and only if $A + x \in$ Inv[Ψ]. The converse is just as easy.

A similar proposition to Proposition 3.49 holds for τ-closings.

Now suppose that Ψ is an algebraic opening. A class $\mathcal{B}$ of images is said to be a *basis* for Inv[Ψ] if Inv[Ψ] is the class generated by $\mathcal{B}$ under translations and infinite unions.

Note that every invariant set of a τ-opening has a basis: Inv[Ψ] is a basis for itself. Proposition 3.49 guarantees that Inv[Ψ] is closed under translations. But it is also closed under unions. Indeed, if $\{A_k\} \subset$ Inv[Ψ], $\Psi(A_k) = A_k$ for all k. Moreover,

$$\Psi[\bigcup_k A_k] \supset \Psi(A_k) = A_k$$

and thus,

$$\bigcup_k A_k \subset \Psi[\bigcup_k A_k]$$

The reverse inclusion follows from antiextensivity, and consequently the union is an element of Inv[Ψ].

Example 3.24.

Suppose $\Psi(A) = O(A, B)$. Then $\{B\}$ is a basis for Inv[Ψ]. To see this, let A be a set generated by $\{B\}$. Then A is a union of translates of B, say

$$A = \bigcup_{x \in S} B + x = B \oplus S.$$

By Proposition 3.25, A is B-open, i.e., $A \in$ Inv[Ψ]. Conversely, if $A \in$ Inv[Ψ], then

Sec. 3.11 Algebraic Openings and Closings of Images

A is B-open, and the same proposition implies that there exists some set E such that

$$A = E \oplus B = \bigcup_{x \in E} B + x$$

which says that A is in the class generated by the single set B.

Theorem 3.8 (Matheron). An image-to-image mapping Ψ is a τ-opening if and only if there exists a class of sets $\mathcal{B}$ such that

$$\Psi(A) = \cup \{O(A, B): B \in \mathcal{B}\}$$

Moreover, $\mathcal{B}$ is a basis for $\text{Inv}[\Psi]$.

Proof. We first show that if Ψ is given by the above union, then Ψ is a τ-opening.

(i) Ψ is antiextensive:

$$\Psi(A) = \cup \{O(A, B): B \in \mathcal{B}\} \subset \cup \{A: B \in \mathcal{B}\} = A$$

(ii) Ψ is increasing: $E \subset F$ implies that

$$\Psi(E) = \cup \{O(E, B): B \in \mathcal{B}\} \subset \cup \{O(F, B): B \in \mathcal{B}\} = \Psi(F)$$

(iii) Ψ is idempotent: Since Ψ is antiextensive, $\Psi[\Psi(A)] \subset \Psi(A)$. As for the reverse inclusion, if $x \in \Psi(A)$, then there exists $B_0 \in \mathcal{B}$ and $y \in R^2$ such that

$$x \in B_0 + y \subset O(A, B_0) \subset \cup \{O(A, B): B \in \mathcal{B}\} \subset A$$

and this implies that

$$x \in \cup \{O[\cup \{O(A, B): B \in \mathcal{B}\}, B]: B \in \mathcal{B}\} = \Psi[\Psi(A)]$$

(iv) Ψ is translation compatible:

$$\Psi(A) + x = [\cup \{O(A, B): B \in \mathcal{B}\}] + x$$
$$= \cup \{O(A, B) + x: B \in \mathcal{B}\}$$
$$= \cup \{O(A + x, B): B \in \mathcal{B}\} = \Psi(A + x)$$

To see that $\mathcal{B}$ is a basis for $\text{Inv}[\Psi]$, suppose $A \in \text{Inv}[\Psi]$ and $x \in A$. Since $A = \cup \{O(A, B): B \in \mathcal{B}\}$, there exists z_x and $B_x \in \mathcal{B}$ such that $x \in B_x + z_x \subset A$. Consequently,

$$A = \cup \{B_x + z_x: x \in A\}$$

and hence, $\text{Inv}[\Psi]$ is generated by $\mathcal{B}$.

We must now show that any τ-opening is of the form specified in the statement of the theorem. To that end, let Ψ be a τ-opening. Since Ψ is idempotent,

$\Psi(A) \in \text{Inv}[\Psi]$. Let
$$Q = \cup \{B + x : x \in R^2, B \in \text{Inv}[\Psi], B + x \subset A\}$$
Since $\Psi[A] + \bar{0} = \Psi[A] \subset A$, $\Psi[A] \subset Q$. Now suppose $z \in Q$. Then there exists $B' \in \text{Inv}[\Psi]$ and $x' \in R^2$ such that $z \in B' + x' \subset A$. Since Ψ is increasing and compatible with translation,
$$\Psi(A) \supset \Psi(B' + x') = \Psi(B') + x' = B' + x'$$
where the last equality follows from the fact that $B' \in \text{Inv}[\Psi]$. Therefore, $z \in \Psi(A)$ and $Q \subset \Psi(A)$. Hence, by Theorem 3.2,
$$\Psi(A) = Q = \bigcup_{B \in \text{Inv}[\Psi]} [\cup \{B + x : x \in R^2, B + x \subset A\}]$$
$$= \bigcup_{B \in \text{Inv}[\Psi]} O(A, B)$$
which is precisely what is to be proved since $\text{Inv}[\Psi]$ is a basis for itself.

Proposition 3.50. If Ψ is a τ-opening, then the dual of Ψ has the representation
$$\Psi^*(A) = \cap \{C(A, B) : B \in \mathcal{B}\}$$
where $\mathcal{B}$ is a basis for $\text{Inv}[\Psi]$.

Proof. Applying Theorem 3.8, together with duality, we have
$$\Psi^*(A) = [\Psi(A^c)]^c = [\cup \{O(A^c, B) : B \in \mathcal{B}\}]^c$$
$$= \cap \{O(A^c, B)^c : B \in \mathcal{B}\}$$
$$= \cap \{C(A, B) : B \in \mathcal{B}\}$$

Example 3.25

Suppose one wishes to filter from an image (i.e., deactivate) any pixel that does not have two adjacent activated neighbors. One might wish to engage in such a process under the assumption that the activation of any such pixel is likely due to noise. Let Ψ be the mapping that provides the desired output. For instance, if

$$S = \begin{pmatrix} * & 1 & * & 1 & 1 & 1 \\ * & 1 & 1 & 1 & * & 1 \\ \circledast & 1 & * & 1 & * & 1 \\ 1 & * & 1 & * & * & * \end{pmatrix}$$

then

$$\Psi(S) = \begin{pmatrix} * & 1 & * & 1 & 1 & 1 \\ * & 1 & 1 & 1 & * & 1 \\ \circledast & 1 & * & 1 & * & * \\ * & * & * & * & * & * \end{pmatrix}$$

Sec. 3.12   Algebraic Granulometric Characterization   163

What is of consequence is that such a morphological filtering operation is a τ-opening. Hence, the discrete form of Theorem 3.8 applies. First, it should be clear that $\text{Inv}[\Psi]$ consists of all "discrete" images (including unbounded ones) only possessing activated pixels that have at least two adjacent neighbors. It can be shown that a basis for $\text{Inv}[\Psi]$ is given by

$$B_1 = \begin{pmatrix} 1 & 1 \\ \circledast & 1 \end{pmatrix} \quad B_2 = \begin{pmatrix} 1 & 1 \\ \textcircled{1} & * \end{pmatrix} \quad B_3 = \begin{pmatrix} 1 & * \\ \textcircled{1} & 1 \end{pmatrix} \quad B_4 = \begin{pmatrix} * & 1 \\ \textcircled{1} & 1 \end{pmatrix}$$

Consequently, by Theorem 3.8,

$$\Psi(S) = \vee \{\text{OPEN}(A, B): B \in \mathcal{B}\}$$

$$= \bigvee_{k=1}^{4} \text{OPEN}(A, B_k)$$

$$= \begin{pmatrix} * & * & * & * & 1 & 1 \\ * & * & 1 & 1 & * & 1 \\ \circledast & * & * & 1 & * & * \end{pmatrix} \vee \begin{pmatrix} * & * & * & 1 & 1 & * \\ * & 1 & 1 & 1 & * & * \\ \circledast & 1 & * & * & * & * \end{pmatrix}$$

$$\vee \begin{pmatrix} * & 1 & * & * & * & * \\ * & 1 & 1 & * & * & * \\ \circledast & * & * & * & * & * \end{pmatrix} \vee \begin{pmatrix} * & * & * & 1 & * & * \\ * & * & 1 & 1 & * & * \\ \circledast & * & * & * & * & * \end{pmatrix}$$

which, if the $\vee$ operations are carried out, will result in exactly the image given previously for $\Psi(S)$.

If one wishes to filter geometric patterns in a space-invariant manner (translation-invariant manner), a good methodology is to operate by a τ-opening, for certainly antiextensitivity, increasing monotonicity, and idempotence are most desirable qualities. It is therefore apparent that Theorem 3.8 is among the most basic results in image processing: it completely characterizes an important class of morphological filters.

3.12 ALGEBRAIC GRANULOMETRIC CHARACTERIZATION

The granulometries $t \rightarrow O(A, tB)$ were introduced in Section 3.4. In this section, we shall discuss a more general theory of granulometries developed by G. Matheron. The main result will be Matheron's Representation Theorem for a class of granulometries known as *Euclidean granulometries*.

The notion of a granulometric mapping, as conceived by Matheron, has to do with the process of *sieving*. Imagine a wire mesh sieve into which particles of different diameters are tossed. Those with a diameter less than the mesh size, say t, will fall through, while those with a diameter larger than the mesh size will remain within the sieve. As was demonstrated in Figure 3.22 for the opening, the action is like a filter. The smaller particles (diameter $< t$) are lost, while the larger particles (diameter $\geq t$) remain.

The situation might best be illustrated by considering two sets, each a disjoint union of connected components. Let

$$A = A_1 \cup A_2 \cup \cdots \cup A_m$$

and

$$B = B_1 \cup B_2 \cup \cdots \cup B_n$$

where $m \leq n$ and $A_k \subset B_k$ for $k = 1, 2, \ldots, m$. Moreover, suppose that the diameter of A_k is equal to r_k, with $r_1 \leq r_2 \leq \cdots \leq r_m$, and that the diameter of B_k is equal to s_k, with $s_1 \leq s_2 \leq \cdots \leq s_n$. (See Figure 3.42.) Several sieving properties are evident. Let Ψ_t denote the function which outputs that part of the input set which remains in the sieve of mesh size t. ($\Psi_t(A)$ is called the *t-oversize* of A.) If $r_k < t \leq r_{k+1}$, then

$$\Psi_t(A) = A_{k+1} \cup A_{k+2} \cup \cdots \cup A_m$$

Consequently, $\Psi_t(A) \subset A$, i.e., Ψ_t is antiextensive. Moreover, since $A_k \subset B_k$ for all k, $\Psi_t(A) \subset \Psi_t(B)$. Next, note that if t' is a different mesh size, then the effect of sieving by both mesh sizes t and t' is the same as sieving by only the greater of the two sizes. In other words,

$$\Psi_t[\Psi_{t'}(A)] = \Psi_{t'}[\Psi_t(A)] = \Psi_{\max(t,t')}(A)$$

Matheron has used the preceding three properties of sieving to give the following axiomatic formulation of a granulometric process.

A *granulometry on X*, the collection of two-dimensional Euclidean images, is a family of mappings $\Psi_t: X \to X$, $t > 0$, such that

(i) $\Psi_t(A) \subset A$ for any $t > 0$ (Ψ_t is antiextensive)

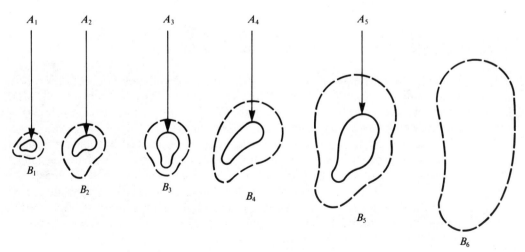

Figure 3.42 Particles to Be Sieved

Sec. 3.12 Algebraic Granulometric Characterization

(ii) $A \subset B$ implies that $\Psi_t(A) \subset \Psi_t(B)$ (Ψ_t is increasing)
(iii) $\Psi_t \circ \Psi_{t'} = \Psi_{t'} \circ \Psi_t = \Psi_{\max(t,t')}$ for all $t, t' > 0$

where "$\circ$" denotes function composition. It should be noted at once that a mapping Ψ_t within a granulometry $\{\Psi_t\}$ satisfies the first two requirements of an algebraic opening. Note also that the granulometries $t \to O(A, tB)$ of Section 3.4 satisfy the requirements of a general granulometry. Finally, for the special case $t = 0$, we define $\Psi_0(A) = A$.

The first result of this section gives a property of granulometries which says, in terms of sieving, that the greater the size of the mesh, the less of the original set which remains within the sieve.

Proposition 3.51. Suppose $\{\Psi_t\}$ is a granulometry. If $r \geq s$, then $\Psi_r(A) \subset \Psi_s(A)$.

Proof. By axiom (i) of a granulometry, $\Psi_r(A) \subset A$. Therefore, applying (iii) and (ii) in that order,

$$\Psi_r(A) = \Psi_s[\Psi_r(A)] \subset \Psi_s(A)$$

The next proposition gives an equivalent formulation of general granulometries in terms of algebraic openings and invariant classes. Notice the intuitive nature of its condition (ii'): for $r \geq s$, a set that is unaffected by the sieving function Ψ_r is also unaffected by the sieving function Ψ_s.

Proposition 3.52. $\{\Psi_t\}$ is a granulometry if and only if

(i') Ψ_t is an algebraic opening for all $t > 0$.
(ii') If $r \geq s > 0$, then $\mathrm{Inv}[\Psi_r] \subset \mathrm{Inv}[\Psi_s]$.

Proof. First assume that $\{\Psi_t\}$ is a granulometry. Then the granulometry axioms (i) and (ii) hold, and all we need do to show that Ψ_t is an algebraic opening is show idempotence. But this follows from axiom (iii):

$$\Psi_t[\Psi_t(A)] = \Psi_{\max(t,t)}(A) = \Psi_t(A)$$

For the proof of (ii'), suppose $A \in \mathrm{Inv}[\Psi_r]$ and $r \geq s$. Then

$$\Psi_s(A) = \Psi_s[\Psi_r(A)] = \Psi_{\max(s,r)}(A) = \Psi_r(A) = A$$

Hence, $A \in \mathrm{Inv}[\Psi_s]$.

For the proof of the converse of the proposition, suppose that Ψ_t is an opening. Then, immediately, the first two requirements of a granulometry hold. To prove requirement (iii), suppose $r \geq s > 0$. By idempotence and (ii'),

$$\Psi_r(A) \in \mathrm{Inv}[\Psi_r] \subset \mathrm{Inv}[\Psi_s]$$

Therefore, $\Psi_s[\Psi_r(A)] = \Psi_r(A)$. Consequently,

$$\Psi_r(A) = \Psi_r(\Psi_r((A))) = \Psi_r(\Psi_s(\Psi_r(A))) \subset \Psi_r(\Psi_s(A)) \subset \Psi_r(A)$$

where the two inclusions hold because Ψ_t is both antiextensive and increasing for any $t > 0$. Accordingly, it follows that

$$\Psi_r(A) \subset \Psi_r[\Psi_s(A)] \subset \Psi_r(A)$$

and hence,

$$\Psi_r[\Psi_s(A)] = \Psi_r(A) = \Psi_{\max(s,r)}(A)$$

However, in almost exactly the same way,

$$\Psi_r(A) = \Psi_r(\Psi_r(A)) = \Psi_s(\Psi_r(\Psi_r(A))) \subset \Psi_s(\Psi_r(A)) \subset \Psi_r(A)$$

where the last inclusion follows directly from the antiextensivity of Ψ_s. But then, it follows that

$$\Psi_s[\Psi_r(A)] = \Psi_r(A) = \Psi_{\max(r,s)}(A)$$

and requirement (iii) of a granulometry is satisfied.

If the following two axioms are added to those for a general granulometry, we obtain a *Euclidean granulometry*:

(iv) For any $t > 0$, Ψ_t is compatible with translation.

(v) For any $t > 0$ and image A, $\Psi_t(A) = t\Psi_1\left(\frac{1}{t}A\right)$.

Note that condition (iv), together with Proposition 3.52, asserts that Ψ_t is a τ-opening. Intuitively, it says that the sieving operation is independent of the position of A in the plane R^2. Practically speaking, the mesh of the sieve is uniform throughout its extent. Condition (v) states that scaling an image by $1/t$, sieving by Ψ_1, and then rescaling by t is the same as sieving by Ψ_t. The condition appears most intuitive if one considers the usual mesh-type sieving. Furthermore, it is satisfied by the granulometries $t \to O(A, tB)$ of Section 3.4. Indeed, for $\Psi_t = O(\cdot, tB)$, application of Propositions 3.12 and 3.13 yields

$$t\Psi_1\left(\frac{1}{t}A\right) = tO\left(\frac{1}{t}A, B\right) = t\left[\left(\frac{1}{t}A \ominus (-B)\right) \oplus B\right]$$

$$= \left[t\left(\frac{1}{t}A \ominus (-B)\right)\right] \oplus tB$$

$$= [A \ominus (-tB)] \oplus tB = O(A, tB) = \Psi_t(A)$$

It is the Euclidean granulometries that Matheron has characterized. To understand his characterization, some preliminary properties must be set out. To

Sec. 3.12 Algebraic Granulometric Characterization

begin with, note that, according to Proposition 3.49, Ψ_t is a τ-opening if and only if Inv$[\Psi_t]$ is closed under translation.

Proposition 3.53. Let $\{\Psi_t\}$ be a granulometry for which axiom (iv) holds. Then axiom (v) is equivalent to Inv$[\Psi_t] = t \cdot$Inv$[\Psi_1]$, which means that $A \in$ Inv$[\Psi_t]$ if and only if $(1/t)A \in$ Inv$[\Psi_1]$.

Proof. Assuming (v) to hold, $A \in$ Inv$[\Psi_t]$ if and only if $\Psi_t(A) = A$, which means precisely that

$$\Psi_1\left(\frac{1}{t}A\right) = \frac{1}{t}\Psi_t(A) = \frac{1}{t}A$$

i.e., $(1/t)A \in$ Inv$[\Psi_1]$. For the converse, since (iv) holds, Ψ_t is a τ-opening. Hence, the mapping $\Phi_t(A) = t\Psi_1((1/t)A)$ is also a τ-opening; indeed, that Φ is antiextensive and increasing follows at once from the corresponding properties of Ψ_t. As for idempotence,

$$\Phi_t[\Phi_t(A)] = t\Psi_1\left[\frac{1}{t}\left(t\Psi_1\left(\frac{1}{t}A\right)\right)\right]$$

$$= t\Psi_1\left[\Psi_1\left(\frac{1}{t}A\right)\right] = t\Psi_1\left(\frac{1}{t}A\right) = \Phi_t(A)$$

Moreover, compatibility with translation follows from

$$\Phi_t(A + x) = t\Psi_1\left[\frac{1}{t}(A + x)\right] = t\left[\Psi_1\left(\frac{1}{t}A + \frac{x}{t}\right)\right]$$

$$= t\left[\Psi_1\left(\frac{1}{t}A\right) + \frac{x}{t}\right]$$

$$= t\Psi_1\left(\frac{1}{t}A\right) + x = \Phi_t(A) + x$$

Now, $A \in$ Inv$[\Phi_t]$ if and only if $(1/t)A \in$ Inv$[\Psi_1]$, which, by the hypothesis of the proposition, means precisely that $A \in$ Inv$[\Psi_t]$. Hence, Φ_t and Ψ_t are τ-openings having the same invariant sets. But, it is an immediate consequence of Theorem 3.8 that τ-openings with identical invariant sets are themselves identical.

Not just any collection of sets can be the invariant class of Ψ_1, where $\{\Psi_t\}$ is some Euclidean granulometry. The next proposition characterizes the appropriate collections.

Proposition 3.54. Let $\mathcal{B}$ be a collection of subsets of R^2. Then there exists a Euclidean granulometry $\{\Psi_t\}$ such that $\mathcal{B} =$ Inv$[\Psi_1]$ if and only if $\mathcal{B}$ satisfies the closure conditions

(i) If $A_i \in \mathcal{B}$ for $i \in I$, then $\cup_i A_i \in \mathcal{B}$.
(ii) If $A \in \mathcal{B}$, then $A + x \in \mathcal{B}$ for all $x \in R^2$.
(iii) If $A \in \mathcal{B}$, then $tA \in \mathcal{B}$ for all $t \geq 1$.

Proof. Suppose there exists a Euclidean granulometry such that $\mathcal{B} = \text{Inv}[\Psi_1]$. To prove (i), consider a collection of sets $A_j \in \text{Inv}[\Psi_1]$. Since Ψ_1 is an algebraic opening, it is antiextensive. Hence,

$$\Psi_1 (\cup_j A_j) \subset \cup_j A_j$$

On the other hand, Ψ_1 is increasing, so that

$$\cup_k A_k = \cup_k \Psi_1(A_k) \subset \cup_k [\Psi_1(\cup_j A_j)] = \Psi_1(\cup_j A_j)$$

the last equality following from the redundancy of the outer union. Therefore, $\cup_j A_j \in \text{Inv}[\Psi_1]$, and $\text{Inv}[\Psi_1]$ is closed under arbitrary unions.

Next, note that property (ii), closure under translations, holds by Proposition 3.49. To prove property (iii), suppose that $A \in \text{Inv}[\Psi_1]$ and $t \geq 1$. Then, by axiom (v), $tA \in \text{Inv}[\Psi_t]$. Therefore, by axiom (iii),

$$\Psi_1(tA) = \Psi_1[\Psi_t(tA)] = \Psi_{\max(1,t)}(tA) = \Psi_t(tA) = tA$$

and $tA \in \text{Inv}[\Psi_1]$. We now proceed to the converse of the proposition. According to Theorem 3.8, Ψ_t defined by

$$\Psi_t(A) = \cup \{O(A, B): B \in t\mathcal{B}\}$$

is a τ-opening with basis $t\mathcal{B}$. Since $\mathcal{B}$ is closed under unions and translations, $\text{Inv}[\Psi_t] = t\mathcal{B}$ and hence $\mathcal{B} = \text{Inv}[\Psi_1]$. In order to show that $\{\Psi_t\}$ is a Euclidean granulometry, it remains to show that (ii') of Proposition 3.52 and axiom (v) hold. For (ii'), suppose $r \geq s > 0$ and $A \in \text{Inv}[\Psi_r]$. Then $A = rB$ for some $B \in \mathcal{B}$. Since $r/s \geq 1$, by the hypothesis of condition (iii) above, $(r/s)B \in \mathcal{B}$, which implies that $rB = sB'$ for some $B' \in \mathcal{B}$, which in turn implies that $A \in s\mathcal{B} = \text{Inv}[\Psi_s]$. Consequently, $\text{Inv}[\Psi_r] \subset \text{Inv}[\Psi_s]$, and (ii') holds. Finally, axiom (v) holds by Proposition 3.53 since, by construction, $\text{Inv}[\Psi_t] = t \, \text{Inv}[\Psi_1]$.

The previous result gives a complete characterization of classes of images which can serve as invariant classes for some sieving function Ψ_1 from a Euclidean granulometry $\{\Psi_t\}$. Taken together with Proposition 3.53, which states that the invariant classes of a Euclidean granulometry are determined by the invariant class for Ψ_1, Proposition 3.54 characterizes the invariant classes of Euclidean granulometries. In fact, much more has been shown within the proof, viz., that $\{\Psi_t\}$ is a Euclidean granulometry if and only if there exists some collection $\mathcal{B}$ of

Sec. 3.12   Algebraic Granulometric Characterization   169

sets such that

$$\Psi_t(A) = \cup \{O(A, B): B \in t\mathcal{B}\}$$
$$= \cup \{O(A, tB): B \in \mathcal{B}\}$$

and in such a case, $\text{Inv}[\Psi_1] = \mathcal{B}$. In other words, Euclidean granulometries themselves have been characterized in terms of the invariant sets of Ψ_1.

Suppose, now, that $\mathcal{B}$ is some collection of sets closed under unions, translations, and scalar multiplications by $t \geq 1$. By Proposition 3.54, $\mathcal{B}$ is the invariant set for Ψ_1 of some Euclidean granulometry. A collection of sets $\mathcal{B}_0$ is called a *generator* of $\mathcal{B}$ if the class closed under arbitrary union, translation, and scalar multiplication by $t \geq 1$ which is generated by $\mathcal{B}_0$ is the class $\mathcal{B}$. If $\{\Psi_t\}$ is the Euclidean granulometry with $\text{Inv}[\Psi_1] = \mathcal{B}$, then we also call $\mathcal{B}_0$ a generator of $\{\Psi_t\}$. Using these concepts, we arrive at Matheron's Representation Theorem for Euclidean granulometries.

Theorem 3.9 (Matheron).   A family of image-to-image mappings $\{\Psi_t\}$, $t > 0$, is a Euclidean granulometry if and only if there exists a class of images $\mathcal{B}_0$ such that

$$\Psi_t(A) = \bigcup_{B \in \mathcal{B}_0} \bigcup_{r \geq t} O(A, rB)$$

Moreover, $\mathcal{B}_0$ is a generator of $\{\Psi_t\}$ and $\text{Inv}[\Psi_t] = t \, \text{Inv}[\Psi_1]$ for all $t > 0$.

Proof. Let $\mathcal{B}$ be the class generated by $\mathcal{B}_0$. Applying Proposition 3.54, let $\{\Psi_t\}$ be the Euclidean granulometry with $\text{Inv}[\Psi_1] = \mathcal{B}$. We need to show that $\Psi_t(A)$ is given by the double union in the statement of the theorem, i.e., that

$$\cup \{O(A, tB): B \in \mathcal{B}\} = \cup \{O(A, t'B'): t' \geq t, B' \in \mathcal{B}_0\}$$

Accordingly, let z be an element of the left-hand union. Then there exists $\overline{B} \in \mathcal{B}$ and $x \in R^2$ such that $z \in t\overline{B} + x \subset A$. Since $\mathcal{B}$ is generated by $\mathcal{B}_0$, there exist some index sets I, J, and K such that

$$\overline{B} = \bigcup_{\substack{i \in I \\ j \in J \\ k \in K}} t_i(B_j + x_k)$$

where $t_i \geq 1$, $B_j \in \mathcal{B}_0$, and $x_k \in R^2$. Consequently, for some i, j, and k,

$$z \in t \cdot t_i(B_j + x_k) + x = (t \cdot t_i)B_j + [(t \cdot t_i)x_k + x] \subset A$$

Since $t \cdot t_i \geq t$, it follows that z is an element of the right-hand union above.

Now suppose z is an element of the right-hand union. Then there exists a $t' \geq t$, $B' \in \mathcal{B}_0$, and x such that $z \in t'B' + x \subset A$. But this can be rewritten as

$$z \in t \cdot \left(\frac{t'}{t}\right) B' + x \subset A$$

Now, since $t'/t \geq 1$, $(t'/t)B' \in \mathcal{B}$ and hence z is an element of the left-hand union. Therefore, the two unions are equal, and one part of the theorem is demonstrated.

For the converse, suppose $\{\Psi_t\}$ is a Euclidean granulometry. Then, by the remarks preceding the theorem,

$$\Psi_t(A) = \cup \{O(A, tB): B \in \text{Inv}[\Psi_1]\}$$

Certainly, $\text{Inv}[\Psi_1]$ is a generator of itself, and hence the representation of $\Psi_t(A)$ as a double union given in the statement of the theorem holds by simply applying the already completed part of the theorem to Ψ_t.

Actually, more has been shown in the proof than has been stated in the theorem, viz., if $\{\Psi_t\}$ is a Euclidean granulometry and $\mathcal{B}_0$ is *any* generator of $\text{Inv}[\Psi_1]$, then

$$\Psi_t(A) = \bigcup_{B \in \mathcal{B}_0} \bigcup_{r \geq t} O(A, rB)$$

Thus, not only does the proof of the theorem guarantee the existence of some generating set, which trivially can be considered to be $\text{Inv}[\Psi_1]$ itself, but any generator will also do.

Now suppose $\mathcal{B}_0 = \{B\}$, a set with a single element. Then

$$\Psi_t(A) = \bigcup_{r \geq t} O(A, rB)$$

defines a Euclidean granulometry. If it happens that B is convex, then, according to Proposition 3.34, $r \geq t$ implies that $O(A, rB) \subset O(A, tB)$, and hence $\{\Psi_t\}$ reduces to the granulometries discussed earlier, where $t \to O(A, tB)$. If B is not convex, then the union cannot be reduced to a single opening. Indeed, according to Theorem 3.5, there exists some $q > 1$ such that qB is not B-open. Accordingly, let $s = qt$. Then $s/t = q$, and

$$O(sB, tB) = t \cdot O\left(\frac{s}{t}B, B\right) \subsetneq t \cdot \left(\frac{s}{t}B\right) = sB = O(sB, sB)$$

where the proper inclusion follows from the fact that $(s/t)B$ is not B-open. Consequently, $O(sB, sB)$ properly contains $O(sB, tB)$, and therefore $O(\cdot, sB)$ cannot be removed from the union representing Ψ_t even though $s > t$. Hence, the union cannot be reduced simply to $O(\cdot, tB)$, as it could be if B were convex. This discussion leads to the following proposition.

Proposition 3.55. Suppose $\text{Inv}[\Psi_1]$ has the singleton generating set $\{B\}$. Then Ψ_t reduces to the granulometry $t \to O(\cdot, tB)$ if and only if B is convex.

Though the details of this section are both abstract and technically difficult, both Theorem 3.9 and Proposition 3.55 are of fundamental practical significance.

From the perspective of filtering, a Euclidean granulometry is a parameterized family of filters that provide increased filtering as $t \to \infty$. Since each Ψ_t is a τ-opening, each Ψ_t filters uniformly over R^2, in that it filters in a manner compatible with translation, and the filtering does not disturb the ordering relation on the class of subsets of R^2. Moreover, axiom (v) in the definition of a Euclidean granulometry implies that the family $\{\Psi_t\}$ behaves well with respect to scaling. Indeed, for positive r and t,

$$\Psi_t(rA) = t\Psi_1\left(\frac{r}{t}A\right) = \left(r\frac{t}{r}\right)\Psi_1\left(\frac{r}{t}A\right) = r\Psi_{t/r}(A)$$

In other words, the output of any sieving function Ψ_t with input rA can be found by inputting A itself into the sieving function $\Psi_{r/t}$. What is important is that A determines the effect of filtering rA. As a totality, the criteria for a Euclidean granulometry yield a family of filters that are highly congruous with human perception, at least insofar as that perception is considered to be Euclidean.

Given the desirable properties of Euclidean granulometries, the importance of Theorem 3.9 is evident: it completely characterizes these granulometries. Thus, if one wishes to construct a family of filters which satisfies the conditions of a Euclidean granulometry (and possibly others), it must be given by a union of openings as in Theorem 3.9. Moreover, should one desire to filter by employing a single generating structuring element, then, according to Proposition 3.55, the union will reduce to a single opening if and only if that structuring element is convex. Furthermore, should one wish to create a finer filter by employing a generator $\mathcal{B}_0 = \{B_1, B_2\}$, then the granulometry will reduce to

$$\Psi_t(\cdot) = O(\cdot, tB_1) \cup O(\cdot, tB_2)$$

if and only if both B_1 and B_2 are convex. A similar remark holds for a generator consisting of some other finite number of elements.

3.13 SAMPLING-ERROR BOUNDS FOR MORPHOLOGICAL OPERATIONS

In this section, we shall consider morphological operations upon compact convex Euclidean images. If $A \subset R^2$ is such an image, then we take as a digitization of that image the set of all grid points $(i/2^k, j/2^k)$ such that the interior of the square centered at $(i/2^k, j/2^k)$ with edge length 2^{-k} intersects A. (See Figure 3.43.) Let $\overline{A}$ denote the digitization of A, and let $\tilde{A}$ denote the union of all grid squares with centers in $\overline{A}$. Assuming B to be a convex compact structuring element, we wish to address the relationship between operating morphologically on A by B and operating on $\overline{A}$ by $\overline{tB}$, where tB is some scalar multiple of B.

If $m(A)$ denotes the Lebesgue measure (area) of A, then the measure of $\tilde{A}$ is given by $m(\tilde{A}) = (1/2^{2k}) \cdot \text{CARD}(\overline{A})$. Consequently, we define a discrete measure

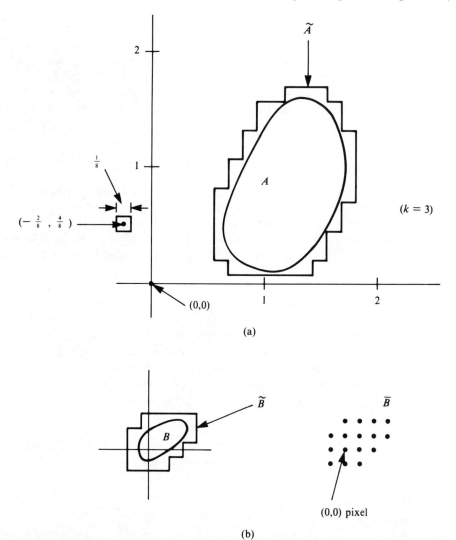

Figure 3.43 Morphological Digitization Scheme

$\overline{m}$ on digital images by $\overline{m}(\overline{A}) = (1/2^{2k}) \cdot \text{CARD}(\overline{A})$. Then, by construction, $\overline{m}(\overline{A}) = m(\tilde{A})$.

Example 3.26

Let $\overline{A}$ be the digitization of A given in Figure 3.43, and let $k = 3$. Then

$$\overline{m}(\overline{A}) = \frac{1}{64} \text{CARD}(\overline{A}) = \frac{99}{64}$$

which is also the area of $\tilde{A}$. (See Figure 3.44.)

Sec. 3.13 Sampling-Error Bounds for Morphological Operations

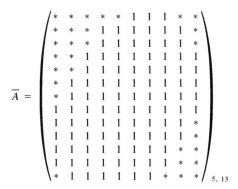

Figure 3.44 Digitization of Image A of Figure 3.43

Suppose one wishes to consider a size distribution $m[\mathscr{E}(A, tB)]$ on a Euclidean image generated by erosion. Suppose, furthermore, that a digital size distribution $\overline{m}[\text{ERODE}(\overline{A}, \overline{tB})]$ is to be computed as an approximation to the Euclidean distribution. We need to get some upper bound on the quantity

$$| m[\mathscr{E}(A, tB)] - \overline{m}[\text{ERODE}(\overline{A}, \overline{tB})] |$$

since this quantity represents the difference between the measure of a direct erosion of A by tB and the corresponding measure of an erosion of a digitization of A by a digitization of tB. For size distributions generated by erosion, this difference denotes the sampling error introduced through digitization.

Example 3.27

Let A be a square of edge length $5\sqrt{2}/4$ with edges at 45° to the coordinate axes. (See Figure 3.45.) Let B be a unit square with center at the origin and sides parallel to the coordinate axes. Then

$$m[\mathscr{E}(A, tB)] = \begin{cases} 2(\tfrac{5}{4} - t)^2 & \text{for } 0 \le t \le \tfrac{5}{4} \\ 0 & \text{for } t > \tfrac{5}{4} \end{cases}$$

Now suppose that a grid with grid squares of edge length $\tfrac{1}{8}$ is placed over A as in Figure 3.45. Note that the exact manner of placement is not determined beforehand. The digitization $\overline{A}$ is given in Figure 3.46. Now let B_k denote the "square" digital image of $2k - 1$ 1s on each side having the center 1 situated at the origin pixel. Then for $0 \le t \le \tfrac{1}{8}$, $\overline{tB} = B_1$, and for $(2k - 3)/8 < t \le (2k - 1)/8$, $\overline{tB} = B_k$, $k = 2, 3,$ .... Consequently, for $(2k - 3)/8 < t \le (2k - 1)/8$,

$$\overline{m}[\text{ERODE}(\overline{A}, \overline{tB})] = \frac{1}{64} \text{CARD}[\text{ERODE}(\overline{A}, B_k)]$$

Denoting this last quantity by $\Phi(k)$, we obtain, by computation, $\Phi(1) = 241/64 = 3.77$; $\Phi(2) = 2.53$; $\Phi(3) = 1.53$; $\Phi(4) = 0.78$; $\Phi(5) = 0.28$; $\Phi(6) = 0.03$, and $\Phi(k) = 0$ for $k > 6$. Using these figures, the graph of $\overline{m}[\text{ERODE}(\overline{A}, \overline{tB})]$, with t as the independent variable, is given in Figure 3.47. Note the manner in which this latter graph approximates $m[\mathscr{E}(A, tB)]$ (See Figure 3.48.)

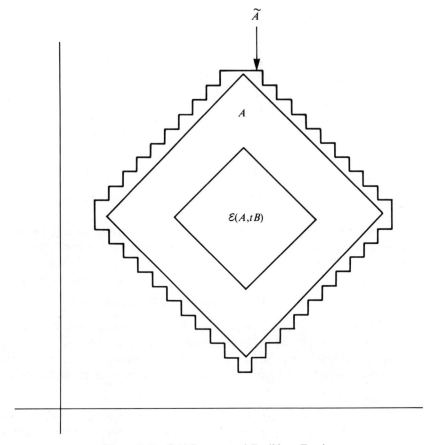

Figure 3.45 Grid Squares and Euclidean Erosion

For a general convex image A and convex structuring element B, it is exactly the manner in which $\overline{m}[\text{ERODE}(\overline{A}, \overline{tB})]$ approximates $m[\mathcal{E}(A, tB)]$ which is of interest. This is the sampling problem for erosion. Theorems will now be stated for both dilation and erosion which give bounds for the sampling error. In both cases, it is assumed that B contains the origin.

Theorem 3.10.[16] Suppose A and B are convex and compact, and the grid is of size 2^{-k}. Then

$$| m[\mathcal{E}(A, tB)] - \overline{m}[\text{ERODE}(\overline{A}, \overline{tB})] | \leq \frac{3\sqrt{2}}{2^{k+1}} \Lambda[A] + \frac{32 + \pi}{2^{2k+1}}$$

where $\Lambda[A]$ denotes the length of the boundary of A.

[16] E. R. Dougherty and C. R. Giardina, "Error Bounds for Morphologically Derived Feature Measurements," *SIAM Journal on Applied Mathematics* (in press).

Sec. 3.13   Sampling-Error Bounds for Morphological Operations

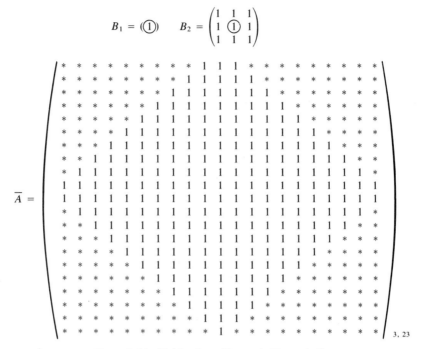

Figure 3.46 Digitization of Image in Figure 3.45

Before illustrating the theorem, we should note that the bars denoting absolute value cannot be removed. For even though $m(A) \leq \overline{m}(\overline{A})$, it is possible that $m[\mathcal{E}(A, tB)]$ is greater than $\overline{m}[\text{ERODE}(\overline{A}, t\overline{B})]$. Moreover, the second summand in the bound has denominator 2^{2k+1} and therefore is very small for large k in comparison to the first summand. Finally, and most importantly, it should be recognized that the bound is uniform in t (i.e., it does not involve the independent variable t) and goes to zero as $k \to \infty$.

Example 3.28

Applying Theorem 3.10 to the preceding example, and using the fact that $\Lambda[A] = 5\sqrt{2}$, one obtains

$$| m[\mathcal{E}(A, tB)] - \overline{m}[\text{ERODE}(\overline{A}, t\overline{B})] | \leq \frac{15}{8} + \frac{32 + \pi}{128}$$

Hence, the maximum difference between the two graphs can be quite large in comparison with the overall area of the image. The problem is that k is not sufficiently large. Indeed, suppose we were to change the problem and ask, Given A and B, how large should k be in order that the bound in Theorem 3.10 be less than 0.1? In other words, given some maximum tolerable error, say 0.1, how fine a grid must be employed? This problem is solved by finding a k such that

$$\frac{3\sqrt{2}}{2^{k+1}}(5\sqrt{2}) + \frac{32 + \pi}{2^{2k+1}} < 0.1$$

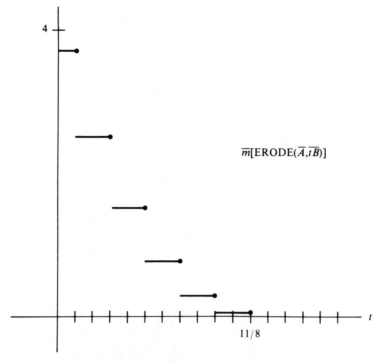

Figure 3.47 Digital Size Distribution from Erosion

The smallest possible solution for k is $k = 8$, in which case the grid will have squares of edge length $\frac{1}{256}$. Each small square will have area 2^{-16}. Of consequence is the fact that the graphs of the two size distributions would differ by no more than 0.1 at any value of t.

Theorem 3.11.[17] Let A and B be compact and convex, let the grid size be 2^{-k}, and let $r = \max_{x \in B} |x|$. Then

$$| m[\mathscr{D}(A, tB)] - \overline{m}[\text{DILATE}(\overline{A}, \overline{tB})] | \leq \frac{3\sqrt{2}}{2^{k+1}} (\Lambda[A] + 2\pi rt) + \frac{9\pi}{2^{2k+1}}$$

The nature of the bound in the theorem for dilation is quite different than that in the theorem for erosion. The latter bound was independent of t, i.e., it was uniform in t. For dilation, there is no such uniform bound in general. However, if t is kept less than some fixed value t_0, then the bound is uniform. In any event, for any given t, the bound goes to zero as $k \to \infty$.

---

[17] Ibid.

Sec. 3.13   Sampling-Error Bounds for Morphological Operations   177

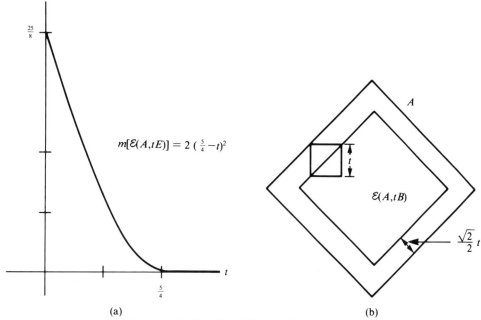

Figure 3.48 Euclidean Size Distribution from Erosion

Example 3.29

Consider Example 3.27, except let $k = 8$. Since B is a unit square centered at the origin, its greatest distance from the origin is $r = \sqrt{2}/2$. Hence, according to Theorem 3.11,

$$| m[\mathcal{D}(A, tB)] - \overline{m}[\text{ERODE}(\overline{A}, \overline{tB})] | \leq \frac{3\sqrt{2}}{512}\left(5\sqrt{2} + 2\pi \frac{\sqrt{2}}{2} t\right) + \frac{9\pi}{2^{17}}$$

which, ignoring the second summand, is approximately equal to $(3\pi t + 15)/256$.

We now turn to the sampling error for size distributions generated by granulometries of the form $t \to O(A, tE)$. In doing so, we shall restrict our attention to the linear structuring element E, where E is a vertical segment of length unity emanating from the origin. (We could just as easily consider a similar horizontal segment of length unity.) The size distribution u is given by $u(t) = m[O(A, tE)]$. Insofar as a digital version of u is concerned, the methodology was indicated in Example 3.19. Employing the digitization technique discussed above, $\overline{tE} = V(1)$ for $0 \leq t \leq 1/2^{k+1}$ and $\overline{tE} = V(n)$ for $(2n - 3)/2^{k+1} < t \leq (2n - 1)/2^{k+1}$, where the $V(n)$ are the structuring elements defined in Section 3.9 for the linear digital granulometries. The digital version of u is then given by $\overline{u}(t) = \overline{m}[\text{OPEN}(\overline{A}, \overline{tE})]$. Using the notation of Section 3.9, $\overline{u}(t) = \omega_1(n)/2^{2k}$ for $(2n - 3)/2^{k+1} < t \leq (2n - 1)/2^{k+1}$, where $\omega_1(n)$ is evaluated relative to the digitization $\overline{A}$.

Unlike the case of erosion, even though the opening will eventually eliminate the input image, and even though for sufficiently large t, $u(t) = \bar{u}(t) = 0$, it is not possible to obtain a uniform bound on $|u(t) - \bar{u}(t)|$ which goes to zero as $k \to \infty$. However, it is possible to obtain an L_1-bound. This is presented in the next theorem.

Theorem 3.12.[18] If A is convex and compact, then

$$\int_0^\infty |u(t) - \bar{u}(t)|\, dt \leq \frac{5\sqrt{2} + 1}{2^{k+2}} (\Lambda[A] + 2^{-k+5})^2$$

Example 3.30

Suppose A is as in Example 3.27 and $k = 8$. Then

$$u(t) = \begin{cases} \frac{25}{8} - (t^2/2) & \text{for } 0 \leq t \leq \frac{5}{2} \\ 0 & \text{for } t > \frac{5}{2} \end{cases}$$

and, according to Theorem 3.12,

$$\int_0^\infty |u(t) - \bar{u}(t)|\, dt \leq \frac{5\sqrt{2} + 1}{1028} \left(5\sqrt{2} + \frac{1}{8}\right)^2 < 0.43$$

Nothing is claimed about the pointwise difference of the functions.

EXERCISES

3.1. Let

$$A = \{(0, 0), (0, 2), (2, 0), (2, 2)\}$$
$$B = \{(0, 0), (1, 0), (2, 0), (3, 0)\}$$
$$C = \{(0, 0), (1, 1), (2, 2)\}$$

Find each of the following sets (in set notation and pictorially):

(a) $(1, 1) + A$
(b) $A \oplus B$
(c) $(A \oplus B) \oplus C$
(d) $A \ominus B$
(e) $B \ominus A$
(f) $\mathscr{E}(A, B)$
(g) $A \oplus (2C)$
(h) $2[\frac{1}{2}A \oplus C]$
(i) $(C \oplus A) \oplus B$
(j) $[A + (-1, 2)] \oplus B$
(k) $A \oplus (B \cup C)$
(l) $A \oplus (B \cap C)$
(m) $(A \oplus B) \cup (A \oplus C)$
(n) $(A \oplus B) \cap (A \oplus C)$
(o) $A \ominus (B \cup C)$
(p) $A \ominus (B \cap C)$
(q) $(A \oplus B) \cap (A \oplus C)$
(r) $(A \ominus B) \ominus C$

[18] Ibid.

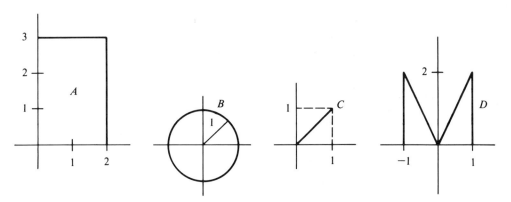

Figure 3.49 Images for Exercise 3.2

3.2. Let

$$A = \{(x, y): 0 \leq x \leq 2, 0 \leq y \leq 3\}$$
$$B = \{(x, y): x^2 + y^2 \leq 1\}$$
$$C = \{(x, y): y = x \text{ and } 0 \leq x \leq 1\}$$
$$D = \{(x, y): 0 \leq y \leq |2x| \text{ and } -1 \leq x \leq 1\}$$

(See Figure 3.49.) Give graphical representations of the following:

(a) $A \oplus B$
(b) $A \oplus C$
(c) $B \oplus C$
(d) $A \ominus B$
(e) $A \ominus C$
(f) $A \ominus \frac{1}{2}B$
(g) $A \ominus \frac{1}{4}C$
(h) $B \oplus \frac{1}{2}B$
(i) $A \oplus \frac{1}{2}B$

(j) $A \oplus \frac{1}{2}C$
(k) $O(A, B)$
(l) $O(A, C)$
(m) $O[B, \frac{1}{2}C]$
(n) $O[A, \frac{1}{2}B]$
(o) $O[A, \frac{1}{4}B]$
(p) $O(D, B)$
(q) $C(D, B)$
(r) $C(D, 2B)$

3.3. Let

$$E = \{(x, y): 0 \leq y \leq 2x \text{ and } 0 \leq x \leq 3\}$$

Find the barycentric coordinates of the point (2, 1).

3.4. The set D of Exercise 3.2 is not convex. Find all boundary points at which there does not exist at least one supporting line.

3.5. Let A and B be the sets given in Exercise 3.2. Find and graph the size distribution generated by the granulometry $t \to O(A, tB)$.

3.6. Let A and C be the sets given in Exercise 3.2. Find and graph the size distribution generated by the granulometry $t \to O(A, tC)$.

3.7. Let E be the set given in Exercise 3.4. The perimeter of E is 9. For all θ, find $\text{Proj}(E, \theta)$ and show that the Cauchy projection theorem holds, i.e., show that the average over $0 \leq \theta \leq 2\pi$ of the projections is equal to $\Lambda(E)/\pi$.

3.8. Using the integral formula that gives the length of a piecewise smooth boundary, show that the boundary length Λ is translationally and rotationally invariant.

3.9. Give an example to show that Λ is not, in general, increasing (does not satisfy Q5), even though it is increasing on convex compact sets. Use the Cauchy projection theorem to show that Λ is increasing on compact convex sets.

3.10. For $n = 1, 2, 3, \ldots$, let

$$A_n = \{(x, y): 0 \leq y \leq \frac{1}{n} x \text{ and } 0 \leq x \leq 1\}$$

and

$$A = \{(x, y): y = 0 \text{ and } 0 \leq x \leq 1\}$$

Show that $\lim_{n \to \infty} A_n = A$ in the Hausdorff metric by actually computing $h(A_n, A)$. Note the extreme sensitivity of the Hausdorff metric to noise by considering $h(B_n, A)$, where $B_n = A_n \cup \{(0, 2)\}$.

3.11. Using the sets B_n from Exercise 3.10, show that $\lim_{n \to \infty} m(B_n) = 0$, where m denotes the Lebesgue measure (area). This result should be expected since m is continuous from above and, using set A of Exercise 3.10,

$$\bigcap_{n=1}^{\infty} B_n = A \cup \{(0, 2)\}$$

But $m[A \cup \{(0, 2)\}] = 0$.

3.12. Let E be the triangle given in Exercise 3.3. In Section 3.6 it was shown that the measure of E can be found morphologically by using a one-sided derivative. (See Example 3.10.) Using that methodology, find the area of E.

3.13. Using the triangle E of Exercise 3.3 and the dilation technique of Section 3.6, find $\text{Proj}(E, \mathbf{u})$, where $\mathbf{u}$ is the unit vector pointing in the $0°$ direction. Repeat for the $90°$ direction.

3.14. Let

$$S = \begin{pmatrix} 1 & * & * & * & * \\ 1 & 1 & * & * & * \\ 1 & 1 & 1 & * & * \\ 1 & ① & 1 & 1 & * \\ 1 & 1 & 1 & 1 & 1 \end{pmatrix}$$

$$T = \begin{pmatrix} ① & * & 1 & * & 1 & 1 & 1 \\ * & 1 & * & 1 & * & 1 & 1 \\ 1 & * & 1 & * & 1 & 1 & 1 \\ * & 1 & * & 1 & * & 1 & * \\ 1 & * & 1 & * & 1 & * & 1 \end{pmatrix}$$

$$E = \begin{pmatrix} 1 & 1 \\ ① & 1 \end{pmatrix}$$

and

$$F = \begin{pmatrix} * & 1 \\ ① & * \end{pmatrix}$$

Find:

(a) ERODE(S, E)
(b) ERODE(S, F)
(c) DILATE(S, E)
(d) DILATE(S, F)
(e) ERODE(T, E)
(f) ERODE(T, F)
(h) DILATE(T, E)
(i) DILATE(T, F)

(j) OPEN(S, E)
(k) OPEN(S, F)
(l) CLOSE(S, E)
(m) CLOSE(S, F)
(n) OPEN(T, E)
(o) OPEN(T, F)
(p) CLOSE(T, E)
(q) CLOSE(T, F)

3.15. Let

$$R = \begin{pmatrix} 1 & 1 & 1 & * & 1 & 1 & 1 & 1 & 1 & 1 \\ 1 & * & * & 1 & * & * & * & * & * & 1 \\ 1 & 1 & 1 & 1 & 1 & 1 & 1 & * & * & * \\ * & * & * & 1 & 1 & 1 & 1 & 1 & 1 & 1 \\ 1 & 1 & 1 & * & 1 & * & 1 & 1 & * & * \\ * & * & * & * & * & 1 & * & 1 & 1 & 1 \\ ① & 1 & 1 & 1 & 1 & 1 & 1 & 1 & 1 & 1 \end{pmatrix}$$

For the image R, find:

(a) $\omega_1(k)$
(b) $\omega_2(k)$
(c) $\mu_1(k)$
(d) $\mu_2(k)$
(e) $\nu_1(k)$
(f) $\nu_2(k)$

3.16. For any digitial constant image S, define HULL(S) in the following manner: a pixel is activated in HULL(S) if it is activated in S or if there exists an activated pixel of S to its left and an activated pixel of S to its right. Find HULL(S) for

$$S = \begin{pmatrix} * & * & * & 1 & 1 \\ 1 & * & * & 1 & * \\ * & ① & 1 & * & 1 \\ * & * & 1 & 1 & * \end{pmatrix}$$

3.17. Develop an algorithm which, given an input constant image S, outputs HULL(S). Specify the algorithm in terms of a block diagram.

3.18. HULL is an increasing τ-mapping. Find the kernel of HULL. Find a basis for the kernel. Using that basis together with the representation of Theorem 3.7, evaluate HULL(S) for the image S of Exercise 3.16.

3.19. Define the mapping Φ in the following manner: pixel (i, j) in S is an element of $\Phi(S)$ if and only if there exists a pixel (u, v) in SQUARE(i, j), besides (i, j) itself, such that (u, v) is in the domain of S.

(a) Show that Φ is a τ-opening.
(b) Find $\Phi(T)$ for the image

$$T = \begin{pmatrix} 1 & * & * & * & 1 & * \\ * & * & * & * & 1 & * \\ 1 & * & * & 1 & 1 & * \\ * & * & 1 & 1 & * & * \\ 1 & ① & 1 & * & * & * \\ * & * & * & * & * & 1 \end{pmatrix}$$

(c) Find Inv[Φ].
(d) Find a basis for Φ.
(e) Apply Theorem 3.8 to find Φ(T).

3.20. Let A be the disk of radius 1. Assume that the sampling grid is overlaid so that the center of A sits at the origin, and assume that the grid squares have sides of length $\frac{1}{16}$. Let E be the vertical structuring element of length unity situated at the origin, i.e.,

$$E = \{(x, y): x = 0 \text{ and } 0 \leq y \leq 1\}$$

(a) Find and graph the Euclidean size distribution $m[\mathscr{E}(A, tE)]$.
(b) Applying the methodology of Section 3.13, find and graph the digital size distribution $\overline{m}[\text{ERODE}(\overline{A}, t\overline{E})]$.
(c) Apply Theorem 3.10 to obtain an upper bound on the difference between the two size distributions.

4

Transform Techniques in Image Processing

4.1 ANALOG (CONTINUOUS) IMAGES

There are numerous types of images to be considered in image processing. For instance, in Chapter 2, digital images are defined in terms of bound-matrix structures, while in Chapter 3, constant Euclidean images are given as subsets of the Euclidean plane. In this chapter we shall discuss *analog*, or continuous, image processing transforms. The definition of an analog or continuous image must be in accordance with the operations and transformations performed on the structure. In particular, since most transform techniques are integral in nature, it is necessary to include some integrability condition in the definition of analog images.

In general, an image f is defined to be a real-valued function defined on a subset of the Euclidean plane; however, this definition is too general for practical purposes and must be restricted. To begin with, if $f: A \to R$ and $g: B \to R$, then $f \oplus g$ is defined by

$$[f \oplus g](x, y) = \begin{cases} f(x, y) + g(x, y) & \text{for } (x, y) \in A \cap B \\ f(x, y) & \text{for } (x, y) \in A - B \\ g(x, y) & \text{for } (x, y) \in B - A \end{cases}$$

Note that in this section $\oplus$ denotes a more general version of EXTADD, and not Minkowski addition. Using it, an image f will be assumed to have the representation

$$f = \left[\bigoplus_{n=1}^{N} g_n \right] \oplus \left[\bigoplus_{m=1}^{M} h_m \right] \oplus \left[\bigoplus_{k=1}^{K} l_k \right]$$

where g_n is a bounded continuous function defined on a bounded domain A_n in R^2, h_m is a continuous function defined on a compact region B_m in R^2, and l_k is defined on a subset C_k of R^2 and is well behaved from some integrability point of view. For instance, l_k might be absolutely integrable or square integrable on C_k, or l_k might be a finite power function on C_k. Note that f need not consist of all three of the above types; indeed, we assume only that $N, M, K \geq 0$. Should $N = M = K = 0$, the null image $\emptyset$ results.

From a modeling point of view, a function g_n in the preceding representation can be thought of as an object of importance in the image f, while a function h_m can be imagined to be an edge, a boundary of some object in f or an object of interest in its own right which happens to include its own boundary. On the other hand, a function l_k has to do with "background" information within an image or with possible distortions or noise within the image. In many cases it will be assumed that $K = 0$, and hence there will be no such functions in the representation of f. Sometimes it is assumed that $K = 1$ and that l_1 is identically zero over all of R^2. This makes the image defined over the whole plane and at the same time gives it a constant background.

In general, an image f has a domain A given by

$$A = \left[\bigcup_{n=1}^{N} A_n \right] \cup \left[\bigcup_{m=1}^{M} B_m \right] \cup \left[\bigcup_{k=1}^{K} C_k \right]$$

If $N = M = K = 0$, then $A = \emptyset$, and the null image is obtained.

It is often the case that further constraints are imposed upon the components of an image. Sometimes further conditions are imposed upon the regions A_n, B_m, and C_k. These constraints may be given in terms of curvature, measurability, shape, or some other geometric property.

Some properties and examples of images using the previously given representation might help to illustrate the significance and universality of the model.

When $K = 0$, the resulting image has a bounded domain and can be enclosed in a (minimal-sized) rectangle. An instance of this type of image is a bound matrix, which can result when $N = K = 0$. In particular, each B_i can be a one-point subset of $Z \times Z \subset R \times R$. A more general situation with $K = 0$ is given in Example 4.2, where $B_i \not\subset Z \times Z$.

Example 4.1

Let

$$f = \bigoplus_{m=1}^{3} h_m = h_1 \oplus h_2 \oplus h_3$$

where $B_1 = \{(0, 0)\}$, $B_2 = \{(1, 0)\}$, $B_3 = \{(2, 1)\}$, $h_1(0, 0) = 2$, $h_2(1, 0) = 3$, and $h_3(2, 1) = 0$. Using the notation of Chapter 2,

$$f = \begin{pmatrix} * & * & 0 \\ 2 & 3 & * \end{pmatrix}_{0,1}$$

Example 4.2

Let

$$f = \bigoplus_{m=1}^{3} h_m$$

where $B_1 = [-\frac{1}{2}, \frac{1}{2}] \times [-\frac{1}{2}, \frac{1}{2}]$, $B_2 = [\frac{1}{2}, 1\frac{1}{2}] \times [-\frac{1}{2}, \frac{1}{2}]$, $B_3 = [-\frac{1}{2}, \frac{1}{2}] \times [1\frac{1}{2}, 2\frac{1}{2}]$, $f = 2$ on B_1, $f = 3$ on B_2, and $f = 0$ on B_3. The image f is depicted in Figure 4.1, where each small "square pixel" is "painted" the appropriate gray value. Notice that $f = 5$ on the "edge" given by $\{(x, y): x = \frac{1}{2}$ and $-\frac{1}{2} \leq y \leq \frac{1}{2}\}$. Depending upon the application, this observation may or may not be of consequence. In any event, one can see that the image fits into the rectangle $[-\frac{1}{2}, 1\frac{1}{2}] \times [-\frac{1}{2}, 2\frac{1}{2}]$.

A type of bounded image that is not digital is the *line-drawing-type* image, for which $K = N = 0$. These images are constant Euclidean images that are defined on sets of Lebesgue measure zero.

Example 4.3

Let

$$B_1 = \{(x, y): x = 0 \text{ and } -1 \leq y \leq 1\}$$

$$B_2 = \{(x, y): x = 2 \text{ and } -1 \leq y \leq 1\}$$

$$B_3 = \{(x, y): y = -1 \text{ and } 0 < x < 2\}$$

$$B_4 = \{(x, y): y = x + 1 \text{ and } 0 < x \leq 1\}$$

$$B_5 = \{(x, y): y = 3 - x \text{ and } 1 < x < 2\}$$

Moreover, let

$$f = \bigoplus_{m=1}^{5} h_m$$

where $h_m = 1$ on its domain of definition. Then f is the *house-type-edge* image depicted in Figure 4.2.

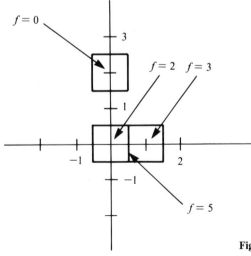

Figure 4.1 Image Not Given by a Bound Matrix

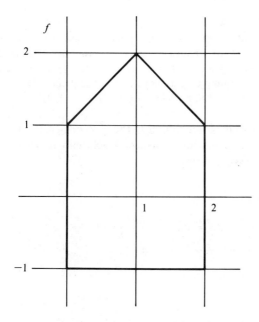

Figure 4.2 Edge Image of House

Various edge images are utilized in the section on Fourier series transforms. They occur as inputs to the Fourier procedure via the Freeman chain code representation. The output of this procedure is also an edge image, which happens to be comprised of a finite number of harmonic components.

When $M = K = 0$, once again the image has a bounded domain; however, in this case, there are no boundaries included in the image.

Example 4.4

Suppose $f(x, y) = g_1(x, y) = xy$ on A_1, where $A_1 = (0, 1) \times (0, 1)$ is a square without a boundary. Then, as illustrated in Figure 4.3, f can be visualized as a surface that is defined only on A_1.

Example 4.5

Let g_1 and A_1 be as in Example 4.4, and let $l_1 = 0$ on $C_1 = R \times R$. If $f = g_1 \oplus l_1$, then the graph of f is similar to that given in Figure 4.3 except that this new image is defined on all of $R \times R$ and has the value zero outside of A_1.

More general types of images will be employed in ensuing sections and manipulated using the Fourier, Hankel, Radon, and other transforms. In the case of Fourier transforms (as well as in numerous others), integrability-type constraints must be imposed on the images under consideration since the Fourier transform operates on images that are assumed to be defined on the entire Euclidean plane.

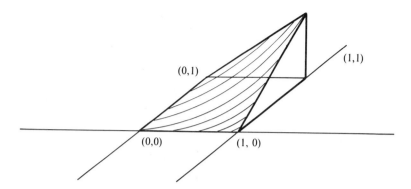

Figure 4.3 Image Given as a Surface

4.2 FOURIER TRANSFORMS OF INTEGRABLE SIGNALS (ONE DIMENSION)

Many properties of the Fourier Transform (FT) in one dimension hold in two or more dimensions; however, these properties are easier to understand and illustrate in the lower dimension. Moreover, the Fourier transform involving one-dimensional signals has direct application in image processing, particularly in the area of tomographical imaging. Consequently, this section will be devoted to the single-variable case.

The simplest class of signals for which the Fourier integral is defined is the class L_1 of Lebesgue integrable signals on $(-\infty, \infty)$. A signal f is in L_1 if it is measurable and absolutely integrable, i.e., if

$$\int_{-\infty}^{\infty} |f(t)|\, dt$$

is finite. If $f \in L_1$, then the areas under both the positive and negative parts of f are finite.

Given a signal f in L_1, the *Fourier transform* of f, $\mathcal{F}(f)$, is defined by the function

$$F(\omega) = \int_{-\infty}^{\infty} f(t)e^{-j\omega t}\, dt,$$

where $j = \sqrt{-1}$. The fact that f is integrable guarantees that $F(\omega)$ exists for all ω in $(-\infty, \infty)$.

The variable ω in the transform $F(\omega)$ is called the (radian) *frequency*. The modulus of $F(\omega)$, $|F(\omega)|$, is called the *frequency spectrum* or *spectral amplitude* of $f(t)$. The symbolism $f(t) \leftrightarrow F(\omega)$ is often employed to mean that $F(\omega)$ is the Fourier transform of $f(t)$. We shall shortly see that, under certain conditions, given an $F(\omega)$, there exists a unique $f(t)$ such that $F(\omega)$ is the Fourier transform of $f(t)$; that is, there is a symmetric relationship between time signals and their frequency representations. The next theorem gives some of the most basic properties of the Fourier transform. As with the other theorems of this chapter, no proof will be provided.

Theorem 4.1. Let $f, g \in L_1$, and let c be a complex number. If $f \leftrightarrow F(\omega)$ and $g \leftrightarrow G(\omega)$, then

(a) $f + g \leftrightarrow F(\omega) + G(\omega)$ (FT is additive.)
(b) $cf \leftrightarrow cF(\omega)$ (FT is homogenous.)
(c) $|F(\omega)| \leq \int_{-\infty}^{\infty} |f(t)|\, dt$ (The spectral amplitude is bounded by the one norm of f.)
(d)[1] $F(\omega)$ is uniformly continuous on $(-\infty, \infty)$.
(e)[2] The Rieman-Lebesgue lemma holds, i.e.,

$$\lim_{|\omega| \to \infty} F(\omega) = 0$$

Note that (a) and (b) of Theorem 4.1 together imply that the Fourier transform is linear, i.e.,

$$bf + cg \leftrightarrow bF(\omega) + cG(\omega)$$

for complex scalars b and c.

Example 4.6

Let $f(t) = e^{-\lambda t} u(t)$, where $\lambda > 0$ and $u(t)$ is the unit step function

$$u(t) = \begin{cases} 1 & \text{if } t \geq 0 \\ 0 & \text{if } t < 0 \end{cases}$$

Then $f \in L_1$ since $\int_{-\infty}^{\infty} |f(t)|\, dt = 1/\lambda$. Moreover,

$$F(\omega) = \int_0^\infty e^{-\lambda t} e^{-j\omega t}\, dt = \left. \frac{e^{-t(\lambda + j\omega)}}{-(\lambda + j\omega)} \right|_0^\infty = \frac{1}{\lambda + j\omega}$$

since $\lim_{t \to \infty} e^{-t(\lambda + j\omega)} = 0$. The latter follows from

$$\lim_{t \to \infty} |e^{-\lambda t} e^{-j\omega t}| = \lim_{t \to \infty} e^{-\lambda t} |\cos(\omega t) - j\sin(\omega t)|$$

$$= \lim_{t \to \infty} e^{-\lambda t} = 0$$

The spectral amplitude is given by

$$|F(\omega)| = \left| \frac{1}{\lambda + j\omega} \right| = (\lambda^2 + \omega^2)^{-1/2}$$

[1] R. R. Goldberg, *Fourier Transforms* (New York: Cambridge University Press, 1965), 6.
[2] Ibid., 7.

Sec. 4.2   Fourier Transforms of Integrable Signals (One Dimension)

Note that $|F(\omega)|$ is uniformly continuous and that $\lim_{|\omega|\to\infty}(\lambda^2 + \omega^2)^{-1/2} = 0$. One normally illustrates spectral properties using the amplitude spectrum rather than $F(\omega)$ itself since the former is real valued.

Example 4.7

Let $f(t) = e^{-\alpha|t|} + 2e^{-\lambda t}u(t)$, where $\lambda > 0$ and $\alpha > 0$. By the result of the previous example and homogeneity,

$$2e^{-\lambda t}u(t) \leftrightarrow \frac{2}{\lambda + j\omega}$$

Moreover, letting $g(t) = e^{-\alpha|t|}$,

$$G(\omega) = \int_{-\infty}^{\infty} e^{-\alpha|t|} e^{-j\omega t} dt$$

$$= \int_{-\infty}^{0} e^{-t(j\omega - \alpha)} dt + \int_{0}^{\infty} e^{-t(\alpha + j\omega)} dt$$

$$= \frac{1}{\alpha - j\omega} + \frac{1}{\alpha + j\omega} = \frac{2}{\alpha^2 + \omega^2}$$

Hence, by linearity, $f(t)$ has Fourier transform

$$F(\omega) = \frac{2\alpha}{\alpha^2 + \omega^2} + \frac{2\alpha}{\lambda + j\omega}$$

Thus, $F(\omega)$ is continuous, $\lim_{|\omega|\to\infty} F(\omega) = 0$ and $|F(\omega)| \leq 2/\alpha + 2/\lambda$.

Theorem 4.2. Suppose $f \in L_1$ and $f \leftrightarrow F(\omega)$. Then

(a) $f(t)e^{-jht} \leftrightarrow F(\omega + h)$, for h real
(b) $f(at) \leftrightarrow (1/|a|) F(\omega/a)$, for a real and $a \neq 0$
(c) $f(t + h) \leftrightarrow e^{jh\omega} F(\omega)$, for h real
(d) If $tf(t) \in L_1$, then the derivative

$$F'(\omega) = \frac{d[F(\omega)]}{d\omega}$$

exists and $-jtf(t) \leftrightarrow F'(\omega)$
(e) If $df/dt \in L_1$, then $df/dt \leftrightarrow j\omega F(\omega)$
(f)[3] If $n > 1$ and the nth derivative $f^{(n)}(t) \in L_1$, then $f', f'', \ldots, f^{(n-1)} \in L_1$ and $f^{(n)}(t) \leftrightarrow (j\omega)^n F(\omega)$.

Example 4.8

Let $f(t) = e^{-t} \cos(bt) u(t)$, for $b > 0$. Since $\cos(bt) = \frac{1}{2}[e^{jbt} + e^{-jbt}]$,

$$f(t) = \frac{1}{2} e^{-t} u(t) e^{jbt} + \frac{1}{2} e^{-t} u(t) e^{-jbt}$$

---
[3] S. Bochner and K. Chandrasekharan, *Fourier Transforms* (New York: Kraus Reprint Corp., 1965), 29.

But, by Example 4.6, $e^{-t}u(t) \leftrightarrow 1/(1 + j\omega)$. Therefore, by part (a) of Theorem 4.2,

$$e^{-t}u(t)e^{jbt} \leftrightarrow \frac{1}{1 + j(\omega - b)}$$

and

$$e^{-t}u(t)e^{-jbt} \leftrightarrow \frac{1}{1 + j(\omega + b)}$$

By linearity,

$$e^{-t}\cos(bt)u(t) \leftrightarrow \frac{1}{2}\left[\frac{1}{1 + j(\omega - b)} + \frac{1}{1 + j(\omega + b)}\right]$$

Consequently,

$$|F(\omega)| = \sqrt{\frac{1 + \omega^2}{(1 + (\omega - b)^2)(1 + (\omega + b)^2)}}$$

Example 4.9

Let $f = I_{(-1/2, 1/2)}(t)$, where $I_A(t)$ is the indicator function,

$$I_A(t) = \begin{cases} 1 & \text{if } t \in A \\ 0 & \text{if } t \notin A \end{cases}$$

Then

$$F(\omega) = \int_{-1/2}^{1/2} e^{-j\omega t} dt = \frac{e^{-j\omega/2} - e^{j\omega/2}}{-j\omega} = \frac{\sin(\omega/2)}{\omega/2}, \text{ for } \omega \neq 0$$

Notice that $F(\omega)$ has a removable discontinuity at $\omega = 0$ since the rightmost expression is not defined at $\omega = 0$ and $\lim_{\omega \to 0} \sin(\omega/2)/\omega/2 = 1$. All removable discontinuities will be removed throughout the chapter without further mention. In this case, we set $F(0) = 1$, which agrees with the result obtained by setting $\omega = 0$ in the Fourier integral.

Next, by property (c) of Theorem 4.2, with $h = -\frac{1}{2}$,

$$I_{(0,1)}(t) = I_{(-1/2,1/2)}\left(t - \frac{1}{2}\right) \leftrightarrow \frac{\sin(\omega/2)}{\omega/2} e^{-j\omega/2}.$$

Next, applying property (b) with scalar $a = 1/d$, where $d > 0$, to $I_{(0,1)}$, we have

$$I_{(0,d)}(t) = I_{(0,1)}\left(\frac{t}{d}\right) \leftrightarrow \frac{d \sin(d\omega/2)}{d\omega/2} e^{-jd\omega/2}$$

Consequently, by property (c) again,

$$I_{(a, a+d)}(t) = I_{(0,d)}(t - a) \leftrightarrow \frac{d \sin(d\omega/2)}{d\omega/2} e^{-jd\omega/2} e^{-ja\omega}$$

Finally, using linearity, the rectangular wave of height A and support in $(a, a + d)$,

Sec. 4.2   Fourier Transforms of Integrable Signals (One Dimension)   191

namely, $AI_{(a,a+d)}(t)$, has Fourier transform given by

$$AI_{(a,a+d)} \leftrightarrow \frac{Ad \sin(d\omega/2)}{d\omega/2} e^{-jd\omega/2} e^{-ja\omega}$$

Pay particular attention in this example to how Theorems 4.1 and 4.2 were utilized in a sequence of steps in order to find the desired transform of an important function.

Example 4.10

Let $f(t) = e^{-\lambda t} u(t)$, for $\lambda > 0$. Notice that $tf(t) \in L_1$. Since $f \leftrightarrow 1/(\lambda + j\omega)$, by part (d) of Theorem 4.2,

$$-jte^{-\lambda t} u(t) \leftrightarrow \frac{d}{d\omega}[1/(\lambda + j\omega)] = \frac{-j}{(\lambda + j\omega)^2}$$

or, by linearity,

$$te^{-\lambda t} u(t) \leftrightarrow (\lambda + j\omega)^{-2}$$

Two points should be noticed. First, $te^{-\lambda t} u(t)$ is continuous for all t, whereas $e^{-\lambda t} u(t)$ has a jump discontinuity at $t = 0$. Second, the high-frequency components of $te^{-\lambda t} u(t)$ fall off like $|\omega|^{-2}$, whereas for the signal $e^{-\lambda t} u(t)$, they fall off like $|\omega|^{-1}$. Indeed, there is a relationship between the smoothness of time signals and the rate of fall of high-frequency components of their spectral representations. This relationship is evidenced in part (f) of Theorem 4.2, where if $f^{(n)} \leftrightarrow G(\omega)$, then $F(\omega) = (j\omega)^{-n} G(\omega)$.

Example 4.11

This example illustrates the calculational aspect of property (e) of Theorem 4.2. Let $g(t) = e^{-t^2}$. By the said property (e),

$$g'(t) = -2te^{-t^2} \leftrightarrow j\omega G(\omega)$$

Applying part (d) of the theorem,

$$-jtg(t) \leftrightarrow G'(\omega)$$

and hence, by linearity,

$$-2te^{-t^2} \leftrightarrow \frac{2}{j} G'(\omega)$$

Since the Fourier transform of a given time function in L_1 is unique, the two representations of the transform of $-2te^{-t^2}$ must be equal. Hence, we obtain the separable differential equation,

$$j\omega G(\omega) = \frac{2}{j} G'(\omega)$$

Separating the variables gives

$$\frac{G'(\omega)}{G(\omega)} = -\omega/2$$

and integrating yields

$$\log[G(\omega)] = \frac{-\omega^2}{4} + C$$

Hence, $G(\omega) = Ae^{-\omega^2/4}$. The constant A can be found by computing $G(0)$ directly from the defining Fourier integral:

$$G(0) = \int_{-\infty}^{\infty} e^{-t^2} dt = 2\int_0^{\infty} e^{-t^2} dt = \frac{2\sqrt{\pi}}{2} = \sqrt{\pi}$$

Hence, $G(\omega) = \sqrt{\pi} e^{-\omega^2/4}$. Using part (b) of Theorem 4.2 gives

$$e^{-t^2/2} \leftrightarrow \sqrt{2\pi} e^{-\omega^2/2}$$

In other words, $e^{-t^2/2}$ acts like an *eigenfunction* for the Fourier transform since, when it is input into the transform, a scalar multiple of it is output.

Theorem 4.3.[4] Suppose $g \in L_1$ and $f(t) = \int_a^t g(u)\, du$ for some fixed point a, where $-\infty \leq a < t$. If $f \in L_1$ and $f \leftrightarrow F$, then $g \leftrightarrow j\omega F(\omega)$.

Example 4.12

Let $f(t) = (t + 1)I_{(-1,0)}(t) + (1 - t)I_{(0,1)}(t)$. (See Figure 4.4.) One could find $F(\omega)$ by integration by parts; however, this would be quite tedious. Instead, note that even though f has no derivative at -1, 0, and 1, so that part (e) of Theorem 4.2 does not apply, we still have $f(t) = \int_{-\infty}^t g(u)\, du$, where $g(u) = I_{(-1,0)}(u) - I_{(0,1)}(u)$. (See Figure 4.5.) Thus, the Fourier transform $G(\omega)$ of g can be found by the methodology employed in Example 4.9. We have

$$G(\omega) = \frac{\sin(\omega/2)}{\omega/2} e^{-j\omega/2} e^{j\omega} - \frac{\sin(\omega/2)}{\omega/2} e^{-j\omega/2}$$

According to Theorem 4.3, $F(\omega) = (j\omega)^{-1} G(\omega)$.

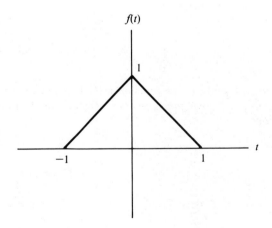

Figure 4.4 Triangular Wave-Type Signal

[4] Ibid., 9.

Sec. 4.2 Fourier Transforms of Integrable Signals (One Dimension)   193

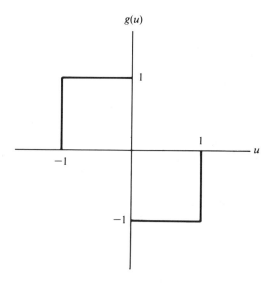

Figure 4.5 Signal to Serve as Integrand

The next operation defined on signals in L_1 plays an important role in image processing, as well as in many other areas of engineering. Suppose $f, g \in L_1$. The *convolution* of f and g is the function h defined by

$$h(t) = \int_{-\infty}^{\infty} f(t - x)g(x)\, dx$$

Notationally, we write $h = f * g$. The following theorem gives some of the fundamental properties of the convolution.

Theorem 4.4. Suppose f, g, and r are signals in L_1. Then

(a)[5] The convolution $[f * g](t)$ exists almost everywhere[6] for $t \in (-\infty, \infty)$ and, furthermore, $f * g \in L_1$.
(b)[7] $\int_{-\infty}^{\infty} |[f * g](t)|\, dt \leq \int_{-\infty}^{\infty} |f(t)|\, dt \int_{-\infty}^{\infty} |g(t)|\, dt$.
(c) $f * g = g * f$ (commutativity)
(d) $(f * g) * r = f * (g * r)$ (associativity)

Example 4.13

Let $f(t) = tI_{(1,2)}(t)$ and $g(t) = I_{(3,4)}(t)$. We wish to find

$$h(t) = [f * g](t) = \int_{-\infty}^{\infty} f(t - x)g(x)\, dx$$

---

[5] Ibid., 6.

[6] A property is said to hold almost everywhere on $(-\infty, \infty)$ if it holds everywhere except on a set of Lebesgue measure zero. For practical purposes regarding integration, sets of measure zero can be ignored. For a complete explanation, see M. E. Munroe, *Measure and Integration*, 2nd ed. (Philippines: Addison-Wesley, 1971), 104.

[7] Bochner, *Fourier Transforms*, 6.

First, notice that

$$f(t - x) = (t - x)I_{(1,2)}(t - x)$$

$$= \begin{cases} t - x & \text{for } t - 2 < x < t - 1 \\ 0 & \text{otherwise} \end{cases}$$

$$= (t - x)I_{(t-2, t-1)}(x)$$

Both $f(t - x)$ and $g(x)$ are illustrated in Figure 4.6. To find the convolution $h(t)$, we must multiply $f(t - x)$, which has support in $(t - 2, t - 1)$, by $g(x)$, which has support in $(3, 4)$. Since the parameter t varies between $-\infty$ and ∞, Figure 4.6 actually represents infinitely many figures. Of course, the only important ones are those for which the intersection of the support regions is nonempty, that is, those for which the product $f(t - x)g(x) \neq 0$. There are several possible cases. If $t - 1 \leq 3$, then the support of $f(t - x)$ is to the left of the support of $g(x)$, and we have $h(t) = 0$. If $3 < t - 1 \leq 4$, or $4 < t \leq 5$, which is the case illustrated in Figure 4.7(a), then

$$h(t) = \int_3^{t-1} (t - x)\, dx = tx - \frac{x^2}{2}\bigg|_3^{t-1} = \frac{t^2}{2} - 3t + 4$$

If $3 < t - 2 \leq 4$, or $5 < t \leq 6$, which is the case illustrated in Figure 4.7(b), then

$$h(t) = \int_{t-2}^4 (t - x)\, dx = tx - \frac{x^2}{2}\bigg|_{t-2}^4 = \frac{-t^2}{2} + 4t - 6$$

Finally, if $4 < t - 2$, then $h(t) = 0$. Consequently, the convolution of f and g is given by

$$h(t) = \left(\frac{t^2}{2} - 3t + 4\right) I_{(4,5]}(t) + \left(\frac{-t^2}{2} + 4t - 6\right) I_{(5,6]}(t)$$

which is integrable (in L_1) and, in fact, is continuous. Note also that part (b) of Theorem 4.4 holds, since

$$\int_{-\infty}^{\infty} |h(t)|\, dt = \frac{3}{2} \leq \int_{-\infty}^{\infty} |f(t)|\, dt \int_{-\infty}^{\infty} |g(t)|\, dt = \frac{3}{2}$$

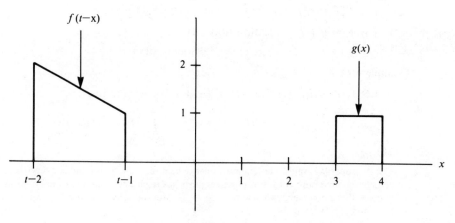

Figure 4.6 Initialization Prior to Convolution

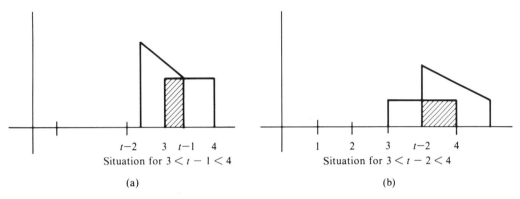

Figure 4.7(a) Situation for $3 < t - 1 < 4$
Figure 4.7(b) Situation for $3 < t - 2 < 4$

The following theorem is one of the prime reasons for the prominence of the convolution in signal processing. In words, it states that multiplication in frequency corresponds to convolution in time.

Theorem 4.5.[8] Suppose $f, g \in L_1$, $f \leftrightarrow F(\omega)$ and $g \leftrightarrow G(\omega)$. Then $f * g \leftrightarrow F(\omega)G(\omega)$.

Example 4.14

Let $f(t) = g(t) = i_{(-1/2, 1/2)}(t)$. Then, in a similar manner to that of the previous example, we can find the convolution $h = f * g$; indeed,

$$h(t) = (t + 1)I_{(-1,0]}(t) + (1 - t)I_{[0,1)}(t)$$

By Theorem 4.5, $H(\omega) = F(\omega)G(\omega)$. But $F(\omega) = G(\omega) = \sin(\omega/2)/(\omega/2)$. Consequently,

$$H(\omega) = \left(\frac{\sin(\omega/2)}{\omega/2}\right)^2$$

Note that there was no need to compute $h(t)$ explicitly in order to find $H(\omega)$.

Theorem 4.6[9] (Uniqueness Theorem). Suppose $f \in L_1$, $f \leftrightarrow F(\omega)$, and $F(\omega) = 0$. Then $f = 0$ almost everywhere.

An immediate consequence of the Uniqueness Theorem (see Figure 4.8) is the fact that if $f \leftrightarrow F(\omega)$, $g \leftrightarrow G(\omega)$, and $F(\omega) = G(\omega)$, then $f = g$ almost everywhere, that is to say, from the point of view of integration, they are the same function. (This is what is meant by uniqueness.) Therefore, knowing $F(\omega)$, once a time function $f(t)$ is found such that $f \leftrightarrow F(\omega)$, we can say that $f(t)$ is *the*

---
[8] Ibid., 6.
[9] Ibid., 11.

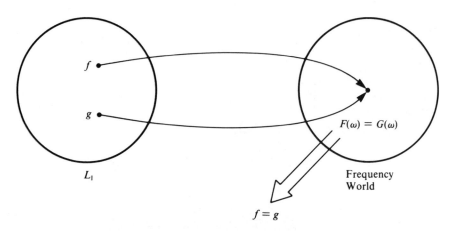

Figure 4.8 Uniqueness Theorem for Fourier Transform

time function satisfying $\mathcal{F}(f) = F(\omega)$. It is this one-to-one relationship between a time function and its Fourier transform which motivates the symmetric symbolism $f \leftrightarrow F(\omega)$. The problem, of course, is to retrieve $f(t)$ when given $F(\omega)$. Under some conditions this can be accomplished.

Given $F(\omega)$, we call the integral (when it exists)

$$[\mathcal{F}^{-1}(F)](t) = \frac{1}{2\pi} \int_{-\infty}^{\infty} F(\omega)e^{j\omega t}\, d\omega$$

the *inverse Fourier transform*. Under appropriate conditions, $\mathcal{F}^{-1}(F) \leftrightarrow F(\omega)$, or equivalently, $\mathcal{F}[\mathcal{F}^{-1}(F)] = F$, which means precisely that operation by $\mathcal{F}^{-1}$ yields a time function $f(t)$ whose Fourier transform is the given function $F(\omega)$. (See Figure 4.9.) The next theorem gives sufficient conditions for the inversion formula to yield a valid inverse.

Theorem 4.7. Suppose $f \in L_1$ and $f \leftrightarrow F(\omega)$. Then

(a)[10] If f is continuous for any $t \in (-\infty, \infty)$, and if $F(\omega) \in L_1$, then

$$f(t) = \frac{1}{2\pi} \int_{-\infty}^{\infty} F(\omega)e^{j\omega t}\, d\omega$$

If f is not continuous, then inversion holds for almost all t (almost everywhere).[11]

(b)[12] If f is bounded and continuous over $(-\infty, \infty)$ and $F(\omega) \ge 0$, then, once again, $f(t) = [\mathcal{F}^{-1}(F)](t)$. Again, if f is not continuous, the relation holds for almost all t.

---

[10] Goldberg, *Fourier Transforms*, 16.
[11] Bochner, *Fourier Transforms*, 19.
[12] Ibid., 66.

Sec. 4.2 Fourier Transforms of Integrable Signals (One Dimension)

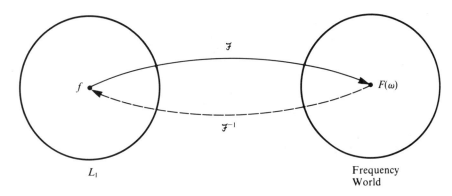

Figure 4.9 Inversion of Fourier Transform

(c)[13] If f is of bounded variation, then

$$\lim_{R \to \infty} \frac{1}{2\pi} \int_{-R}^{R} e^{j\omega t} F(\omega) \, d\omega = \frac{1}{2} [f(t^+) + f(t^-)]$$

In other words, the inversion integral exists in the CPV (Cauchy principle value) sense, and it equals the average value of the left- and right-hand limits of f at t.

In part (c) of the theorem, if f is continuous at t, then the average value of the left- and right-hand limits equals $f(t)$. Moreover, note that the continuity of $f(t)$ in parts (a) and (b) of Theorem 4.7 is usually unimportant, since if $\mathcal{F}^{-1}(F)$ equals $f(t)$ almost everywhere, then they are equal from the point of view of integration.

Example 4.15

Let $f(t) = e^{-|t|}$. In Example 4.7 it was seen that $F(\omega) = 2/(1 + \omega^2)$, which is in L_1. Since $f(t)$ is everywhere continuous, Theorem 4.7 (part a) implies that

$$e^{-|t|} = \frac{1}{2\pi} \int_{-\infty}^{\infty} \frac{2}{1 + \omega^2} e^{j\omega t} \, d\omega$$

for all t. This equality may be verified directly by complex integration.

Example 4.16

Let $f(t) = I_{(-1/2, 1/2)}(t)$. We have seen that $F(\omega) = \sin(\omega/2)/(\omega/2)$ which is not in L_1. Hence, part (a) of Theorem 4.7 is not applicable, and neither is part (b) since $F(\omega)$ is not nonnegative. However, part (c) does apply since the total variation of $f(t)$ over $(-\infty, \infty)$ is 2. As a result,

$$\lim_{R \to \infty} \frac{1}{2\pi} \int_{-R}^{R} \frac{\sin(\omega/2)}{\omega/2} e^{j\omega t} \, d\omega = \begin{cases} 1 & \text{for } |t| < \frac{1}{2} \\ \frac{1}{2} & \text{for } |t| = \frac{1}{2} \\ 0 & \text{for } |t| > \frac{1}{2} \end{cases}$$

This result may be verified using residue theory.

---

[13] Goldberg, *Fourier Transforms*, 12.

We next briefly consider the Fourier transform of those signals which are square integrable, that is, for time functions in L_2. L_2 signals are known as *energy* signals. Since there exist L_2 signals that are not in L_1, one should not expect the same spectral formulas to be used for signals in L_2 that were used for signals in L_1. Nonetheless, in many engineering situations where energy signals are employed, L_2 signals are also in L_1. Therefore, in this section we will limit ourselves to signals that are both integrable and square integrable, i.e., signals in $L_1 \cap L_2$. For such signals, everything that was said for L_1 signals is again applicable. The next theorem, Parseval's theorem, relates the energy in the time domain with the energy in the frequency domain.

Theorem 4.8.[14] If $f \in L_1 \cap L_2$, then $F(\omega) \in L_2$ and

$$\int_{-\infty}^{\infty} |f(t)|^2\, dt = \frac{1}{2\pi} \int_{-\infty}^{\infty} |F(\omega)|^2\, d\omega$$

Example 4.17

Let $f(t) = I_{(-1/2, 1/2)}(t)$. Then $F(\omega) = \sin(\omega/2)/(\omega/2)$. Since $f \in L_1 \cap L_2$ with energy $E = \int_{-\infty}^{\infty} f^2 = 1$, by the previous theorem,

$$\frac{1}{2\pi} \int_{-\infty}^{\infty} \left(\frac{\sin(\omega/2)}{\omega/2}\right)^2 d\omega = 1$$

Example 4.18

Since $e^{-|t|} \leftrightarrow 2/(1 + \omega^2)$, $e^{-|t|} \in L_1 \cap L_2$, and $E = \int_{-\infty}^{\infty} e^{-2|t|}\, dt = 1$, Theorem 4.8 implies that

$$\frac{1}{2\pi} \int_{-\infty}^{\infty} \frac{4}{(1 + \omega^2)^2}\, d\omega = 1$$

4.3 THE FOURIER TRANSFORM OF INTEGRABLE IMAGES

It is possible to extend the theory of Fourier transforms to images. When applied to images, many properties of the Fourier transform are analogous to those for signals, except that for images a double integral is involved. For instance, the class L_1 now consists of measurable functions on R^2 for which the double integral

$$\int_{-\infty}^{\infty} \int_{-\infty}^{\infty} |f(x, y)|\, dx\, dy$$

is finite. As before, images in L_1 are said to be *integrable*.

The Fourier transform of an image $f(x, y)$ in L_1 is the complex-valued function $F(\omega_1, \omega_2)$ of two real variables ω_1, ω_2 in $(-\infty, \infty)$ defined by

$$F(\omega_1, \omega_2) = \int_{-\infty}^{\infty} \int_{-\infty}^{\infty} f(x, y) e^{-j(\omega_1 x + \omega_2 y)}\, dx\, dy$$

---

[14] Ibid., 44.

Sec. 4.3   The Fourier Transform of Integrable Images

Symbolically, one often writes $\mathcal{F}(f) = F$, or $f \leftrightarrow F$. Notice that the defining double integral can be interpreted, and therefore evaluated, as an iterated integral. Hence,

$$F(\omega_1, \omega_2) = \int_{-\infty}^{\infty} \left[\int_{-\infty}^{\infty} f(x, y) e^{-j\omega_1 x} \, dx \right] e^{-j\omega_2 y} \, dy$$

$$= \int_{-\infty}^{\infty} \left[\int_{-\infty}^{\infty} f(x, y) e^{-j\omega_2 y} \, dy \right] e^{-j\omega_1 x} \, dx$$

Theorem 4.9. Suppose $f, g \in L_1$, α and β are complex numbers, $f \leftrightarrow F$, and $g \leftrightarrow G$. Then

(a) $\alpha f + \beta g \leftrightarrow \alpha F + \beta G$ (FT is linear.)
(b) $|F(\omega_1, \omega_2)| \leq \int_{-\infty}^{\infty} \int_{-\infty}^{\infty} |f(x, y)| \, dy \, dx$
   That is, the amplitude spectrum is bounded by the one norm of f.
(c) $F(\omega_1, \omega_2)$ is uniformly continuous on R^2.
(d)[15] The Riemann-Lebesgue lemma holds, i.e.,

$$\lim_{\omega_1^2 + \omega_2^2 \to 0} F(\omega_1, \omega_2) = 0$$

(e) $f(x, y) e^{-j(hx + ky)} \leftrightarrow F(\omega_1 + h, \omega_2 + k)$, for h and k real.
(f) $f(ax, by) \leftrightarrow |ab|^{-1} F(\omega_1/a, \omega_2/b)$, for nonzero real numbers a, b.
(g) $f(x + h, y + k) \leftrightarrow F(\omega_1, \omega_2) e^{j(\omega_1 h + \omega_2 k)}$, for h and k real.

Example 4.19

Let $f(x, y) = e^{-x} I_{[0, \infty) \times (0, 1)}(x, y)$. Then

$$F(\omega_1, \omega_2) = \int_0^{\infty} e^{-x} e^{-j\omega_1 x} \, dx \int_0^1 e^{-j\omega_2 y} \, dy$$

Applying the results of Examples 4.6 and 4.9, we have

$$F(\omega_1, \omega_2) = \frac{1}{1 + j\omega_1} \frac{\sin(\omega_2/2)}{\omega_2/2} e^{-j\omega_2/2}$$

Many of the properties set out in Theorem 4.9 are easily verified. For instance, it can readily be shown that $|F(\omega_1, \omega_2)| \leq 1$ and that $F(\omega_1, \omega_2)$ is uniformly continuous.

As in the signal processing case, convolution is a basic image processing operation. The convolution of two L_1 images f and g is defined by

$$h(x, y) = \int_{-\infty}^{\infty} \int_{-\infty}^{\infty} f(x - u, y - v) g(u, v) \, du \, dv$$

---

[15] Bochner, *Fourier Transforms*, 57.

and is denoted by $h = f * g$. The following theorem gives the major properties of the two-dimensional convolution. Notice the similarities with Theorems 4.4 and 4.5.

Theorem 4.10. For $f, g, r \in L_1$,

(a) $f * g$ exists almost everywhere in R^2.
(b) $\int_{-\infty}^{\infty} \int_{-\infty}^{\infty} |[f * g](x, y)| \, dy \, dx \leq \int_{-\infty}^{\infty} |f| \, dy \, dx \int_{-\infty}^{\infty} |g| \, dy \, dx$.
(c) $f * g = g * f$ (commutativity)
(d) $(f * g) * r = f * (g * r)$ (associativity)
(e) If $f \leftrightarrow F$ and $g \leftrightarrow G$, then $f * g \leftrightarrow F \cdot G$.

Example 4.20

Let $f(x, y) = g(x, y) = I_{(0,1) \times (0,1)}(x, y)$, and let $h = f * g$. Then

$$H(\omega_1, \omega_2) = \frac{\sin(\omega_1/2)}{\omega_1/2} e^{-j\omega_1/2} \frac{\sin(\omega_2/2)}{\omega_2/2} e^{-j\omega_2/2}$$

Though we could have evaluated $f * g$ and then taken the convolution, such a procedure would have been tedious and unnecessary.

As in the case of one variable, any two L_1 functions with the same Fourier transform must be equal almost everywhere and hence, from the point of view of integration, are identical. Theorem 4.11 is the uniqueness theorem in two dimensions. An immediate consequence of it is that $F = G$ implies $f = g$ almost everywhere, exactly as in the one-dimensional case. Thus, when it is possible, inversion yields a "unique" inverse image. The importance of this uniqueness cannot be overemphasized.

Theorem 4.11.[16] If $f \in L_1$ and $f \leftrightarrow 0$, then $f = 0$ almost everywhere.

We now proceed to define the inverse Fourier transform for images in a manner analogous to that used for signals. Given the complex-valued function $F(\omega_1, \omega_2)$, the inverse Fourier transform, when it exists, is given by the formula

$$[\mathcal{F}^{-1}(F)](x, y) = \frac{1}{4\pi^2} \int_{-\infty}^{\infty} \int_{-\infty}^{\infty} F(\omega_1, \omega_2) e^{j(\omega_1 x + \omega_2 y)} \, dy \, dx$$

The following inversion theorem contains analogies to parts (a) and (b) of Theorem 4.7.

Theorem 4.12. Suppose $f(x, y) \in L_1$ and $f \leftrightarrow F$. Then

(a) If $F(\omega_1, \omega_2) \in L_1$, then $\mathcal{F}^{-1}(F) = f$ almost everywhere. If, in addition, f is continuous, then $\mathcal{F}^{-1}(F) = f$ everywhere.

---
[16] Ibid., 59.

Sec. 4.3   The Fourier Transform of Integrable Images

(b) If there exists a constant M such that $|f(x, y)| \leq M$ on R^2 (i.e., if f is bounded), and if $F(\omega_1, \omega_2) \geq 0$, then $\mathscr{F}^{-1}(F) = f$ almost everywhere.[17] The equality holds everywhere if f is continuous.[18]

We have by now seen that many results in the theory of the one-dimensional Fourier transform carry over to the two-dimensional case. For a certain class of images, those which exhibit circular symmetry, more can be said. Specifically, a *radial* image f is one that can be expressed in the form $f(x, y) = g(\sqrt{x^2 + y^2})$ for all $(x, y) \in R^2$. Such functions are invariant under all orthogonal transformations relative to the origin. They arise in image processing by the observation of wave propagation from energy sources that exhibit "natural" symmetry.

The Fourier transform of a radial image also has circular symmetry. Note that

$$F(\omega_1, \omega_2) = \int_{-\infty}^{\infty} \int_{-\infty}^{\infty} f(x, y) e^{-j(\omega_1 x + \omega_2 y)} \, dx \, dy$$

$$= \int_{r=0}^{\infty} \int_{\theta=0}^{2\pi} g(r) e^{-j(\omega_1 r \cos\theta + \omega_2 r \sin\theta)} r \, dr \, d\theta$$

where the second integral is now expressed in polar coordinates. If we subsequently let $\omega_1 = R \cos \phi$ and $\omega_2 = R \sin \phi$, we obtain, after some trigonometric manipulations,

$$F(\omega_1, \omega_2) = \int_{r=0}^{\infty} \int_{\theta=0}^{2\pi} g(r) e^{-jRr\cos(\theta - \phi)} r \, d\theta \, dr$$

Changing variables gives

$$F(\omega_1, \omega_2) = \int_{r=0}^{\infty} \int_{\theta=0}^{2\pi} g(r) e^{-jRr\cos\theta} r \, d\theta \, dr,$$

which is only a function of $R = (\omega_1^2 + \omega_2^2)^{1/2}$. Hence, one can write $F(\omega_1, \omega_2) = G(R)$.

The next theorem formalizes the preceding discussion, expressing the last representation of $F(\omega_1, \omega_2)$ in terms of $J_0(u)$, the zero-order Bessel function of the first kind, which is defined by

$$J_0(u) = \frac{1}{2\pi} \int_0^{2\pi} e^{-ju\cos\theta} \, d\theta$$

Theorem 4.13.[19]   Suppose $f \in L_1$ is a radial function with $f(x, y) = g(r)$, $r = (x^2 + y^2)^{1/2}$. Then the Fourier transform of f depends only on $R =$

---

[17] Ibid., 66.
[18] Ibid., 66.
[19] Ibid., 69.

$(\omega_1^2 + \omega_2^2)^{1/2}$, i.e., $F(\omega_1, \omega_2) = G(R)$, where

$$G(R) = 2\pi \int_0^\infty g(r)rJ_0(rR)\,dr.$$

The transformation $g(r) \leftrightarrow G(R)$ is called the *Hankel transform of zero order*. The following theorem provides a sufficient condition for the inversion of the Hankel transform.

Theorem 4.14.[20]   Let $g(r)$ be continuous for all r in $[0, \infty)$, and suppose that

$$\int_0^\infty |g(r)|\,r\,dr < \infty$$

If $G(R)$ is the Hankel transform of zero order of $g(r)$ and

$$\int_0^\infty |G(R)|\,R\,dR < \infty$$

then for all r,

$$g(r) = \frac{1}{2\pi} \int_0^\infty G(R)J_0(Rr)R\,dR$$

Example 4.21

Let $f(x, y) = I_D(x, y)$, where D is the closed unit disk centered at the origin. Since f is a radial image, its Fourier transform is given by the Hankel transform, where $g(r) = I_{[0,1]}(r)$. We have

$$G(R) = 2\pi \int_0^\infty g(r)rJ_0(rR)\,dr = 2\pi \int_0^1 rJ_0(rR)\,dr$$

But Bessel functions of the first kind satisfy the identity

$$\frac{d}{dz}[z^n J_n(z)] = z^n J_{n-1}(z)$$

Applying this identity with $z = rR$ and $n = 1$ gives the identity

$$d[rRJ_1(rR)] = rRJ_0(rR)\,d(rR)$$

Substituting this into the expression for $G(R)$ yields

$$G(R) = 2\pi \int_0^1 \frac{1}{R} d[rJ_1(rR)] = \frac{2\pi J_1(R)}{R}$$

where $J_1(R)$ is the first-order Bessel function of the first kind and $J_1(0) = 0$.

---

[20] Ibid., 75.

Example 4.22

Let $f(x, y) = e^{-(x^2+y^2)^{1/2}} = g(r)$, $r = (x^2 + y^2)^{1/2}$. Then the Fourier transform also has circular symmetry and is given by the Hankel transform

$$G(R) = \int_0^\infty e^{-r} r \left[\int_0^{2\pi} e^{-jRr\cos\theta}\, d\theta \right] dr$$

where the integral representation of $J_0(rR)$ has been directly substituted into the integral. Since the iterated integrals are absolutely convergent, the order of integration can be reversed, so that

$$G(R) = \int_0^{2\pi} \int_0^\infty r e^{-r[1+jR\cos\theta]}\, dr\, d\theta$$

$$= \int_0^{2\pi} \frac{1}{[1 + jR\cos\theta]^2}\, d\theta$$

This last integral can be evaluated easily, but tediously, using standard complex variable integration. The result is $G(R) = 2\pi(1 + R^2)^{-3/2}$.

4.4 THE PLANCHEREL THEOREM (FOURIER TRANSFORM ON L_2)

At the end of Section 4.2, we briefly discussed signals that are in $L_1 \cap L_2$. In this section, we consider "pure" energy signals, i.e., those that are in L_2 but not in L_1. We extend these considerations to images in L_2 but not in L_1 at the end of the section. The approach requires the use of function sequences that converge in the L_2 norm. Those not familiar with such covergence can either omit the section or read it lightly. Nothing herein will play a role in the rest of the text.

The Fourier transform of a pure energy signal is defined by employing a sequence of functions obtained from f by restricting its support region. For $n = 1, 2, \ldots$, let $f_n(t) = f(t) \cdot I_{[-n,n]}(t)$, the restriction of f to the closed interval $[-n, n]$. Then $f_n \in L_1 \cap L_2$. Hence, by the discussion at the close of Section 4.2, f_n has a Fourier transform $F_n(\omega)$ in L_2. The intention is to define the Fourier transform of $f(t)$ to be the limit of the sequence $\{F_n(\omega)\}$. In fact, such a limit does exist, but not necessarily in the pointwise sense. Instead, it must be taken in the sense of the L_2 norm. The following theorem, due to Plancherel, is one of the most notable in the theory of transforms, stating numerous propositions of importance. Among these propositions is that the sequence $\{F_n(\omega)\}$ converges in the L_2 norm to an L_2 function $F(\omega)$, which is defined to be the Fourier transform of the signal $f(t)$. Moreover, if $f(t)$ happens to be in $L_1 \cap L_2$, then this limit transform $F(\omega)$ agrees with the one directly obtained from the Fourier integral. Finally, an interesting symmetry relation holds by letting $F^{[n]}(\omega)$ denote the restriction of $F(\omega)$ to the closed interval $[-n, n]$ and by letting $f^{[n]} = \mathcal{F}^{-1}(F^{[n]})$, which is defined since it happens that $F^{[n]} \in L_1 \cap L_2$. More will be said on this point following the theorem.

Theorem 4.15[21] **(Plancherel).** Let $f \in L_2$, let f_n be the restriction of f to $[-n, n]$, and let $f_n \leftrightarrow F_n(\omega)$. Then

(a) There exists a function $F(\omega)$ in L_2 such that F_n converges to F in the L_2 norm, i.e.,

$$\lim_{n \to \infty} \int_{-\infty}^{\infty} |F(\omega) - F_n(\omega)|^2 \, d\omega = 0$$

(b) For any $F(\omega) \in L_2$, there exists a unique $f(t) \in L_2$ such that $f \leftrightarrow F(\omega)$, where the double arrow denotes the L_2 Fourier transform.

(c) If $f \in L_1 \cap L_2$, then $F(\omega) = \lim_{n \to \infty} F_n(\omega)$ pointwise, and $F(\omega)$ agrees with the original definition of the Fourier transform for L_1 functions.

(d) For any $f \in L_2$, the Parseval identity holds, i.e.,

$$\int_{-\infty}^{\infty} |f(t)|^2 \, dt = \frac{1}{2\pi} \int_{-\infty}^{\infty} |F(\omega)|^2 \, d\omega$$

(e) If $f^{[n]} = \mathcal{F}^{-1}(F^{[n]})$, then $f^{[n]}$ converges to f in the L_2 norm, i.e.,

$$\lim_{n \to \infty} \int_{-\infty}^{\infty} |f(t) - f^{[n]}(t)|^2 \, dt = 0$$

Part (e) of the Plancherel theorem deserves special attention. Because of the manner in which f_n and $F^{[n]}$ have been defined as restrictions, it follows that

$$F_n(\omega) = \int_{-n}^{n} f(t) e^{-j\omega t} \, dt$$

and

$$f^{[n]}(t) = \frac{1}{2\pi} \int_{-n}^{n} F(\omega) e^{j\omega t} \, d\omega$$

Hence, there is a symmetry (up to a multiple of $1/2\pi$) between parts (a) and (e) of the theorem. Part (e) is sometimes called the L_2 *inversion theorem*. Indeed, in terms of limits in the L_2 sense, (e) can be rewritten as $\lim_{n \to \infty} f^{[n]} = f$, and $f^{[n]}$ can be computed directly from $F(\omega)$.

For those familiar with Hilbert space theory, the mapping $f \leftrightarrow F(\omega)$ is an isometry of L_2 onto L_2.

Example 4.23

Let $f(t) = \sum_{k=1}^{\infty} (1/k) I_{(k-1, k]}(t)$. We would like to find $F(\omega)$. Since f is in L_2 but not in L_1, we must use the functions

$$f_n(t) = f(t) \cdot I_{[-n, n]}(t) = \sum_{k=1}^{n} \frac{1}{k} I_{(k-1, k]}(t)$$

[21] Goldberg, *Fourier Transforms*, 46–51.

We obtain

$$F_n(\omega) = \int_{-\infty}^{\infty} f_n(t)e^{-j\omega t}\,dt = \sum_{k=1}^{n} \frac{1}{k}\int_{k-1}^{k} e^{-j\omega t}\,dt$$

$$= \sum_{k=1}^{n} \frac{2e^{-(k+1/2)\omega j}\sin(\omega/2)}{k\omega}$$

$$= \frac{\sin(\omega/2)}{\omega/2} e^{-j\omega/2} \sum_{k=1}^{n} \frac{e^{-jk\omega}}{k}$$

To find $F(\omega)$, we must find the L_2 limit of the $F_n(\omega)$. This can be shown to be

$$F(\omega) = \frac{e^{-j\omega/2}\sin(\omega/2)}{\omega/2} \log(1 - e^{-j\omega})$$

Example 4.24

Let $F(\omega) = I_{(-1/2,1/2)}(\omega)$. Since $F(\omega)$ is in L_2, it must be the transform of some function f in L_2. We need to find $f(t)$. We have

$$f^{[n]}(t) = \frac{1}{2\pi}\int_{-n}^{n} I_{(-1/2,1/2)}(\omega)e^{j\omega t}\,d\omega$$

$$= \frac{1}{2\pi}\int_{-1/2}^{1/2} e^{j\omega t}\,d\omega$$

$$= \frac{1}{2\pi}\frac{\sin(t/2)}{t/2}$$

for all n. Hence, $f(t) = (1/2\pi)\sin(t/2)/(t/2)$.

The result of the preceding exercise could have been determined directly from the defining equations for $F_n(\omega)$ and $f^{[n]}(t)$. Indeed, suppose $f \in L_2$ and $f \leftrightarrow F(\omega)$. Moreover, suppose F is an even function, i.e., $F(-\omega) = F(\omega)$. Then

$$f^{[n]}(t) = \frac{1}{2\pi}\int_{-n}^{n} F(\omega)e^{j\omega t}\,d\omega$$

Changing variables by letting $v = -\omega$ yields

$$f^{[n]}(t) = \frac{1}{2\pi}\int_{-n}^{n} F(v)e^{-jvt}\,dv$$

which is exactly $(1/2\pi)\tilde{F}_n(t)$, where $\tilde{F}(\omega)$ denotes the L_2 Fourier transform of F, which makes sense since F is in L_2. By part (a) of the Plancherel theorem, $\tilde{F}_n(t)$ converges to $\tilde{F}(t)$ in the L_2 norm. Hence, $f(t) = (1/2\pi)\tilde{F}(t) = (1/2\pi)[\mathcal{F}(F)](t)$. In Example 4.24, the Fourier transform of F, evaluated at t, is $\sin(t/2)/(t/2)$ (from Example 4.9), which, divided by 2π, is the result obtained. Similar considerations apply when F is not an even function, but the result is not quite as simple: in general, one obtains $2\pi f(t) = [\mathcal{F}(F^*)](t)$, where F^* is given by $F^*(t) = F(-t)$.

The convolution of two L_2 signals is defined just as in the L_1 case; however, the resulting signal is "better behaved," in that if f and g are in L_2 and

$$h(t) = [f * g](t) = \int_{-\infty}^{\infty} f(t - x)g(x)\, dx$$

then $h(t)$ is bounded, continuous, and converges to zero as t goes to infinity.[22]

Insofar as images are concerned, an image $f(x, y)$ defined on R^2 is in L_2 if it is measurable and if

$$\int_{-\infty}^{\infty} \int_{-\infty}^{\infty} |f(x, y)|^2\, dy\, dx$$

is finite. If $f(x, y)$ is a pure energy image, i.e., if $f(x, y)$ is in L_2 but not in L_1, then one must define the Fourier transform in the Plancherel sense by defining the restricted images

$$F_{m,n}(x, y) = I_{[-m,m] \times [-n,n]}(x, y) f(x, y)$$

and by taking the two-dimensional Fourier transforms of these restricted images. A Plancherel-type theorem then applies in that there exists a function $F(\omega_1, \omega_2)$ in L_2 such that $F_{m,n}(\omega_1, \omega_2)$ converges in the L_2 sense to $F(\omega_1, \omega_2)$, where $f_{m,n} \leftrightarrow F_{m,n}(\omega_1, \omega_2)$. In terms of spectral representations, this means that

$$\lim_{m,n \to \infty} \int_{-\infty}^{\infty} \int_{-\infty}^{\infty} |F(\omega_1, \omega_2) - F_{m,n}(\omega_1, \omega_2)|^2\, d\omega_1\, d\omega_2 = 0$$

where

$$F_{m,n}(\omega_1, \omega_2) = \int_{-n}^{n} \int_{-m}^{m} f(x, y) e^{-j(\omega_1 x + \omega_2 y)}\, dx\, dy$$

Moreover, if $f \leftrightarrow F(\omega_1, \omega_2) \in L_2$, $F^{[m,n]}$ is the restriction to the closed rectangle $[-m, m] \times [-n, n]$, and if $f^{[m,n]}$ is the two-dimensional inverse $\mathcal{F}^{-1}(F^{[m,n]})$, then part (e) of the Plancherel theorem applies, i.e., $\lim_{m,n \to \infty} f^{[m,n]} = f$ in the two-dimensional L_2 sense. Finally, the Parseval identity holds, i.e.,

$$\int_{-\infty}^{\infty} \int_{-\infty}^{\infty} |f(x, y)|^2\, dx\, dy = \frac{1}{4\pi^2} \int_{-\infty}^{\infty} \int_{-\infty}^{\infty} |F(\omega_1, \omega_2)|^2\, d\omega_1\, d\omega_2$$

4.5 SINE AND COSINE TRANSFORMS

A signal $f(t)$ is said to be *even* if $f(-t) = f(t)$ and *odd* if $f(-t) = -f(t)$. Examples are the cosine wave and sine wave, respectively. Other examples of even functions are the even powers of t, given by $f(t) = t^{2n}$ for $n = 1, 2, \ldots$. The odd powers of t, given by $f(t) = t^{2n+1}$, are odd functions.

---

[22] N. I. Akhiezer, *Theory of Approximation* (New York: Frederick Ungar, 1956), 105.

Sec. 4.5 Sine and Cosine Transforms

A simple calculation shows that if $f(t)$ is even, then its Fourier transform, if it exists, is given by

$$F(\omega) = 2 \int_0^\infty f(t) \cos(\omega t)\, dt$$

Similarly, an odd signal $f(t)$ has a Fourier transform (if it exists) given by

$$F(\omega) = 2 \int_0^\infty f(t) \sin(\omega t)\, dt$$

The preceding expressions motivate the definitions of the Fourier cosine transform, $F_c(\omega)$, and the Fourier sine transform, $F_s(\omega)$, both of which are defined for signals that are not necessarily even or odd. The Fourier cosine transform and its inverse are respectively given by

$$F_c(\omega) = 2 \int_0^\infty f(t) \cos(\omega t)\, dt$$

and

$$f(t) = \frac{1}{\pi} \int_0^\infty F_c(\omega) \cos(\omega t)\, dt, \; t \geq 0$$

The Fourier sine transform and its inverse are respectively given by

$$F_s(\omega) = 2 \int_0^\infty f(t) \sin(\omega t)\, dt$$

and

$$f(t) = \frac{1}{\pi} \int_0^\infty F_s(\omega) \sin(\omega t)\, d\omega, \; t \geq 0$$

Existence conditions for these transforms using an L_1 or an L_2 condition follow directly from the corresponding discussion in regard to the Fourier transform itself. Moreover, recovery of the original signal using the inverse transforms is immediately obtainable from the corresponding discussion of Fourier transform inversion. (See Theorem 4.7.) Note, however, that inversion holds only for $t \geq 0$, and, as before, only almost everywhere.

Example 4.25

Suppose $f(t) = I_{[0,1]}(t)$. Then direct calculation shows that $F_c(\omega) = 2\sin(\omega)/\omega$. Application of the inverse cosine transform formula yields

$$\frac{1}{\pi} \int_0^\infty \frac{2 \sin(\omega)}{\omega} \cos(\omega t)\, d\omega = I_{(-1,1)}(t)$$

where the integration can be done by the use of residue theory, a route we shall not pursue here. In any event, the inversion holds for all $t \geq 0$ except $t = 1$. As is assured, inversion retrieves the original signal almost everywhere for nonnegative t.

Useful formulas for obtaining the Fourier cosine and sine transforms from the Fourier transform $F(\omega)$ are

$$F_c(\omega) = F(\omega) + F(-\omega)$$

and

$$F_s(\omega) = \frac{1}{j}[F(\omega) - F(-\omega)]$$

Example 4.26

The Fourier transform of $f(t) = e^{-t}u(t)$ is given by $F(\omega) = (1 + j\omega)^{-1}$. Hence, by the preceding identities,

$$F_c(\omega) = \frac{1}{1 + j\omega} + \frac{1}{1 - j\omega} = \frac{2}{1 + \omega^2}$$

and

$$F_s(\omega) = \frac{1}{j}\left(\frac{1}{1 + j\omega} - \frac{1}{1 - j\omega}\right) = \frac{-2\omega}{1 + \omega^2}$$

The corresponding inverse transformations give $e^{-|t|}$, which equals $f(t)$ for $t \geq 0$.

4.6 MELLIN TRANSFORM OF SIGNALS

The Mellin transform is useful in image processing for simplifying the scaling of images. Whereas the Fourier transform converts convolution (sum kernel) integrals into products of transforms, the Mellin transform converts product kernel integrals into products of transforms. Before considering the Mellin transform, we shall briefly discuss the Laplace transform and the generalized Fourier transform.

Let $f(t)$ be a signal with support in $[0, \infty)$ such that $e^{-kt}f(t) \in L_1$ for some real number k. The *Laplace transform* of $f(t)$ is defined for all complex numbers s such that the real part of s, $\text{Re}(s)$, is greater than k. It is given by

$$F(s) = \int_0^\infty f(t)e^{-st}\, dt$$

and is denoted by $\mathcal{L}(f)$. Notice that when s is real, the Laplace integral can be obtained from the Fourier integral by using $\omega = -js$. By the assumption concerning the integrability of $e^{-kt}f(t)$, $F(s)$ is defined at least in the right half of the complex plane, that half being defined by $\text{Re}(s) > k$; in fact, it can be shown that $F(s)$ is actually analytic there.

When the inversion conditions for the Fourier transform hold (see Theorem 4.7), we also have an inversion for the Laplace transform given by

$$f(t) = \frac{1}{2\pi j}\int_{a-j\infty}^{a+j\infty} e^{ts}F(s)\, ds$$

for $t > 0$, where a is any real number greater than k, and k is such that $e^{-kt}f(t) \in L_1$.[23] The meaning of the integral is that the integration is to be done in the complex plane along any vertical line $z = a$ such that $a > k$. In practice, the integral is calculated around a contour in the complex plane by the use of residue theory, and limits are taken. We shall not pursue the matter any further in this text. Fortunately, for many important applications, the inversion can be done without recourse to the preceding *Bromwich integral formula*. These algebraic-type inversion techniques are thoroughly discussed in any elementary course in differential equations. Notationally, we write $\mathcal{L}^{-1}$ for the inverse Laplace transform.

The following theorem summarizes many of the basic properties of the Laplace transform. We temporarily use the notation $f \leftrightarrow F$ to denote that F is the Laplace transform of f.

Theorem 4.16. Suppose $f(t)$ and $g(t)$ have support in $[0, \infty)$, $e^{-kt}f(t)$ and $e^{-ht}g(t)$ are in L_1, $f \leftrightarrow F$, and $g \leftrightarrow G$. Then

(a) $f(t) + g(t) \leftrightarrow F(s) + G(s)$, for $\text{Re}(s) > \max(k, h)$.
(b) $af(t) \leftrightarrow aF(s)$, for a real and $\text{Re}(s) > k$.
(c) $(-1)^n t^n f(t) \leftrightarrow F^{(n)}(s)$, for $\text{Re}(s) > k$.
(d) If $f(t)$ is continuous on $(0, \infty)$ and possesses a derivative $f'(t)$ such that $e^{-kt}f'(t) \in L_1$, then $f'(t) \leftrightarrow sF(s) - f(0^+)$, for $\text{Re}(s) > k$.
(e) If $f \subset L_1$, then $\int_0^t f(u)\, du \leftrightarrow F(s)/s$, for $\text{Re}(s) > 0$.
(f) $f(at) \leftrightarrow (1/a)F(s/a)$, for $a > 0$ and $\text{Re}(s) > ak$.
(g) $f(t - a)I_{(a, \infty)}(t) \leftrightarrow e^{-as}F(s)$, for $a > 0$ and $\text{Re}(s) > k$.
(h) $e^{bt}f(t) \leftrightarrow F(s - b)$, for complex b and $\text{Re}(s) > k + \text{Re}(b)$.

The convolution theorem for the Laplace transform is similar to that for the Fourier transform; however, in the case of the Laplace transform, the convolution integral is given by

$$h(t) = \int_0^t f(u)g(t - u)\, du$$

Under the assumption that $f(t)$ and $g(t)$ have support in $[0, \infty)$, $g(t - u) = 0$ for $t < u$.

Theorem 4.17.[24] Let f and g be in L_1 on $[0, R)$. Then the Laplace-type convolution $h(t)$ exists almost everywhere in $[0, R)$. If this is true for all R,

---
[23] E. C. Titchmarsh, *Introduction to the Theory of Fourier Integrals* (London: Clarendon Press, 1962), 6.
[24] David Vernon Widder, *The Laplace Transform* (Princeton: Princeton University Press, 1966), 91.

and if there exists a k such that $e^{-kt}f(t) \in L_1$ and $e^{-kt}g(t) \in L_1$, then $[\mathscr{L}(f)](s)[\mathscr{L}(g)](s) = [\mathscr{L}(h)](s)$ for $\text{Re}(s) > k$.

Example 4.27

Let $h(t) = \int_0^t e^{-u}(t - u)^{-a} du$, where $0 < a < 1$. A simple integration shows that $\mathscr{L}(e^{-t}) = 1/(s + 1)$. It is also well known that $\mathscr{L}(t^{-a}) = s^{a-1}\Gamma(1 - a)$, where Γ denotes the gamma function. By Theorem 4.17,

$$\mathscr{L}(h) = s^{a-1}\frac{\Gamma(1 - a)}{s + 1}$$

Example 4.28

Suppose $h(t) = \int_0^t J_0(u)J_0(t - u) du$, where J_0 is the zero-order Bessel function of the first kind. It is well known that $\mathscr{L}(J_0) = (1 + s^2)^{-1/2}$. By Theorem 4.17, $\mathscr{L}(h) = (1 + s^2)^{-1}$, which is precisely the Laplace transform of $\sin(t) \cdot u(t)$. Hence, $h(t) = \sin(t) \cdot u(t)$, the sine signal on the nonnegative real axis.

For certain classes of signals which do not have support in $[0, \infty)$, a *bilateral* Laplace transform can be obtained. It is defined by

$$F(s) = \int_{-\infty}^{\infty} f(t)e^{-st} dt$$

and is denoted by $\mathscr{L}_{11}(f)$. Whereas the Laplace integral converges in half-planes $\text{Re}(s) > k$, the bilateral Laplace integral converges in infinite strips $k_1 < \text{Re}(s) < k_2$. It, too, has an inverse obtained from the Bromwich integral formula, but in this case the vertical line over which the complex integral is taken must lie in the strip $k_1 < \text{Re}(s) < k_2$.

Example 4.29

Let $f(t) = e^{at}I_{(0,\infty)}(t)$, $g(t) = -e^{bt}I_{(-\infty,0)}(t)$, and $h(t) = f(t) + g(t)$. Then $\mathscr{L}_{11}(f) = (s - a)^{-1}$ for $\text{Re}(s) > a$, $\mathscr{L}_{11}(g) = (s - b)^{-1}$ for $\text{Re}(s) < b$, and $\mathscr{L}_{11}(h) = (s - a)^{-1} + (s - b)^{-1}$, for $a < \text{Re}(s) < b$, assuming that $a < b$ to begin with. Notice that if $a = b$, $\mathscr{L}_{11}(f)$ and $\mathscr{L}_{11}(g)$ appear to be the same but in fact are not, since their domains are different. Furthermore, $\mathscr{L}_{11}(h)$ does not exist in this case.

A transform closely related to the bilateral Laplace transform is the *generalized Fourier transform*. Suppose $e^{-k|t|}f(t) \in L_1$ on $(-\infty, \infty)$ for some $k > 0$. We define

$$F_+(\omega) = \int_0^{\infty} e^{-j\omega t} f(t) dt$$

and

$$F_-(\omega) = \int_{-\infty}^0 e^{-j\omega t} f(t) dt$$

Together, $F_+(\omega)$ and $F_-(\omega)$ form the generalized Fourier transform of f. $F_+(\omega)$

Sec. 4.6   Mellin Transform of Signals   211

is defined and is analytic for $\text{Im}(\omega) < -k$, while $F_-(\omega)$ is defined and is analytic for $\text{Im}(\omega) > k$.

Example 4.30

Let $f(t) = e^{|t|}$. Then $e^{-k|t|}f(t) \in L_1$ for $k > 1$. Therefore, the generalized Fourier transform is given by

$$F_+(\omega) = \int_0^\infty e^{-j\omega t} e^t \, dt = -(1 - j\omega)^{-1}$$

for $\text{Im}(\omega) < -1$, and

$$F_-(\omega) = \int_{-\infty}^0 e^{-j\omega t} e^{-t} \, dt = -(1 + j\omega)^{-1}$$

for $\text{Im}(\omega) > 1$.

We are now in a position to define the Mellin transform. If $t^{\text{Re}(s)-1}f(t) \in L_1$ on $(0, \infty)$, the *Mellin transform* is given by the formula

$$F(s) = \int_0^\infty t^{s-1} f(t) \, dt$$

Like the other transforms thus far introduced in this section, the Mellin transform can be obtained from the Fourier transform by a change of variables. In fact, if we write

$$F(s) = \int_0^1 t^{s-1} f(t) \, dt + \int_1^\infty t^{s-1} f(t) \, dt$$

then the first integral corresponds to a generalized Fourier integral F_+, while the second corresponds to F_-. (This can be seen by letting $t = e^{-x}$ and $j\omega = s$.) Consequently, under suitable integrability conditions, the first integral represents an analytic function in a right half-plane $\text{Re}(s) > s_1$, and the second represents an analytic function in a left half-plane $\text{Re}(s) < s_2$. (We henceforth assume that $s_1 < s_2$.)

Though we will not go into detail, the Mellin inverse formula is given by the Bromwich integral-type formula

$$f(t) = \frac{1}{2\pi j} \int_{a-j\infty}^{a+j\infty} t^{-s} F(s) \, ds$$

where $s_1 < a < s_2$.[25]

Example 4.31

Let $f(t) = I_{[0,a)}(t)$. Then

$$F(s) = \int_0^a t^{s-1} dt = \frac{a^s}{s}$$

which holds for $\text{Re}(s) > 0$ since $t^{\text{Re}(s)-1} f(t) \in L_1$ for $\text{Re}(s) > 0$.

[25] Titchmarsh, *Fourier Integrals*, 6.

Example 4.32

Let $f(t) = e^{-t}I_{(0,\infty)}(t)$. Then, for $\operatorname{Re}(s) > 0$,

$$F(s) = \int_0^\infty e^{-t}t^{s-1}dt = \Gamma(s)$$

since the integral is precisely the definition of the gamma function.

Example 4.33

Let $f(t) = (1 - t)^{a-1}I_{(0,1)}(t)$ for $a > 0$. Then

$$F(s) = \int_0^1 t^{s-1}(1 - t)^{a-1}dt$$

for $\operatorname{Re}(s) > 0$. But this integral is precisely the definition of the beta function $B(s, a)$, and it is well known that $B(s, a) = \Gamma(s)\Gamma(a)/\Gamma(s + a)$, which is therefore the value of $F(s)$.

Example 4.34

Let $f(t) = (1 + t)^{-b}I_{(0,\infty)}(t)$ for $b > 0$. Then

$$F(s) = \int_0^\infty t^{s-1}(1 + t)^{-b}dt$$

$$= \int_0^1 x^{b-1-s}(1 - x)^{s-1}dx = B(b - s, s) = \frac{\Gamma(b - s)\Gamma(s)}{\Gamma(b)}$$

where the change of variable $x = (1 + t)^{-1}$ has been made to obtain the second integral from the first. Note that the relationship holds for $0 < \operatorname{Re}(s) < b$, since

$$\int_0^\infty t^{\operatorname{Re}(s)-1}(1 + t)^{-b}dt$$

converges when $0 < \operatorname{Re}(s) < b$. As a special case, when $b = 1$ we obtain $F(s) = \pi/\sin(\pi s)$.

The convolution for the Mellin transform involves a product-type kernel and is given by

$$h(t) = \int_0^\infty f(x)g(t/x)x^{-1}dx$$

for $t > 0$. We now state, for the Mellin transform, the analogue of Theorem 4.5.

Theorem 4.18.[26] Suppose $t^{\operatorname{Re}(s)-1}f(t)$ and $t^{\operatorname{Re}(s)-1}g(t)$ are in L_1 on $(0, \infty)$, $F(s)$ and $G(s)$ are the respective Mellin transforms of $f(t)$ and $g(t)$, and $h(t)$ is the aforementioned Mellin convolution. Then $t^{\operatorname{Re}(s)-1}h(t)$ is in L_1 on $(0, \infty)$ and $H(s) = F(s) \cdot G(s)$, where $H(s)$ is the Mellin transform of $h(t)$.

We conclude the section with a brief look at two other transforms that are

[26] Ibid., 60.

useful in image processing, especially for radial images. The Abel and Weyl transforms are respectively given by $A^\alpha(f)$ and $W^\alpha(f)$, where

$$[A^\alpha(f)](x) = \frac{1}{\Gamma(\alpha)} \int_a^x (x - t)^{\alpha - 1} f(t)\, dt$$

and

$$[W^\alpha(f)](x) = \frac{1}{\Gamma(\alpha)} \int_x^\infty (t - x)^{\alpha - 1} f(t)\, dt$$

for $\mathrm{Re}(\alpha) > 0$. If f is in L_1, the existence of $A^\alpha(f)$ and $W^\alpha(f)$ is assured. We therefore make that assumption.

Both the Abel and Weyl transforms are related to fractional integration. For $\alpha = n$, a nonnegative integer,

$$[A^n(f)](x) = \int_a^x \int_a^{x_1} \cdots \int_a^{x_{n-1}} f(x_n)\, dx_n \cdots dx_1$$

where this representation can be shown to be equal to the defining integral by a change in the order of integration. Furthermore,

$$A^\alpha(A^\beta(f)) = A^{\alpha + \beta}(f)$$

and

$$W^\alpha(W^\beta(f)) = W^{\alpha + \beta}(f)$$

whenever $\mathrm{Re}(\alpha) > 0$ and $\mathrm{Re}(\beta) > 0$. If, in addition, f is continuous, then $A^0(f) = f$ and $W^0(f) = f$.

Example 4.35

By the preceding comments, with $a = 0$,

$$A^{1/2}(A^{1/2}(f)) = A^1(f) = \int_0^x f(t)\, dt$$

4.7 HILBERT TRANSFORM

The Hilbert transform pair consists of two functions g and f related by the expressions

$$g(x) = \frac{1}{\pi} \int_{-\infty}^\infty \frac{f(t)}{t - x}\, dt$$

and

$$f(t) = \frac{-1}{\pi} \int_{-\infty}^\infty \frac{g(x)}{x - t}\, dx$$

The integrals are interpreted in the Cauchy principal value (CPV) sense at

$t = x$. The first integral yields the Hilbert transform of f, $\mathcal{H}(f)$, and the second is often termed the inverse Hilbert transform. Some formal manipulations will be given in several examples.

Example 4.36

Let $f(t) = I_{(-a,a)}(t)$, for $a > 0$. Then the Hilbert transform $g(x)$ is given by

$$g(x) = \frac{1}{\pi} \text{CPV} \left(\int_{-a}^{a} (t-x)^{-1} \, dt \right)$$

For $-a < x < a$, this integral gives, in the CPV sense,

$$g(x) = \frac{1}{\pi} \lim_{r \to 0} \int_{-a}^{x-r} \frac{dt}{t-x} + \int_{x+r}^{a} \frac{dt}{t-x}$$

$$= \frac{1}{\pi} [-\log|-x-a| + \log|a-x|]$$

$$+ \frac{1}{\pi} \lim_{r \to 0} [\log|-r| - \log|r|]$$

$$= \frac{1}{\pi} \log \left| \frac{x-a}{x+a} \right|$$

since the terms within the limit cancel. (Note that if the CPV had not been employed, there would have been no cancellation and the limit would not have existed.) For $x < -a$ and $x > a$, the integral can be evaluated directly without limits, and the same value results. For $x = -a$ and $x = a$, the evaluation of the appropriate improper integral does not converge. Consequently, the Hilbert transform is not defined at these two values. For instance, at $x = a$ we obtain

$$\frac{1}{\pi} \lim_{r \to 0} \int_{-a}^{a-r} \frac{dt}{t-a} = \frac{1}{\pi} \lim_{r \to 0} [\log|-r| - \log|-2a|]$$

which does not exist since $\lim_{r \to 0} \log|-r| = -\infty$.

Note that residue integration could be performed to show the inversion

$$f(t) = \frac{-1}{\pi^2} \int_{-\infty}^{\infty} \frac{\log(|x-a|/|x+a|)}{x-t} \, dx$$

Example 4.37

Let $f(t) = \cos t$. Then $g(x) = -\sin x$, which is the cosine transform of $(t-x)^{-1}$. This can be verified by residue integration.

We now give rigorous conditions under which the Hilbert transform pair is valid. Some of these conditions are easy to apply in image reconstruction, a matter which will be discussed in subsequent sections. The following theorem holds for L_2 signals.

Theorem 4.19. Suppose $f(t)$ is in L_2 on $(-\infty, \infty)$. Then (1) the Hilbert integral exists almost everywhere, and hence $\mathcal{H}(f)$ is defined almost everywhere;

(2) $g(x) = \mathcal{H}(f)(x)$ is in L_2 on $(-\infty, \infty)$ and the inversion integral equals $f(t)$ almost everywhere;[27] and (3) the L_2 norms of f and g are identical,[28] i.e.,

$$\int_{-\infty}^{\infty} |f(t)|^2 \, dt = \int_{-\infty}^{\infty} |g(x)|^2 \, dx$$

Example 4.38

From Example 4.36, since $f(t)$ is in L_2,

$$\frac{1}{\pi^2}\int_{-\infty}^{\infty} [\log|(x-a)/(x+a)|]^2 \, dx = 2a$$

Theorem 4.20
(a) The Hilbert transform of a constant is zero.
(b) If f is in L_2, then $\mathcal{H}(\mathcal{H}(f)) = -f$.
(c)[29] If f is in L_2, then f and $\mathcal{H}(f)$ are orthogonal, i.e.,

$$\lim_{r \to \infty} \int_{-r}^{r} [f \mathcal{H}(f)](u) \, du = 0$$

Part (a) shows the many-to-one property of the Hilbert transform: if $f(t)$ and $g(t)$ differ by a constant, e.g., $g(t) = f(t) + c$, then they have the same Hilbert transform.

Example 4.39

To illustrate part (c) of Theorem 4.20, consider $f(t) = I_{(-a,a)}(t)$, as in Example 4.36. By (c),

$$\lim_{r \to \infty} \frac{1}{\pi} \int_{-r}^{r} I_{(-a,a)}(t) \log|(t-a)/(t+a)| \, dt = 0$$

Recognizing the support of $I_{(-a,a)}(t)$, and realizing that the CPV must be employed, this means precisely that

$$\frac{1}{\pi} \int_{-a}^{a} \log|(t-a)/(t+a)| \, dt = 0$$

which can be checked by using integration by parts.

We next explore some properties of the Hilbert transform, as well as the relation of the Hilbert transform to the Fourier transform. To begin with, notice that $\mathcal{H}(f)$ is a convolution of $f(t)$ with $-1/\pi t$, that is, $\mathcal{H}(f) = f * (-1/\pi t)$. But the

---

[27] Ibid., 121.
[28] Ibid., 123.
[29] Ibid., 138.

Fourier transform of $-1/(\pi t)$ is given by $j\,\text{sgn}(\omega)$, where sgn, the *sign* function, is given by

$$\text{sgn}(\omega) = \begin{cases} 1 & \text{for } \omega > 0 \\ 0 & \text{for } \omega = 0 \\ -1 & \text{for } \omega < 0 \end{cases}$$

Using this last fact, the following theorem can be proven.

Theorem 4.21. If $f(t)$ with Fourier transform $F(\omega)$ is in L_2, then the Fourier transform of $\mathcal{H}(f)$ is equal to $j\,\text{sgn}(\omega)F(\omega)$ almost everywhere.

Example 4.40

Referring again to Example 4.36, since

$$I_{(-a,a)}(t) \leftrightarrow \frac{2\sin(a\omega)}{\omega}$$

we have

$$\frac{1}{\pi}\log|(t-a)/(t+a)| \leftrightarrow \frac{j2\sin(a\omega)}{\omega}\,\text{sgn}(\omega)$$

The Hilbert transforms of signals in $L_1 \cap L_2$ are usually not well behaved from the integrability point of view. Indeed, suppose f is in $L_1 \cap L_2$ and $f \leftrightarrow F(\omega)$. Then, according to Theorem 4.1, $F(\omega)$ is uniformly continuous. Now, we know from Theorem 4.19 that $\mathcal{H}(f)$ is in L_2, but it will usually not be in L_1. If it were in L_1, then $jF(\omega)\,\text{sgn}(\omega)$ would have to be uniformly continuous, which would imply that $F(0) = 0$. But $F(0) = \int_{-\infty}^{\infty} f(t)\,dt$. Hence, the area under f must be zero. Accordingly, this is a necessary condition for the Hilbert transform of a function in $L_1 \cap L_2$ to be in L_1. Given a particular $f(t)$, this condition is unlikely to be satisfied.

Conditions other than being in L_2 can be imposed on signals in order to obtain well-behaved Hilbert transforms. Among the simplest conditions is that given in the next theorem.

Theorem 4.22. If f satisfies the *Lipshitz condition*

$$|f(t+h) - f(t)| \leq m|h|$$

and the integral of f is uniformly bounded, that is, there exists an $M > 0$ such that

$$\left|\int_a^b f(t)\,dt\right| < M$$

for all real a and b, then $\mathcal{H}(f)$ exists almost everywhere,[30] and

$$|\mathcal{H}(f)| \leq \frac{4}{\pi}\log(2) \times (Mm)^{1/2}$$

---

[30] B. Logan, "Hilbert Transform of a Function Having a Bounded Integral and a Bounded Derivative," *SIAM Jour. Math Anal.*, Vol. 14, No. 2 (1983).

Sec. 4.8 Radon Transform

Example 4.41

It was seen in Example 4.37 that $f(t) = \cos t$ has the Hilbert transform $g(x) = -\sin x$. Since $|f'(t)| \leq 1$, $f(t)$ satisfies a Lipshitz condition with $m = 1$. Moreover, $|\int_a^b \cos t \, dt| \leq 2$ for all a and b. Hence,

$$|\mathcal{H}(f)| \leq \frac{4\sqrt{2} \log(2)}{\pi}$$

Example 4.42

Let $f(t) = \sin(t)/t$. It can be shown that $|f'(t)| \leq 1$, which results in a Lipshitz condition with $m = 1$, and that

$$\left| \int_a^b \sin(t)/t \, dt \right| \leq 2\left(\pi + \frac{2}{\pi}\right)$$

for all a and b. Consequently,

$$|\mathcal{H}(f)| \leq \frac{4}{\pi} \log 2 \cdot \sqrt{2\left(\pi + \frac{2}{\pi}\right)}$$

Since f is in L_2, $\mathcal{H}(f) \leftrightarrow j \operatorname{sgn}(\omega) F(\omega)$, where

$$F(\omega) = \pi I_{(-1,1)}(\omega)$$

is the Fourier transform of $f(t)$. Using the inverse Fourier transform yields

$$[\mathcal{H}(f)](t) = \frac{1}{2\pi} \left[\int_{-1}^{0} -j\pi e^{j\omega t} d\omega + \int_{0}^{1} j\pi e^{j\omega t} d\omega \right]$$

$$= \frac{\cos t - 1}{t}$$

4.8 RADON TRANSFORM

Methodology employing the Radon transform has been successfully developed using computer tomography in medical imaging. The Radon transform has also proven useful in radio astronomy and in nondestructive testing of materials.

The Radon transform involves the evaluation of integrals of a given image over lines in the plane. Figure 4.10 presents the situation graphically; the image $f(x, y)$ is imagined as a surface, and the "area directly under" that surface and above the line L is to be evaluated. More generally, for functions of more than two variables, the Radon transform is computed by integrating functions of n variables over $(n - 1)$-dimensional hyperplanes. For instance, for a function $f(x, y, z)$ of three variables, one would integrate over planes in R^3.

In practice, the Radon transform of a given image is obtained empirically by sensors, and the original image is to be reconstructed. Consequently, the practical mathematical problem concerns inversion of the transform. In other words, the so-called *projections*, i.e., integrals over lines, are observed, and a *back projection*, i.e., an inversion of the Radon transform, must be performed. Theoreti-

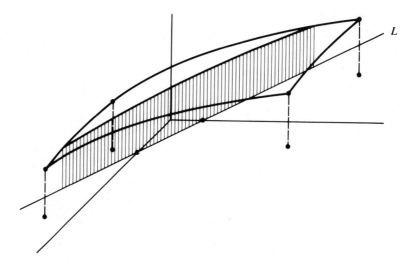

Figure 4.10 Evaluation of Radon Transform in Two Dimensions

cally, one inverts the transform by using all the projections; nonetheless, in practice, an approximate inversion must be performed from a finite number of observations of the transform values. This approximation involves the theories of sampling, interpolation, and extrapolation and will not be considered in this chapter.

Given an image $f(x, y)$ defined on R^2, the Radon transform of f, $R(p, \theta)$, is found by integrating f along the line L given by

$$x \cos \theta + y \sin \theta = p$$

which is the normal-form equation of the line. As usual, θ is the angle between the abscissa and the perpendicular from the origin to the line L ($0 \leq \theta < 2\pi$), and $p \geq 0$ is the length of that perpendicular. (See Figure 4.11.) In parametric form, L is given by

$$\frac{x - p \cos \theta}{-\sin \theta} = \frac{y - p \sin \theta}{\cos \theta} = z$$

Solving for x and y, we have

$$x = p \cos \theta - z \sin \theta$$

$$y = p \sin \theta + z \cos \theta$$

Therefore, the Radon transform $R(p, \theta)$ is given by the integral

$$R(p, \theta) = \int_{-\infty}^{\infty} f(p \cos \theta - z \sin \theta, p \sin \theta + z \cos \theta) \, dz$$

Extensions for values of p and θ other than those given will sometimes be used. These are given by

$$R(p, \theta) = R(-p, \theta + \pi) = R(p, \theta + 2\pi)$$

As are the other transforms studied thus far, the Radon transform is linear. Symbolically, if R_f denotes the Radon transform of image f, then $R_{f+g} = R_f + R_g$.

If the image $f(x, y)$ is given in polar form as $g(r, \theta)$, then

$$R(p, \theta) = \begin{cases} \int_{-\infty}^{\infty} g(\sqrt{p^2 + z^2}, \theta + \arctan(z/p)) \, dz & \text{if } p \neq 0 \\ \\ \int_{-\infty}^{\infty} g\left(z, \theta + \frac{\pi}{2}\right) dz & \text{if } p = 0 \end{cases}$$

The last form of the Radon transform will be utilized in the ensuing discussion. Various examples will now be given to illustrate the aforementioned definitions.

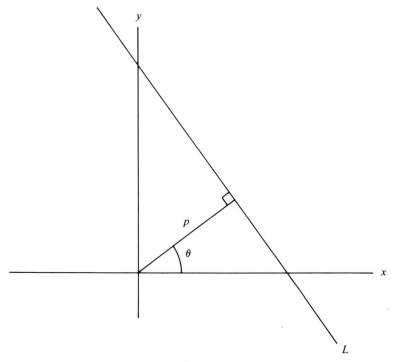

Figure 4.11 Normal Representation of Line

Example 4.43

Let $f(x, y) = e^{-(x^2+y^2)}$. The polar form of f is given by $g(r, \theta) = e^{-r^2}$, and the Radon transform is

$$R(p, \theta) = \int_{-\infty}^{\infty} e^{-(p^2+z^2)} dz = \pi^{1/2} e^{-p^2}$$

Example 4.44

Let $f(x, y) = I_D(x, y)$, where D is the closed unit disk centered at the origin. In polar form,

$$g(r, \theta) = \begin{cases} 1 & \text{for } 0 \leq r \leq 1 \\ 0 & \text{otherwise} \end{cases}$$

Since on D the function is the constant 1, the integral over any line is simply the length of the intersection of the line with the support D. Consequently, $R(p, \theta) = 0$ for $p \geq 1$. Other values can be obtained with the aid of Figure 4.12. We obtain

$$R(p, \theta) = \begin{cases} 2(1 - p^2)^{1/2} & \text{for } 0 \leq p \leq 1 \\ 0 & \text{otherwise} \end{cases}$$

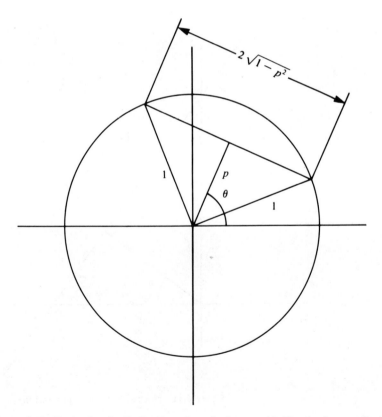

Figure 4.12 Evaluating the Radon Transform for Image with Circular Support Region

Example 4.45

For $\alpha > 0$, let

$$f(x, y) = \begin{cases} (1 - x^2 - y^2)^{\alpha-1} & \text{for } x^2 + y^2 \leq 1 \\ 0 & \text{otherwise} \end{cases}$$

The polar representation of this image is

$$g(r, \theta) = \begin{cases} (1 - r^2)^{\alpha-1} & 0 \leq r \leq 1 \\ 0 & \text{otherwise} \end{cases}$$

Since the image is zero outside the unit disk, $R(p, \theta) = 0$ for $p \geq 1$. Otherwise, the image is radial, so that the integral along L will be independent of the angle θ. Therefore, we integrate along a horizontal line $\theta = \pi/2$. Using the original nonpolar definition of the Radon transform, we obtain, for $p \leq 1$ and all θ,

$$R(p, \theta) = \int_{-(1-p^2)^{1/2}}^{(1-p^2)^{1/2}} (1 - p^2 - z^2)^{\alpha-1} \, dz$$

Making the substitution $z = (1 - p^2)^{1/2} u$ in this integral gives

$$R(p, \theta) = \int_{-1}^{1} (1 - p^2)^{\alpha - 1/2} (1 - u^2)^{\alpha-1} \, du$$

$$= (1 - p^2)^{\alpha - 1/2} \int_{-1}^{1} (1 + u)^{\alpha-1}(1 - u)^{\alpha-1} \, du$$

Substituting $u = 2t - 1$ in the last integral changes that integral to

$$2^{2\alpha-1} \int_0^1 t^{\alpha-1}(1 - t)^{\alpha-1} \, dt = 2^{2\alpha-1} B(\alpha, \alpha) = 2^{2\alpha-1} \frac{\Gamma^2(\alpha)}{\Gamma(2\alpha)}$$

where B denotes the beta function. (See Example 4.33.) Hence, for $0 \leq p \leq 1$ and all θ,

$$R(p, \theta) = 2^{2\alpha-1}(1 - p^2)^{\alpha - 1/2} \frac{\Gamma^2(\alpha)}{\Gamma(2\alpha)}$$

Example 4.46

Let $f(x, y) = I_Q(x, y)$, where Q is the unit square $[0, 1] \times [0, 1]$ in the first quadrant. Since $f(x, y) = 1$ on the domain of f, $R(p, \theta)$ is the length of the intersection of the line L determined by p and θ with Q. In attacking this problem, it should be recognized that the set of all lines intersecting Q in more than one point can be broken down into six classes, each one determined by the manner in which a line intersects two of the four edges of Q. Representative lines, labeled L_k, are shown in Figure 4.13. We shall refer to that labeling scheme throughout this example.

In evaluating $R(p, \theta)$, it is convenient to proceed in eight steps, each step concerning a region of θ, namely,

$$\frac{(n - 1)\pi}{4} < \theta < \frac{n\pi}{4}$$

for $n = 1, 2, \ldots, 8$.

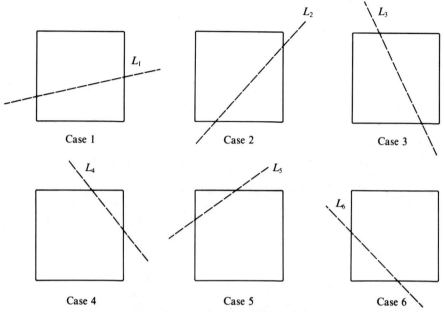

Figure 4.13 Labeling Scheme for Radon Transform

Step 1. For $0 < \theta \leq \pi/4$, Figure 4.14(a) illustrates the possibilities for crossing according to the six classes of Figure 4.13. First consider a line of the type L_6. The computation of its length is illustrated in Figure 4.15. In the figure,

$$R(p, \theta) = \overline{CB} + \overline{BA} = \overline{OC} \cos \theta + \overline{OA} \sin \theta$$

$$= \frac{p}{\sin \theta} \cos \theta + \frac{p}{\cos \theta} \sin \theta$$

$$= \frac{p}{\sin \theta \cos \theta}$$

Now consider a line of the type L_3. Figure 4.16 illustrates the computation of its length. In the figure,

$$R(p, \theta) = \overline{BC} = \frac{\overline{AC}}{\cos \theta} = \frac{1}{\cos \theta}$$

Type L_4 is considered in Figure 4.17. There,

$$\overline{FE} = \frac{\overline{FD}}{\cos \theta} = \frac{1 - \overline{FA}}{\cos \theta} = \frac{1 - p \sin \theta}{\cos \theta}$$

$$\overline{FC} = \frac{1 - p \cos \theta}{\cos \theta}$$

$$\overline{FB} = \overline{BC} \cos \theta = \frac{\overline{FC}}{\sin \theta} \cos \theta = \frac{1 - p \cos \theta}{\sin \theta}$$

Sec. 4.8 Radon Transform

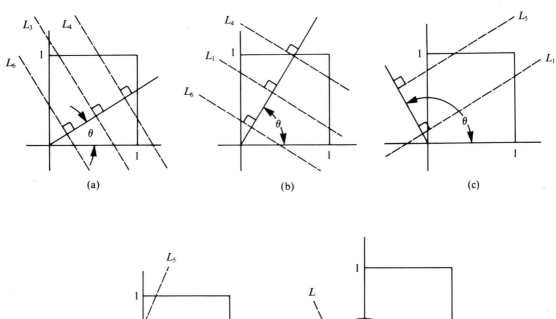

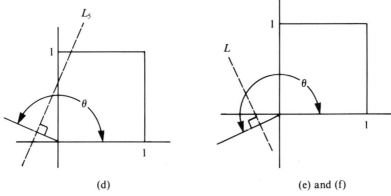

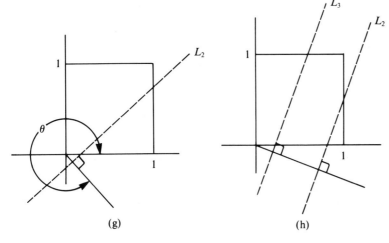

Figure 4.14 Eight Steps in Evaluation of Radon Transform for Image with Square Support Region

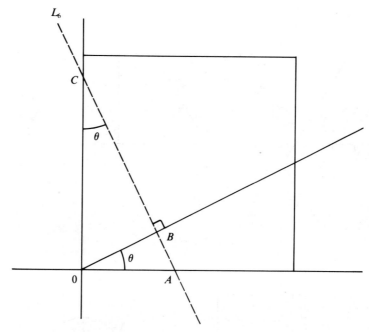

Figure 4.15 Evaluation of Radon Transform on L_6

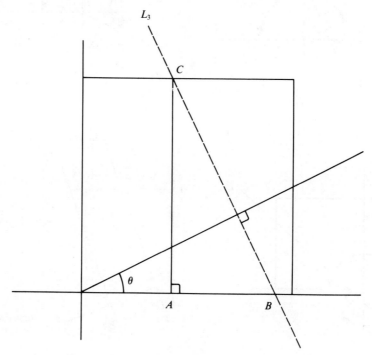

Figure 4.16 Evaluation of Radon Transform on L_3

Sec. 4.8 Radon Transform

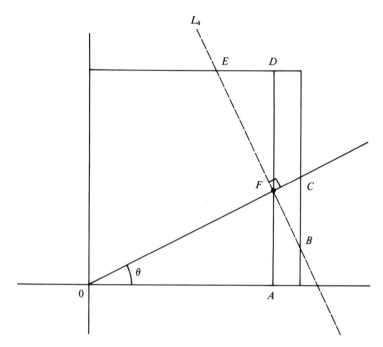

Figure 4.17 Evaluation of Radon Transform on L_4

and

$$R(p, \theta) = \overline{FE} + \overline{FB} = \frac{\cos \theta + \sin \theta - p}{\cos \theta \sin \theta}$$

All possibilities in Step 1 are now completed. In terms of the variables p and θ, L_6 holds for $0 \leq p \leq \sin \theta$, L_3 holds for $\sin \theta \leq p \leq \cos \theta$, L_4 holds for $\cos \theta \leq p \leq \sin \theta + \cos \theta$, and $R(p, \theta) = 0$ for $p > \sin \theta + \cos \theta$ because in such an instance the line does not intersect Q.

Step 2. For $\pi/4 \leq \theta < \pi/2$, Figure 4.14(b) illustrates the possibilities. We will not go through all the details, but for a line of type L_4,

$$R(p, \theta) = (\cos \theta + \sin \theta - p)/(\cos \theta \sin \theta)$$

and the constraint on p is given by $\sin \theta \leq p \leq \sin \theta + \cos \theta$. For a line of type L_1, $R(p, \theta) = \csc \theta$, and the constraint is $\cos \theta \leq p \leq \sin \theta$. For integration along L_6, $R(p, \theta) = p/(\cos \theta \sin \theta)$ with constraint $0 \leq p \leq \cos \theta$. Finally, if $p \geq \sin \theta + \cos \theta$, $R(p, \theta) = 0$.

Steps 3 through 8. The remaining steps can be completed in a similar fashion. They are illustrated in Figures 4.14(c) through 4.14(h), respectively. Notice that the values $\theta = 0$, $\pi/2$, π, and $3\pi/2$ have been skipped. Calculations for them must be done independently.

4.9 INVERSION OF THE RADON TRANSFORM

In this section, we give a rigorous treatment of some inversion theorems for the Radon transform for a particular class of images denoted by S. For those interested in practical results, attention should be focused on the construction of hyperplane functional averages and the actual statements of the inversion theorems themselves.

Thus far the Radon transform has been presented only for images $f(x, y)$ defined on R^2. And while it might appear that a simple generalization from integration over lines to integration over planes would move the entire discussion into functions $f(x, y, z)$ of three variables in a completely straightforward manner, such is not the case. In fact, in terms of inversion, the situation is less complicated for functions of three variables than for functions of two. Consequently, we shall proceed to define the n-dimensional Radon transform, where n can be taken to be either 2 or 3. (Nowhere in the discussion will n actually be restricted to 2 or 3, and hence the entire approach is actually n-dimensional.) Moreover, whereas in the previous section, explicit representations were given in terms of $\cos \theta$ and $\sin \theta$, the approach here will be entirely geometric (except for examples that refer back to the last section). Surface integrals will be presented as integrals evaluated with respect to surface measure, and that measure will not be explicitly presented in a parameterized form.

We begin by restricting our attention to functions that possess all possible partial derivatives, the so-called *infinitely differentiable*, or C^∞, functions. Further, of these, only functions in the class $S(R^n)$ of *rapidly decreasing* functions will be allowed.[31] A function $f(x)$ is in $S(R^n)$ if and only if for any nonnegative integer k and any linear combination of partial derivatives of $f(x)$, say $[Pf](x)$,

$$\sup_{x \in R^n} \big| \, \|x\|^k \, [Pf](x) \, \big| < \infty$$

where $\|x\|$ denotes the Euclidean norm of $x \in R^n$. Note that the class of C^∞ functions with compact support, C_c^∞, is in $S(R^n)$. (For those familiar with the theory of distributions, C_c^∞ is dense in $S(R^n)$ and the class of tempered distributions consists of those linear forms on $S(R^n)$ which are continuous.) It is precisely the functions in $S(R^n)$ for which the inversion theorems will be stated.

Example 4.47

Let $f(u, v) = e^{-(u^2+v^2)}$. Then $f \in S(R^2)$. Indeed, all partial derivatives produce terms with powers of u and v multiplied by $e^{-(u^2+v^2)}$. Multiplication by $\|x\|^k = (u^2 + v^2)^{k/2}$ still leaves the negative exponential intact, and in the limit, as $\|x\| \to \infty$, $\|x\|^k [Pf](x) \to 0$, thereby assuring that the defining supremum is finite.

---

[31] M. J. Lighthill, *Introduction to Fourier Analysis and Generalized Functions* (London: Cambridge University Press, 1962), 15.

Suppose now, that $f(x)$ is integrable over all hyperplanes in R^n, i.e., for all lines in R^2 or for all planes in R^3. Let $H^{(n)}$ denote the class of all hyperplanes in R^n. For any hyperplane $P \in H^{(n)}$, we define $R_f(P)$ to be the integral of $f(x)$ over the hyperplane P. That is,

$$R_f(P) = \int_P f(x) \, dm_P(x)$$

where $dm_P(x)$ refers to $(n-1)$ dimensional measure as it is induced on P.[32]

R_f is thus a real-valued function on $H^{(n)}$ and is called the *Radon transform* of f. We write $\mathcal{R}: f \to R_f$ or $\mathcal{R}(f) = R_f$.

Given a point $x \in R^n$, let $H^{(n)}[x]$ denote the set of all hyperplanes through x. (See Figure 4.18.) In the reconstruction of $f(x)$ from its Radon transform $R_f(P)$, the average value of $R_f(P)$ over the set of hyperplanes $H^{(n)}[x]$ plays a crucial role. Note that for each unit vector $\mathbf{w}$, there is exactly one hyperplane through x. Let $\boldsymbol{\phi} = (\phi_1, \phi_2, \ldots, \phi_{n-1})$ be a polar representation of S^n, the unit sphere in R^n. When restricted to $H^{(n)}[x]$, R_f is actually a function of $\boldsymbol{\phi}$, so we shall write $R_f\langle x; \boldsymbol{\phi}\rangle$. The average value of $R_f(P)$ over $H^{(n)}[x]$ is, in fact, a surface integral over S^n. If $d\sigma$ denotes area on S^n, the desired average value is given by

$$A_n(R_f; x) = \frac{1}{s_n} \int_{S^n} R_f\langle x; \boldsymbol{\phi}\rangle \, d\sigma$$

where s_n denotes the total surface area of S^n.

It is useful to recognize that $A_n(R_f; x)$ can be expressed directly in terms of f, the original function. Indeed, for $f \in S(R^n)$, it can be shown that

$$A_n(R_f; x) = \frac{s_{n-1}}{s_n} \int_{R^n} \frac{f(y)}{\|x - y\|} \, dy$$

For $n = 2$, this reduces to

$$A_2(R_f; (u, v)) = \frac{2}{2\pi} \int_{-\infty}^{\infty} \int_{-\infty}^{\infty} \frac{f(\xi, \zeta)}{[(u - \xi^2) + (v - \zeta^2)]^{1/2}} d\xi \, d\zeta$$

At this point we can state an inversion theorem for R^3. Let Δ denote the Laplacian

$$\Delta f = f_{11} + f_{22} + \cdots + f_{nn}$$

where f_{ii} denotes the second partial derivative of f with respect to the ith variable.

---

[32] Recall that a hyperplane is defined by the dot product $\mathbf{x} \cdot \mathbf{w} = p$, where $\mathbf{w}$ is a unit vector and p gives the perpendicular distance from the hyperplane to the origin. For instance, in Section 4.8, where R^2 was under consideration,

$$p = \mathbf{x} \cdot \mathbf{w} = (u, v) \cdot (\cos \theta, \sin \theta) = u \cos \theta + v \sin \theta$$

As is customary, we do not distinguish between the point x and the vector $\mathbf{x}$.

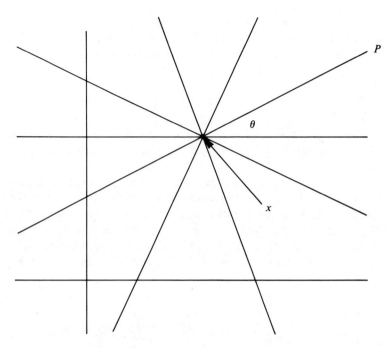

Figure 4.18 Set of All Hyperplanes in $H^{(1)}[x]$.

Theorem 4.23.[33] Supppose $f \in S(R^3)$. Then

$$f(x) = -\frac{1}{2\pi} \Delta [A_3(R_f;x)]$$

Before investigating the Radon inversion formula on R^2, it is necessary to introduce a generalization of the Laplacian. For $f \in S(R^n)$, we define the *Riesz potential* I^a by

$$[I^a(f)](x) = \frac{1}{H_n(a)} \int_{R^n} \| x - y \|^{a-n} f(y) \, dy$$

where $H_n(a)$ is the dimensional constant

$$H_n(a) = 2^a \pi^{n/2} \frac{\Gamma(a/2)}{\Gamma[(n-a)/2]}$$

I^a is closely related to the Laplacian Δ. In fact, it extends Δ since $-\Delta f = I^{-2}(f)$. In general, for $f \in S(R^n)$,

$$I^a \Delta f = \Delta I^a f = -I^{a-2} f$$

---

[33] Sigurdur Helgason, *The Radon Transform* (Boston: Birkhauser, 1980), 20.

Though we shall not go into detail, the Riesz potentials have been well studied and play an important role in the theory of potentials.

We can now state the main form of the inversion theorem in terms of Riesz potentials for R^n. In the case $n = 3$, it reduces to Theorem 4.23. Its complicated form in R^2 was foreshadowed at the beginning of the section, where it was mentioned that inversion in R^2 is more problematic than in R^3.

Theorem 4.24.[34] Suppose $f \in S(R^n)$. Then

$$f(x) = (-1)^{n-1} \lambda_n I^{1-n}[A_n(R_f;x)]$$

where λ_n is the dimensional constant

$$\lambda_n = (4\pi)^{(1-n)/2} \frac{\sqrt{\pi}}{\Gamma(n/2)}$$

For $n = 3$, $\lambda_3 = 1/(2\pi)$ and $I^{-2} = -\Delta$, resulting in Theorem 4.3. For $n = 2$, we obtain $\lambda_2 = 1/2$; hence,

$$f(x) = \frac{1}{2} I^{-1}[A_2(R_f;x)]$$

We now state a different inversion theorem. In this theorem, although the hypothesis still has the original image in $S(R^n)$, it gives the inversion in a different form. Once again, the situation in R^3 will be less complicated than in R^2. First of all, note that the preceding discussion was geometric in content and that, in fact, $R_f(P) = R_f(p, \boldsymbol{\theta})$, where p is the perpendicular distance from the origin to P and $\boldsymbol{\theta} = (\theta_1, \theta_2, \ldots, \theta_{n-1})$ is the vector of $n - 1$ angles determining the unit vector perpendicular to P. Consequently, R_f can be partially differentiated with respect to p. We shall write $\mathcal{H}_p$ to denote the Hilbert transform applied with respect to the variable p, and $R_f^{(k)}$ to denote the kth derivative of R_f with respect to p, i.e.,

$$\frac{d^k}{dp^k} R_f(p, \boldsymbol{\theta})$$

Theorem 4.25.[35] Let $f \in S(R^n)$ and let λ_n be the dimensional constant given in Theorem 4.24. Then

(i) For n odd, $f(x) = (-1)^{(n-1)/2} \lambda_n A_n(R_f^{(n-1)};x)$
(ii) For n even, $f(x) = (-1)^{n/2} \lambda_n A_n(\mathcal{H}_p(R_f^{(n-1)});x)$

For $n = 2$, Theorem 4.25 gives a method for reconstruction of $f(x)$; however, the method is extremely difficult. Letting $n = 2$ in part (ii), $f(x) = -\lambda_2 A_2 (\mathcal{H}_p(R'_f);x)$.

[34] Ibid, 20.
[35] Ibid, 25.

Hence, the methodology is:

1. Find the derivative $R'_f(p, \theta)$.
2. Find the Hilbert transform of $R'_f(p, \theta)$ as applied with respect to the variable p.
3. Write $\mathcal{H}_p[R'_f(p, \theta)]$ as $[\mathcal{H}_p(R'_f)]\langle x;\phi\rangle$, i.e., as a function of x and ϕ, which means that it is now defined, for each x, on the set of lines $H^{\langle 1 \rangle}[x]$.
4. Integrate $(1/2\pi)[\mathcal{H}_p(R'_f)]\langle x;\phi\rangle$ over the unit circle, and multiply by the dimensional constant $-\lambda_2$.

For $n = 3$, there are only three steps:

1. Find the second derivative $R''_f(p, \theta_1, \theta_2)$ with respect to p.
2. Write $R''_f(p, \theta_1, \theta_2)$ as $R''_f\langle x;\phi_1, \phi_2\rangle$, which means that R''_f is now defined, for each x, on the set of planes $H^{\langle 2 \rangle}[x]$.
3. Integrate $(1/4\pi)R''_f\langle x;\phi_1, \phi_2\rangle$ over the unit sphere and multiply by the dimensional constant $-\lambda_3$.

Both forms of the inversion theorem, Theorems 4.24 and 4.25, are highly problematic, though not as much so in R^3 as in R^2. To begin with, the set of hyperplanes $H^{\langle n \rangle}$ is parameterized in two different manners: first by (p, θ), where p gives the perpendicular distance from the origin to a hyperplane and θ determines the hyperplane orientation by specifying an orthogonal vector, and second by $\langle x;\phi\rangle$, where x is contained in the hyperplane and ϕ determines the orientation by specifying a subspace of dimension $n - 1$ which is parallel to the given hyperplane. Figure 4.19 shows the situation in R^2, where $\theta = \theta$ and $\phi = \phi$. The relationship between the variables is given by $\theta = \phi + \pi/2$ and $p = \|x\| \cos(\phi - \psi + \pi/2)$, where $\psi = \arctan(v/u)$. Consequently, the relation between $R_f\langle x;\phi\rangle$ and $R_f(p, \theta)$ is given by

$$R_f\langle x;\phi\rangle = R_f\left(\|x\| \cos\left(\phi - \psi + \frac{\pi}{2}\right), \phi + \frac{\pi}{2}\right)$$

Applied to R^2, Theorem 4.24 then reduces to

$$f(x) = \frac{1}{2} I^{-1} \left(\frac{1}{2\pi} \int_0^{2\pi} R_f\left(\|x\| \cos\left(\phi - \psi + \frac{\pi}{2}\right), \phi + \frac{\pi}{2}\right) d\phi \right)$$

One must not forget that I^{-1} is itself a troublesome integral in this equation.

Let us next apply Theorem 4.25 in R^2, following the steps outlined after the theorem. We first find $R'_f(p, \theta)$ and then apply the negative of the Hilbert transform with respect to the variable p, the minus one in front of the constant being

Sec. 4.9   Inversion of the Radon Transform

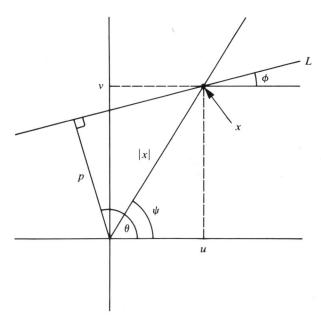

Figure 4.19 Parameterization for Inversion of Radon Transform in R^2

applied now rather than later. This gives

$$h(p, \theta) = \frac{1}{\pi} \int_{-\infty}^{\infty} \frac{\frac{d}{dt}[R_f(t, \theta)]}{p - t} \, dt$$

We then find $h\langle x; \phi \rangle$:

$$h\langle x; \phi \rangle = \frac{1}{\pi} \int_{-\infty}^{\infty} \frac{\frac{d}{dt}\left[R_f\left(t, \phi + \frac{\pi}{2}\right)\right]}{\|x\| \cos\left(\phi - \psi + \frac{\pi}{2}\right) - t} \, dt$$

Due to symmetry, instead of averaging over $[0, 2\pi]$, we can average over $[0, \pi]$ and then multiply by $\lambda_2 = 1/2$. We obtain

$$f(x) = \frac{1}{2\pi^2} \int_0^\pi \int_{-\infty}^{\infty} \frac{\frac{d}{dt}\left[R_f\left(t, \phi + \frac{\pi}{2}\right)\right]}{\|x\| \cos\left(\phi - \psi + \frac{\pi}{2}\right) - t} \, dt \, d\phi$$

The last expression can be simplified by a change of variables, letting $\theta = \phi + \pi/2$. The averaging integral would then run from $\pi/2$ to $3\pi/2$; however, due to the symmetry of the hyperplane orientations, the limits could be returned to 0 to π.

Example 4.48

Using the image f of Example 4.43, we have $R_f(p, \theta) = \sqrt{\pi}\, e^{-p^2}$. Applying the remarks of the preceding paragraph and recognizing that $R'_f(p, \theta) = -2\sqrt{\pi}\, p e^{-p^2}$, inversion yields

$$-(\pi)^{-3/2} \int_0^\pi \int_{-\infty}^\infty \frac{t e^{-t^2}}{\|x\| \cos(\theta - \psi) - t}\, dt\, d\theta$$

Though inversion is less complicated in R^3, we have concentrated mainly on R^2 since the subject matter herein concerns primarily images of the form $f(x, y)$. In any event, it should be recognized that, although the theorems of this section provide a theoretical basis for inversion of the Radon transform, they do not give practical manipulative algorithms.

4.10 IMAGE CREATION FROM PROJECTIONS VIA THE FOURIER TRANSFORM

The purpose of this section is to demonstrate how Fourier transforms can be exploited in the inversion of the Radon transform. The methodology depends upon the relationship $\mathcal{F}_1[\mathcal{R}(\cdot)] = \mathcal{F}_2(\cdot)$, where $\mathcal{F}_1$ and $\mathcal{F}_2$ represent the one- and two-dimensional Fourier transforms, respectively. This relationship will be demonstrated formally, in the sense that no strict conditions will be imposed upon the initial image $f(x, y)$ in order to make the demonstration mathematically rigorous. It should be recognized, however, that our intent is to employ the inverse Radon transform $\mathcal{R}^{-1}$, and consequently, the methodology of the section holds at least whenever f is in $S(R^2)$.

Let the image f be given in polar form, i.e., $g(r, \phi)$. Then, as shown in Section 4.3, the polar form of the two-dimensional Fourier transform of f is given by

$$G(h, \theta) = \int_{r=0}^\infty \int_{\phi=0}^{2\pi} g(r, \phi) e^{-jhr\cos(\phi - \theta)}\, d\phi\, dr$$

On the other hand, the Radon transform of $g(r, \phi)$ is

$$R(p, \theta) = \int_{-\infty}^\infty g[(p^2 + z^2)^{1/2}, \theta + \arctan(z/p)]\, dz$$

where it is assumed that $p \neq 0$. Taking the Fourier transform of $R(p, \theta)$ with respect to p gives

$$F(h, \theta) = \int_{-\infty}^\infty R(p, \theta) e^{-jph}\, dp$$

$$= \int_{-\infty}^\infty \int_{-\infty}^\infty g[(p^2 + z^2)^{1/2}, \theta + \arctan(z/p)] e^{-jph}\, dz\, dp$$

Sec. 4.10   Image Creation from Projections via the Fourier Transform

Going to polar coordinates (that is, letting $r = (p^2 + z^2)^{1/2}$ and $\phi = \theta + \arctan(z/p)$), solving for p, and employing the Jacobian of the transformation yields precisely $G(h, \theta)$. Hence, it has been formally demonstrated that

$$\mathcal{F}_1[\mathcal{R}(g)] = \mathcal{F}_2[g]$$

Finally, taking the inverse two-dimensional Fourier transform gives the desired practical method for inverting the Radon transform:

$$\mathcal{R}^{-1} = \mathcal{F}_2^{-1} \mathcal{F}_1$$

Figure 4.20 gives a commuting diagram illustrating this relation.

Example 4.49

Let $f(x, y) = (x^2 + y^2)e^{-(x^2+y^2)}$. Then the corresponding polar version of f is given by $g(r, \phi) = r^2 e^{-r^2}$, and its Radon transform is

$$R(p, \theta) = \int_{-\infty}^{\infty} (p^2 + z^2)e^{-(p^2+z^2)} \, dz$$

$$= p^2 e^{-p^2} \int_{-\infty}^{\infty} e^{-z^2} \, dz + e^{-p^2} \int_{-\infty}^{\infty} z^2 e^{-z^2} \, dz$$

$$= \sqrt{\pi}\, p^2 e^{-p^2} + \frac{\sqrt{\pi}}{2} e^{-p^2}$$

(It should be noted that $f \in S(R^2)$ and hence either inversion theorem, 4.24 or 4.25, applies. However, our intent is to demonstrate the relationship $\mathcal{F}_2[g] = \mathcal{F}_1[R]$.) Together, Theorem 4.2 and Example 4.11 can be used to show that

$$[\mathcal{F}_1(R)](\omega, \theta) = \pi e^{-\omega^2/4} - \frac{\pi \omega^2 e^{-\omega^2/4}}{4}$$

The two-dimensional Fourier transform of the original image is most expediently

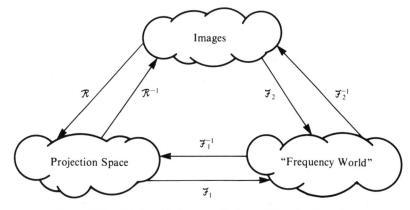

Figure 4.20 Relationship between Radon and Fourier Transforms

found using Cartesian coordinates, since x and y are separable in the original image. We obtain

$$F(\omega_1, \omega_2) = \int_{-\infty}^{\infty} \int_{-\infty}^{\infty} (x^2 + y^2) e^{-(x^2+y^2)} e^{-j(\omega_1 x + \omega_2 y)} \, dx \, dy$$

$$= \int_{-\infty}^{\infty} \left(e^{-y^2} \int_{-\infty}^{\infty} x^2 e^{-x^2} e^{-j\omega_1 x} \, dx + y^2 e^{-y^2} \right.$$

$$\left. \times \int_{-\infty}^{\infty} e^{-x^2} e^{-j\omega_1 x} \, dx \right) e^{-j\omega_2 y} \, dy$$

$$= \int_{-\infty}^{\infty} \left[e^{-y^2} \sqrt{\pi} \left(\frac{e^{-\omega_1^2/4}}{2} - \frac{\omega_1 e^{-\omega_1^2/4}}{4} \right) \right.$$

$$\left. + y^2 e^{-y^2} \sqrt{\pi} \, e^{-\omega_1^2/4} \right] e^{-j\omega_2 y} \, dy$$

$$= \pi \frac{e^{-\omega_2^2/4} e^{-\omega_1^2/4}}{2} - \frac{\omega_1^2 e^{-\omega_1^2/4} e^{-\omega_2^2/4}}{4} \pi$$

$$+ \frac{\pi e^{-\omega_1^2/4} e^{-\omega_2^2/4}}{2} - \frac{\omega_2^2 e^{-\omega_1^2/4} e^{-\omega_2^2/4}}{4} \pi$$

$$= \pi e^{-(\omega_1^2 + \omega_2^2)/4} - \frac{\pi}{4} (\omega_1^2 + \omega_2^2) e^{-(\omega_1^2 + \omega_2^2)/4}$$

$$= G(\sqrt{\omega_1^2 + \omega_2^2})$$

Letting $\omega = \sqrt{\omega_1^2 + \omega_2^2}$ and writing $F(\omega_1, \omega_2)$ in polar coordinates gives the desired result: $[\mathcal{F}_1(R)](\omega, \theta) = G(\omega, \theta)$.

4.11 CONVOLUTIONAL METHOD OF IMAGE CREATION FROM PROJECTIONS

Fourier methods are important for providing an alternative method of inverting the Radon transform. They also provide a background for understanding the convolutional method of inversion. If one employs the R^2 methodology outlined after Theorem 4.25, it is necessary to perform a differentiation followed by a Hilbert transformation. In practice, however, it is the Radon transform that is observed. In other words, an observed function $R(p, \theta)$ is given, and it is necessary to find an image $f(x, y)$ such that $\mathcal{R}(f) = R$. Hence, it is the observed image that must be differentiated. Differentiation of observed quantities is a process that is not recommended due to its sensitivity to noise and other disturbances. The convolutional method bypasses the differentiation and Hilbert transformation steps.

Let $R(p, \theta)$ be the observed projection function. Then the first two steps of the methodology of Theorem 4.25 yield

$$g(x, \theta) = \frac{1}{\pi} \int_{-\infty}^{\infty} \frac{\frac{d}{dt}[R(t, \theta)]}{t - x} \, dt$$

Proceeding formally, let the Fourier transform of $R(t, \theta)$ with respect to the first variable be given by $F(\omega, \theta)$. Then

$$\frac{d}{dt}[R(t, \theta)] \leftrightarrow j\omega F(\omega, \theta)$$

Once again formally, since no L_2 condition has been verified, application of Theorem 4.21 gives the Fourier transform of $g(x, \theta)$ as

$$G(\omega, \theta) = \omega F(\omega, \theta)(-1)\,\text{sgn}(\omega)$$

Rewriting this as $G(\omega, \theta) = F(\omega, \theta)(-1)\omega\,\text{sgn}(\omega)$ motivates the expression of g as a convolution of f with the inverse Fourier transform of $(-1)\omega\,\text{sgn}(\omega)$. Thus, we write

$$g(x, \theta) = R(x, \theta)\mathcal{F}^{-1}[(-1)\omega\,\text{sgn}(\omega)]$$

Intuitively,

$$\mathcal{F}^{-1}[(-1)\omega\,\text{sgn}(\omega)] = \frac{-1}{\pi}\frac{d}{dt}\left(\frac{1}{t}\right) = \frac{1}{\pi t^2}$$

However, there is trouble with this approach since the Fourier transform of $1/t^2$ does not exist in a strict sense. More than that, many of the preceding manipulations have been carried out without proper justification. Nevertheless, insight has been obtained, and, under certain conditions on R, a slight modification of the method will make the approach rigorous.

Suppose, then, that $R(t, \theta)$ is *bandlimited* in the variable t, i.e., its Fourier transform has compact support. Then $F(\omega, \theta)$ is identically zero for $|\omega| > a$, for some $a > 0$. In this case, $G(\omega, \theta)$ is also zero for $|\omega| > a$. Letting

$$Q(\omega) = (-1)\omega\,\text{sgn}(\omega)I_{(-a,a)}(\omega)$$

we can write $G(\omega, \theta) = F(\omega, \theta)Q(\omega)$, which is exactly what was given previously. However, since $R(\cdot, \theta)$ is in L_2, it is given by $\mathcal{F}^{-1}[F(\omega, \theta)]$. Hence, Fourier inversion gives

$$g(t, \theta) = R(t, \theta) * q(t)$$

where

$$q(t) = \frac{-1}{2\pi}\int_{-a}^{a}\omega\,\text{sgn}(\omega)e^{j\omega t}\,d\omega$$

$$= \frac{-1}{\pi}\int_{0}^{a}\omega\cos(\omega t)\,d\omega$$

$$= \frac{-1}{\pi}\left(\frac{a\sin(at)}{t} + \frac{\cos(at) - 1}{t^2}\right)$$

which is also in L_2, thus justifying the preceding convolution.

In sum, when the projections are bandlimited, the first two steps of the inversion methodology of Theorem 4.25 (in R^2) can be replaced by a convolution involving the projection $R(t, \theta)$ itself and the preceding function $q(t)$. The reparameterization in terms of $\langle x;\phi\rangle$, the averaging, and the multiplication by $-\lambda_2$ should then be carried out as usual.

4.12 HOUGH TRANSFORM

In this section, we describe a particular adaptation of the Radon transform known as the *Hough transform*. This adaptation is often used for feature extraction, compression, and image recognition. The images to which it is applied are usually composed of straight lines, and the output is an image of points. The output image is a compressed version of the original and is often more suitable for image recognition. A pair of examples should serve to illustrate the Hough methodology.

Example 4.50

Consider the boundary-type image illustrated in Figure 4.21(a). This image represents the boundary of the image given in Example 4.46 and is defined by $f = \oplus_{m=1}^{4} h_m \oplus l_1$, where $h_m = 1$ on B_m and $l_1 = 0$ on C_1, with the domain sets given by

$$B_1 = \{(x, y): x = 0 \text{ and } 0 \leq y \leq 1\}$$

$$B_2 = \{(x, y): x = 1 \text{ and } 0 \leq y \leq 1\}$$

$$B_3 = \{(x, y): y = 0 \text{ and } 0 < x < 1\}$$

$$B_4 = \{(x, y): y = 1 \text{ and } 0 < x < 1\}$$

$$C_1 = R^2$$

Strictly speaking, this image is not a line-drawing image since its domain consists of more than just lines in the plane; however, it would be a line drawing if the "background" l_1 were not included. Taking the Radon transform of f yields the point image $R(p, \theta)$, where, for $p \geq 0$ and $0 \leq \theta < 2\pi$,

$$R(p, \theta) = \begin{cases} 1 & \text{for } (p, \theta) \in \{(0, 0), (1, 0), (0, \pi/2), \\ & (1, \pi/2), (0, \pi), (0, 3\pi/2)\} \\ 0 & \text{elsewhere.} \end{cases}$$

For all other p and θ, $R(p, \theta)$ is defined by extension, as described in Section 4.8; however, no useful information is provided by extending the domain of θ outside $[0, 2\pi)$. $R(p, \theta)$ is illustrated in Figure 4.21(b).

Given a Radon transform consisting of only a finite number of points in the region $p \geq 0$ and $0 \leq \theta < 2\pi$, the inversion procedure lacks uniqueness in that there exist numerous images that could have produced the point image. To see this, for each point in $R(p, \theta)$, simply draw a line described in normal form using the corresponding p and θ, ignoring overlapping lines.

Sec. 4.12   Hough Transform

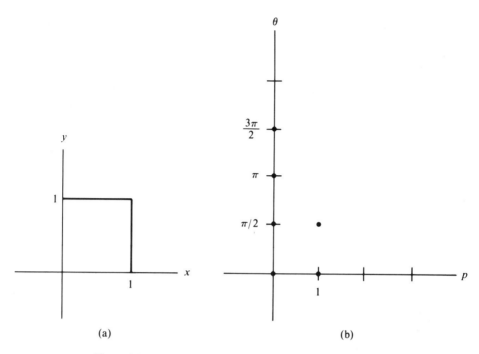

Figure 4.21 Representation of Image Using Hough Transform

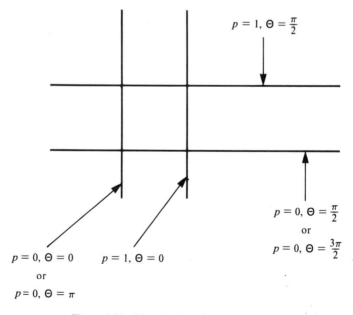

Figure 4.22 Lines Using Hough Transform

Example 4.51

Let $R(p, \theta)$ be as given in Figure 4.21(b). Then four distinct lines correspond to points within that figure. (See Figure 4.22.) Not only is $R(p, \theta)$ the transform of f given in Example 4.49, but it also is the transform of any image g that is zero everywhere on R^2 except for four distinct unit-length line segments, one on each of the four lines in Figure 4.22. In practical applications, other information concerning the original image must be registered together with the Hough point features.

EXERCISES

4.1. Consider the image

$$f = \begin{cases} 1 & \text{if } x^2 + y^2 \leq 1 \\ \dfrac{1}{|xy|} & \text{if } x^2 + y^2 > 1 \end{cases}$$

(a) Use the image representation model discussed in Section 4.1 to identify possible subimages g_n, h_m, and l_k comprising the image f.
(b) For the representation of the image f given in part (a), is l_k in L_1 (wherever it is defined)? Is l_k in L_2?
(c) Show that f is not in L_1 but is in L_2.

4.2. Let $f(t) = te^{-\lambda t}u(t)$, where $\lambda > 0$. Calculate the Fourier transform $F(\omega)$ of $f(t)$. Show directly that $F(\omega)$ is uniformly continuous and $\lim_{|\omega|\to\infty} F(\omega) = 0$. Compare this with Theorem 4.1.

4.3. By using Theorem 4.2, show that $f(t) = t^2 e^{-\lambda t}u(t)$, where $\lambda > 0$, has a Fourier transform $F(\omega)$ such that $\lim_{|\omega|\to\infty} \omega^2 F(\omega) = 0$.

4.4. Refer to Example 4.11 and determine whether $f(t) = te^{-t^2}$ is an eigenfunction for the Fourier transform.

4.5. Determine the convolution h of $f(t) = I_{[0,2]}(t)$ and $g(t) = e^{-t}u(t)$, and then illustrate the conclusions given in Theorem 4.4 on the result h.

4.6. Apply Theorem 4.7 to $f(t) = e^{-t}u(t)$ and discuss in what sense the Fourier inversion integral exists.

4.7. Illustrate the Fourier inversion integral for the following signals:
(a) $f(t) = e^{-|t|}$
(b) $f(t) = e^{-t}u(t)$

4.8. For the signals $f(t)$ given in Exercise 4.7, show that Parseval's Theorem (Theorem 4.8) holds true.

4.9. Let the image

$$f(x, y) = \begin{cases} x & \text{if } 0 < x < y < 1 \\ 0 & \text{otherwise} \end{cases}$$

Find the corresponding Fourier transform $F(\omega_1, \omega_2)$.

4.10. Using the image $f(x, y)$ given in Exercise 4.9, illustrate parts (b), (c), (d), (e), (f), and (g) of Theorem 4.9.

4.11. Find the convolution image $h(x, y)$ of the image $f(x, y)$ given in Exercise 4.9 and the image $g(x, y) = I_{(0,1) \times (0,1)}(x, y)$.

4.12. Show that Theorem 4.10 holds for the image $h(x, y)$ given in Exercise 4.11.

4.13. Let the image

$$f(x, y) = \begin{cases} 1 & \text{if } x^2 + y^2 < 1 \\ 2 & \text{if } 1 \le x^2 + y^2 < 2 \\ 0 & \text{otherwise} \end{cases}$$

Show that $f(x, y)$ is a radial function, and find the corresponding Hankel transform.

4.14. Explain and illustrate why the Riemann-Lebesgue lemma need not hold for signals in L_2.

4.15. Must the Fourier transform of an L_2 signal be continuous?

4.16. Find the Fourier sine and cosine transforms of $f(t) = I_{[-1,2]}(t)$.

4.17. Find the Mellin-type convolution of $f(t) = g(t) = I_{(0,a)}(t)$ and use this result to illustrate Theorem 4.18.

4.18. Use the definition of the Abel transform to show that

$$A^3(f) = \int_0^x \int_0^{x_1} \int_0^{x_2} f(x_3) \, dx_3 \, dx_2 \, dx_1$$

4.19. Determine the Hilbert transforms of $f(t) = \sin(3t)$ and of $g(t) = 2 + \sin(3t)$. Verify that Theorem 4.22 holds for the Hilbert transform of f.

4.20. Find the Radon transform of the image $f(x, y) = (x^2 + y^2)^2 e^{-(x^2+y^2)}$. Is $f(x, y)$ an element of S? Show all work.

4.21. Can the convolutional method of image creation from projections be carried out for a Radon transform

$$R(t, \theta) = \frac{\sin t}{t} q(\theta)$$

where $q(\theta)$ is well behaved?

4.22. Find the Hough transform of the image given in the following figure.

5

Projection Methods in Image Processing

5.1 PROJECTIONS IN EUCLIDEAN SPACE

In Chapter 4 we discussed the so-called projections $R(p, \theta)$ which result from integrating a function over hyperplanes in Euclidean space. We now turn to the usual vector-type projections, which play a fundamental role throughout mathematics and engineering. In image processing, they are of particular importance for compression, image creation, and feature parameter generation. We shall begin by examining the genesis of the problem as it exists in Euclidean space R^n.

Let $u_1, u_2, \ldots, u_n$ be a collection of n vectors in R^n which form a coordinate system. For instance, in R^3 they might be $\mathbf{i}$, $\mathbf{j}$, and $\mathbf{k}$, but not necessarily. By a *coordinate system* we simply mean that each u_k is of length unity and each pair u_k and u_l are *orthogonal* (perpendicular). Mathematically, $\| u_k \| = 1$ and the dot product $u_k \cdot u_l = 0$. Given a coordinate system, each vector x can be written as a linear combination of the u_k. This representation always takes the form

$$x = \hat{x}(u_1)u_1 + \hat{x}(u_2)u_2 + \cdots + \hat{x}(u_n)u_n$$

where $\hat{x}(u_k) = x \cdot u_k$. These n dot products with the coordinate system elements are known as the *Fourier coefficients* of the vector x with respect to the coordinate system $u_1, u_2, \ldots, u_n$. They represent the coordinates of x with respect to the system $u_1, u_2, \ldots, u_n$.

Example 5.1

In R^3, the vectors $\mathbf{i}$, $\mathbf{j}$, and $\mathbf{k}$ form a coordinate system. Moreover, given the vector $x = (x_1, x_2, x_3)$, $\hat{x}(\mathbf{i}) = x \cdot \mathbf{i} = x_1$, $\hat{x}(\mathbf{j}) = x \cdot \mathbf{j} = x_2$, and $\hat{x}(\mathbf{k}) = x \cdot \mathbf{k} = x_3$. Thus, the choice of $\mathbf{i}$, $\mathbf{j}$, and $\mathbf{k}$ for the coordinate system leads to Fourier coefficients of x which are simply its usual coordinates.

Example 5.2

Once again consider R^3, but this time let the coordinate system be given by $u_1 = (\frac{\sqrt{2}}{2}, \frac{\sqrt{2}}{2}, 0)$, $u_2 = (-\frac{\sqrt{2}}{2}, \frac{\sqrt{2}}{2}, 0)$, and $u_3 = (0, 0, 1)$. This system represents the usual **i, j, k** coordinate system rotated 45° in the horizontal x-y plane. The Fourier coefficients of the vector $x = (4, 2, -5)$ with respect to this system are $\hat{x}(u_1) = x \cdot u_1 = 3\sqrt{2}$, $\hat{x}(u_2) = x \cdot u_2 = -\sqrt{2}$, and $\hat{x}(u_3) = x \cdot u_3 = -5$. A simple calculation shows that $x = \hat{x}(u_1)u_1 + \hat{x}(u_2)u_2 + \hat{x}(u_3)u_3$. This equation represents x in terms of the given coordinate system.

The problem we wish to address is that of *compression*. Suppose one desires to represent a vector x in terms of a coordinate system without utilizing all the Fourier coefficients of x. If $w_1, w_2, \ldots, w_m$ represent a subcollection of the entire coordinate system, we would like to represent x as a linear combination of the w_k. Geometrically, we are looking for a vector in the subspace of R^n determined by the subsystem. Needless to say, if x is not in that subspace, then the best we can hope for is a "best approximation." But how is that best approximation to be determined? The answer is straightforward: we want to find a vector y in the subspace determined by the w_k such that $\|x - y\|$ is minimized. In fact, since the square root in the determination of $\|x - y\|$ plays no role in the minimization, we wish to minimize $\sum_{k=1}^{n} [x^{(k)} - y^{(k)}]^2$, where $x^{(k)}$ and $y^{(k)}$ are the usual Euclidean coordinates of x and y, respectively. This gives the so-called best *least-squares approximation*.

Example 5.3

Let u_1, u_2, and u_3 be the coordinate system given in Example 5.2, and let x be the vector given in that example. Suppose we wish to best approximate x by some vector lying in the subspace determined by the vectors u_1 and u_2. Geometrically, that subspace is the horizontal plane. The solution to the problem is simply to take y as the projection of x into the horizontal plane. (See Figure 5.1.) But then, in terms of u_1 and u_2, y has precisely the same coordinates as x, namely,

$$y = \hat{x}(u_1)u_1 + \hat{x}(u_2)u_2 = 3\sqrt{2}\, u_1 - \sqrt{2}\, u_2$$

In other words, the best least-squares approximation is provided by the utilization of the Fourier coefficients of x.

Example 5.3 provides the key to best subspace approximation relative to the norm $\|\cdot\|$. The solution is always given by the projection. If x is a vector in R^n, and $w_1, w_2, \ldots, w_m$ is the coordinate system of a subspace P in R^n, then x is best approximated by the vector

$$x_P = \hat{x}(w_1)w_1 + \hat{x}(w_2)w_2 + \cdots + \hat{x}(w_m)w_m$$

which is the *orthogonal projection* of x onto the subspace. Notice that the original coordinate system representation of the vectors $w_1, w_2, \ldots, w_m$ and x plays no role. This point is absolutely crucial: all that matters is that the w_k themselves form a coordinate system for some subspace. Everything, including the repre-

Sec. 5.1   Projections in Euclidean Space   243

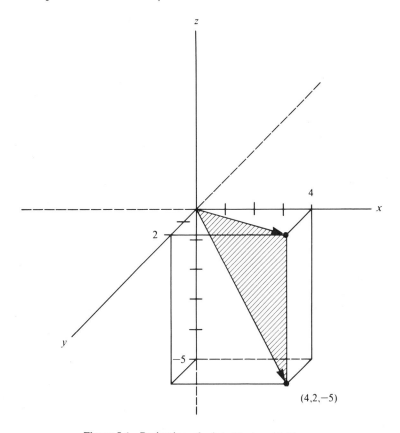

Figure 5.1 Projection of x into Horizontal Plane

sentation of y, the approximating vector, is in terms of the dot product. Moreover, note that in R^3 a subspace determined by a coordinate system with two vectors is a plane through the origin, and one determined by a coordinate system with a single vector is simply a straight line through the origin. The following theorem formalizes this discussion; we use the notation $d(x, P)$ to denote the perpendicular distance from x to the subspace P.

Theorem 5.1.[1]   Suppose $w_1, w_2, \ldots, w_m$ form the coordinate system for a subspace P of R^n and x is a vector in R^n. If $z = t_1 w_1 + t_2 w_2 + \cdots + t_m w_m$ is any vector in P, where the t_k are the coordinates of z relative to the w_k, then

(i) $\|x - x_P\| \leq \|x - z\|$.
(ii) $d(x, P)^2 = \|x\|^2 - \sum_{k=1}^{m} |\hat{x}(w_k)|^2$.

---

[1] Walter Rudin, *Real and Complex Analysis* (New York: McGraw-Hill, 1966), 84.

Part (i) of the theorem states that, in terms of the Euclidean norm, x is closer to x_P than it is to z. (See Figure 5.2.) As for part (ii), a particularly interesting point concerning the norm of x can be made. Suppose that $w_1, w_2, \ldots, w_m$ actually form a coordinate system for R^n. Then $m = n$ and $x = \sum_{k=1}^{m} \hat{x}(w_k) w_k$. Consequently, $x = x_P$ and $d(x, P) = 0$. Hence, by (ii), $\| x \|^2 = \sum_{k=1}^{m} | \hat{x}(w_k) |^2$, a fact that could be directly, but tediously, verified by computing the norm of the linear combination giving x in terms of the w_k.

Put concisely, (i) states that x_P is the best least-squares approximation in the subspace P and (ii) quantifies the perpendicular distance from x to P.

Example 5.4

Let x, u_1, u_2, and u_3 be as in Example 5.3. Then $x_P = 3\sqrt{2}\, u_1 - \sqrt{2}\, u_2$. Now let z be any other vector in the subspace spanned (determined) by u_1 and u_2. By Theorem 5.1, since $\| x - x_P \| = 5$,

$$5 \leq \sqrt{(3\sqrt{2} - t_1)^2 + (-\sqrt{2} - t_2)^2 + (-5 - 0)^2}$$

and the value 5 is attained if and only if $z = x_P$. Note how the norm was calculated in terms of the coordinate system u_1, u_2, u_3. This was easy since x was given in that coordinate system and z was in the subspace spanned by u_1 and u_2 ($\hat{z}(u_3) = 0$).

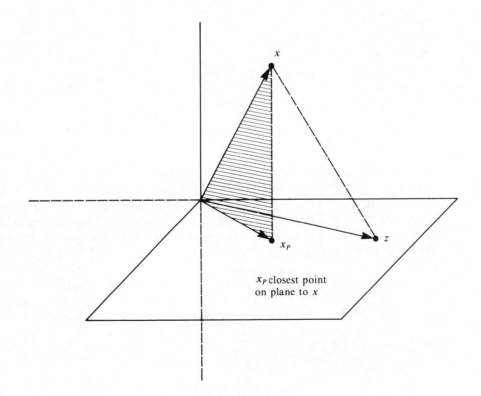

Figure 5.2 Perpendicular Projection as Closest Point to x in Subspace P

Sec. 5.2   Inner-Product Spaces

Finally, applying part (ii) of the theorem,
$$d(x, P)^2 = \|x\|^2 - [(3\sqrt{2})^2 + (-\sqrt{2})^2] = 45 - 20 = 25$$

5.2 INNER-PRODUCT SPACES

What is most notable about the discussion of projections in the preceding section is the generality of the essentials. Accordingly, this section describes a general theory of orthogonal projections based upon those essentials. Throughout the presentation, one should keep the Euclidean case in mind and also stay close to the examples. In the preceding section we utilized the algebraic properties of Euclidean vectors together with the properties of the dot product. Here we shall isolate these properties in order to extend the Euclidean results.

There are two fundamental algebraic operations on vectors in R^n: vector addition, and multiplication of a vector by a (real-number) scalar. These operations satisfy some basic properties, among which are the following:

L1. $x + y = y + x$.   (commutativity of vector addition)
L2. $x + (y + z) = (x + y) + z$.   (associativity of vector addition)
L3. There exists a unique zero vector **0** such that $x + \mathbf{0} = x$ for all x.   (additive identity element)
L4. Every vector x has an additive inverse $-x$ such that $x + (-x) = \mathbf{0}$.
L5. $t(sx) = (ts)x$ for any scalars s and t.
L6. $1x = x$.   (multiplicative identity element)
L7. $t(x + y) = tx + ty$.   (distributivity)
L8. $(t + s)x = tx + sx$.   (distributivity)

Note that although Euclidean vectors satisfy L1 through L8, so do numerous other structures. Similarly, although R^n utilizes scalars that are real numbers, there is no reason why we cannot extend scalar multiplication to encompass complex numbers.

We define a *linear space* to be a collection, say, X, of objects (1) upon which there exists a binary operation $+$, (2) for which there is a scalar multiplication operation, and (3) for which these two operations satisfy L1 through L8. Though the notion of a linear space might at first seem abstract, in fact it is most natural. Many common structures are linear spaces.

Example 5.5

Let $\mathcal{B}_{r,t}^{m;n}$ denote the collection of saturated m by n bound matrices with upper left entry situated at (r, t). Elements of X are of the form $(a_{pq})_{rt}$, where $a_{pq} \neq *$, and where there are m rows and n columns. $\mathcal{B}_{r,t}^{m;n}$ is a linear space, with $+$ being given by the operator ADD, $f + g = \text{ADD}(f, g)$, and scalar multiplication being given by $tf = \text{SCALAR}(t; f)$. The zero element is $(z_{pq})_{rt}$, where $z_{pq} = 0$ for all p and q. Note that we have assumed saturated bound matrices: there are no star values. Also, all scalars are assumed to be real. Consequently, $\mathcal{B}_{r,t}^{m;n}$ is called a *real* linear space.

Example 5.6

Let l_1 denote the collection of sequences of complex numbers $\{c_k\}$ for which $\sum_{k=1}^{\infty} |c_k|$ converges. We define addition in this linear space by $\{c_k\} + \{d_k\} = \{c_k + d_k\}$; in other words, sequences are added componentwise. We define scalar multiplication by $t\{c_k\} = \{tc_k\}$, where t is a complex scalar. Thus, l_1 is simply the set of sequences of complex numbers for which the corresponding infinite series are absolutely convergent (summable).

Example 5.7

Let the linear space l_2 be defined exactly as was l_1, except that we assume that $\sum_{k=1}^{\infty} |c_k|^2$ is convergent. Then l_2 is the linear space of square summable complex sequences.

Example 5.8

The collection of integrable signals over some set S forms a linear space. We use the usual notation $L_1(S)$ for these signals. Addition is then simply function addition, and the usual scalar multiplication is allowed. The square integrable signals form the linear space $L_2(S)$. If we choose S to be a subset of the plane, then $L_1(S)$ is the linear space of integrable images with domain S.

It should be clear from the examples thus far presented that the notion of a linear space is quite natural and that any seeming abstractness simply results from the articulation of the eight properties L1 through L8.

Now that the basic algebraic properties of Euclidean vectors have been isolated, we next consider the fundamental properties of the dot product:

I1. $x \cdot x \geq 0$.
I2. $x \cdot x = 0$ if and only if $x = \mathbf{0}$.
I3. $x \cdot y = \overline{y \cdot x}$, where the bar denotes complex conjugation.
I4. $x \cdot (y + z) = x \cdot y + x \cdot z$.
I5. $(tx) \cdot y = t(x \cdot y)$ for scalar t.

Just as the eight properties defining a linear space are not unique to R^n, neither are the above five unique to the dot product in R^n. (Obviously, the use of conjugation in I3 does not apply in R^n, where we simply have $x \cdot y = y \cdot x$.) We define the term *inner product* to mean any binary mapping on a linear space X which outputs a complex or real number and which satisfies I1 through I5. The Euclidean dot product is an example of an inner product. Any linear space on which an inner product is defined is called an *inner-product space*. The concept is among the most significant in both signal processing and image processing, as well as in many other areas of engineering.

Sec. 5.2 Inner-Product Spaces

Example 5.9

Consider the linear space of saturated bound matrices $\mathcal{B}_{r,t}^{m,n}$. This space is a real inner-product space with inner product defined by $f \cdot g = \text{DOT}(f, g)$. Note that DOT is always defined because f and g have the same actual domain due to the assumption of saturation.

Example 5.10

The linear space l_2 is an inner-product space under the inner product $\{c_k\} \cdot \{d_k\} = \sum_{k=1}^{\infty} c_k \bar{d}_k$, where the bar denotes complex conjugation. Notice how this is a generalization of the dot product in R^n. Indeed, if we define C^n to be the set of complex valued n-vectors, then C^n is a complex version of R^n and this dot product is an infinite extension of the dot product in C^n.

Observe that whereas l_2 is an inner-product space under the inner product $\{c_k\} \cdot \{d_k\}$, no such claim has been made for l_1. The reason is because the infinite series defining the said inner product can diverge when the sequences are in l_1. This cannot happen when they are in l_2, a matter of no little importance in the theory of trigonometric Fourier series.

Example 5.11

$L_2(S)$ is an inner-product space under the inner product $f \cdot g = \int_S f(t) \overline{g(t)} \, dt$. If we are dealing only with real-valued functions, we simply omit the conjugation. Like l_1, $L_1(S)$ is not an inner-product space under the given definition of inner product. Similar reasoning to l_1 applies. To see why, consider the signal $f(t) = t^{-1/2}$ on the interval $(0, 1]$. Then $f \in L_1(0, 1]$ since $\int_0^1 |f(t)| \, dt = 2$; however, f is not in $L_2(0, 1]$. Note that the inner product $f \cdot f$ does not exist since

$$\int_0^1 f(t)^2 \, dt = \int_0^1 \frac{1}{t} \, dt$$

does not exist. This problem cannot arise for L_2 functions since, by the Cauchy-Schwarz inequality for integrals,

$$\int_a^b |f(t)| \, |g(t)| \, dt \leq \left[\int_a^b |f(t)|^2 \, dt \right]^{1/2} \left[\int_a^b |g(t)|^2 \, dt \right]^{1/2}$$

The fact that L_1 is not an inner-product space will result in difficulties when attempting to use orthogonal projections for L_1 signals that are not also in L_2.

Once an inner product has been defined, it is possible to define a norm. The situation is perfectly analogous to Euclidean space, where $\| x \| = (x \cdot x)^{1/2}$. Using the inner product of the space, the norm is defined by the same equation. For instance, in the case of $L_2(S)$ the norm is defined by

$$\| f \| = \left[\int_a^b |f(t)|^2 \, dt \right]^{1/2}$$

where we recall that for complex conjugates, $c\bar{c} = |c|^2$. In l_2,

$$\|\{c_k\}\| = \left[\sum_{k=1}^{\infty} |c_k|^2\right]^{1/2}$$

Finally in $\mathcal{B}_{r,t}^{m;n}$, $\|f\| = [\text{DOT}(f, f)]^{1/2}$.

The next theorem lists two fundamental properties that hold in inner-product spaces.

Theorem 5.2.[2]   In an inner-product space,

(i) $|x \cdot y| \leq \|x\| \cdot \|y\|$.   (Schwarz inequality)
(ii) $\|x + y\| \leq \|x\| + \|y\|$.   (Triangle inequality)

5.3 ORTHONORMAL SYSTEMS

For the moment, consider Euclidean space R^n. The norm $\|x - y\|$ gives the distance between x and y in terms of the Pythagorean formula. Consequently, to say that a sequence of vectors (or points) $\{x_k\}$ converges to x in R^n means that $\lim_{k\to\infty} \|x_k - x\| = 0$. It is well known that this implies that $\lim_{k,l\to\infty} \|x_k - x_l\| = 0$, i.e., that given an $\epsilon > 0$ there exists an N such that $k, l \geq N$ implies $\|x_k - x_l\| < \epsilon$. A sequence that possesses this latter property is called a *Cauchy sequence*. Consequently, what we have thus far stated is that a convergent sequence in R^n must be a Cauchy sequence. What is relevant at this point, however, is the converse, namely, that in R^n every Cauchy sequence must also be convergent. This latter point is a cornerstone of modern mathematical analysis, though we shall not pursue the details here.

One can also define the notion of convergence in an inner-product space X, but in that setting it is the norm derived from the inner product in X which characterizes convergence. Formally, if $\{x_k\}$ is a sequence of elements in an inner-product space X and $x \in X$, then we say that $\{x_k\}$ converges to x and write $\lim_{k\to\infty} x_k = x$ if $\lim_{k\to\infty} \|x_k - x\| = 0$. Although at first glance the notion may seem to be merely an abstract generalization of Euclidean convergence, in fact convergence in inner-product spaces has played a central role in the development of applied mathematics.

Example 5.12

Suppose f_k is a sequence of saturated bound matrices in $\mathcal{B}_{r,t}^{m;n}$. Then it can be shown that $\lim_{k\to\infty} f_k = f$, where f is also an element of $\mathcal{B}_{r,t}^{m;n}$, if and only if $\lim_{k\to\infty} a_{pq}^{(k)} = a_{pq}$, where we have denoted the bound matrix f_k by $(a_{pq}^{(k)})_{rt}$ and the bound matrix f

---

[2] Ibid., 75.

Sec. 5.3   Orthonormal Systems

by $(a_{pq})_{rt}$. For instance, suppose $f_k \in \mathcal{B}_{3;5}^{2;2}$, where

$$f_k = \begin{pmatrix} 1/k & 2 \\ 1/k^2 & (1 + 1/k)^k \end{pmatrix}_{3,5}$$

Then the limit, in the norm generated by the inner product given in Example 5.9, is given by

$$f = \begin{pmatrix} 0 & 2 \\ 0 & e \end{pmatrix}_{3,5}$$

since $\lim_{k \to \infty} 1/k = 0$, $\lim_{k \to \infty} 2 = 2$, $\lim_{k \to \infty} 1/k^2 = 0$, and $\lim_{k \to \infty} (1 + 1/k)^k = e$.

Example 5.13

Consider the inner-product space $L_2(T)$, where T is the interval $[0, 2\pi)$. A sequence of signals $\{f_k\}$ converges in the L_2 inner-product space to $f \in L_2$ if

$$\lim_{k \to \infty} \int_0^{2\pi} |f_k(t) - f(t)|^2 \, dt = 0$$

But this is just the usual notion of L_2 convergence. In other words, L_2 convergence is a special case of inner-product-space convergence. Indeed, it is the most important case!

It was noted earlier, without going into details, that convergence in R^n is equivalent to Cauchy convergence. Does the same principle hold for convergence in inner-product spaces? Unfortunately, the answer is no: while it is true that a convergent sequence in an inner-product space must be Cauchy convergent, i.e., $\lim_{k, l \to \infty} \| x_k - x_l \| = 0$ in the inner-product norm, the converse does not hold in all inner-product spaces. Those for which it does hold are called *Hilbert spaces*. Though it is certainly not our intent to pursue the theory of Hilbert spaces, we briefly mention that the spaces $\mathcal{B}_{r;t}^{m;n}$, l_2, and L_2 are Hilbert spaces.[3] It is important to know these facts since we shall shortly present a number of theorems concerning projections which hold in Hilbert spaces. Though the notions of linear space and inner product are the generalizations required to pursue the theory of approximation by projection, many of the results depend upon the inner-product space under consideration being a Hilbert space. Since the spaces in which we are interested are all Hilbert spaces, the theoretical details concerning Cauchy convergence are of no direct interest, and therefore we have omitted any real discussion of the matter.

We have seen that two vectors are orthogonal (perpendicular) in R^n if and only if their dot product is zero. In an inner-product space (and therefore a Hilbert space) two elements x and y are said to be *orthogonal* if $x \cdot y = 0$, the inner product being the one defined on the space.

Example 5.14

Consider $L_2(T)$, where $T = [0, 2\pi)$. If k and m are different positive integers, then $\sin(kt)$ and $\sin(mt)$ are orthogonal. To see this, we apply the trigonometric product

---
[3] Ibid., 66.

formula for the sine function to obtain

$$[\sin(kt)] \bullet [\sin(mt)] = \int_0^{2\pi} \sin(kt) \sin(mt) \, dt$$

$$= \frac{1}{2} \int_0^{2\pi} \{\sin[(k-m)t] + \sin[(k+m)t]\} \, dt$$

$$= -\frac{1}{2} \left(\frac{\cos[(k-m)t]}{k-m} + \frac{\cos[(k+m)t]}{k+m} \Bigg|_0^{2\pi} \right) = 0$$

Example 5.15

Let f and g in $\mathcal{B}_{0,1}^{2,2}$ be given by

$$f = \begin{pmatrix} 2 & 4 \\ 0 & -2 \end{pmatrix}_{0,1}$$

and

$$g = \begin{pmatrix} 1 & 0 \\ 7 & 1 \end{pmatrix}_{0,1}$$

Then f and g are orthogonal since $\text{DOT}(f, g) = 0$.

Now suppose $u_1, u_2, u_3, \ldots$ are a collection of mutually orthogonal elements in an inner-product space; that is, u_k is orthogonal to u_m if $k \neq m$. Such a collection is called an *orthogonal set*. In Example 5.14, $\{\sin(kt)\}$ was shown to be an orthogonal set of elements in $L_2(T)$. If, furthermore, $\| u_k \| = 1$ for any element in the set, the set is said to be an *orthonormal system*. Looking back to Section 5.1, we see that the coordinate systems of R^n are orthonormal systems. It was with respect to those systems (in R^n) that we defined the notion of orthogonal projection and discussed the role of the Fourier coefficients.

Example 5.16

Consider $\mathcal{B}_{0,1}^{2,2}$. A most natural orthonormal system is given by

$$u_1 = \begin{pmatrix} 1 & 0 \\ 0 & 0 \end{pmatrix}, u_2 = \begin{pmatrix} 0 & 1 \\ 0 & 0 \end{pmatrix}, u_3 = \begin{pmatrix} 0 & 0 \\ 1 & 0 \end{pmatrix}, u_4 = \begin{pmatrix} 0 & 0 \\ 0 & 1 \end{pmatrix}$$

Example 5.17

Referring to Example 5.14, $\{\sin(kt)\}$ is not an orthonormal system since, by the half-angle formula,

$$\| \sin(kt) \|^2 = \int_0^{2\pi} | \sin(kt) |^2 \, dt$$

$$= \frac{1}{2} \int_0^{2\pi} [1 - \cos(2kt)] \, dt$$

$$= \frac{1}{2} \left[t - \frac{\sin(2kt)}{2k} \Bigg|_0^{2\pi} \right] = \pi$$

Sec. 5.3 Orthonormal Systems

Hence, $\| \sin(kt) \| = \sqrt{\pi}$. But it is a general property of the norm that $\| tf \| = | t | \cdot \| f \|$ for any scalar t. Consequently, $\{\sin(kt)/\sqrt{\pi}\}$ is an orthonormal system in $L_2(T)$. More importantly, the system $\{1/\sqrt{2\pi}, \sin(kt)/\sqrt{\pi}, \cos(kt)/\sqrt{\pi}\}$, for $k = 1, 2, \ldots$, called the *trigonometric system*, is an orthonormal system on $L_2(T)$.

Example 5.18

Another orthonormal system on $L_2(T)$ is the system $\{e^{jkt}/\sqrt{2\pi}\}$, for k any integer, nonnegative or negative. To see this, first suppose $k \neq m$. Then, since e^{jkt} is a complex-valued function, the complex conjugate must be employed in the inner product and we obtain

$$\left(\frac{e^{jkt}}{\sqrt{2\pi}}\right) \cdot \left(\frac{e^{jmt}}{\sqrt{2\pi}}\right) = \frac{1}{2\pi} \int_0^{2\pi} e^{jkt} \overline{e^{jmt}} \, dt$$

$$= \frac{1}{2\pi} \int_0^{2\pi} [\cos(kt) + j\sin(kt)][\cos(mt) - j\sin(mt)] \, dt.$$

But this integral just involves products from the trigonometric system, each of which has $k \neq m$, and hence the integral equals zero. If $k = m$, then, using the fact that $z\bar{z} = |z|^2$ together with $|e^{j\theta}| = 1$ gives

$$\left\| \frac{e^{jkt}}{\sqrt{2\pi}} \right\|^2 = \frac{1}{2\pi} \int_0^{2\pi} dt = 1$$

There is a close relationship between the orthonormal system in this example and the trigonometric system of the previous example. $\{e^{jkt}/\sqrt{2\pi}\}$ is known as the *complex trigonometric system*.

Orthonormal systems play the role of coordinate systems in Hilbert spaces. The exact manner in which they do so will now be presented. The discussion parallels the presentation on projections in R^n given in Section 5.1.

Suppose $U = \{u_1, u_2, \ldots, u_m\}$ is a finite orthonormal set in a Hilbert space X. Then U determines a subspace $S\langle U \rangle$ of X in the sense that $S\langle U \rangle$ consists of all linear combinations of the u_k; mathematically, $x \in S\langle U \rangle$ if and only if there exist scalars $t_1, t_2, \ldots, t_m$ such that $x = \sum_{k=1}^{m} t_k u_k$. Recalling the discussion of Section 5.1, it turns out that in R^n if x is in a subspace determined by a finite orthonormal system (coordinate system), then the coefficients with respect to that system have to be the appropriate Fourier coefficients of x. The same is true in a Hilbert space. If we define the Fourier coefficients of x by $\hat{x}(u_k) = x \cdot u_k$, then the following theorem holds.

Theorem 5.3.[4] If $U = \{u_1, u_2, \ldots, u_m\}$ is a finite orthonormal system in a Hilbert space and $x \in S\langle U \rangle$, the subspace spanned by U, then

(i) $x = \hat{x}(u_1)u_1 + \hat{x}(u_2)u_2 + \cdots + \hat{x}(u_m)u_m$
(ii) $\| x \|^2 = \sum_{k=1}^{m} | \hat{x}(u_k) |^2$

---
[4] Ibid., 82.

The situation described in Theorem 5.3 held in R^n in terms of the Euclidean inner product.

Now consider the orthonormal system $\{e^{jkt}/\sqrt{2\pi}\}$ on T for $-n \leq k \leq n$. In other words, let U be only a finite subcollection of the entire complex trigonometric system. In what follows, the full exponential trigonometric system will be denoted by $\mathcal{T}^C$ and the finite subclass U will be denoted by $\mathcal{T}_n^C$. $S\langle \mathcal{T}_n^C \rangle$ consists of all linear combinations of the form

$$p(t) = \frac{1}{\sqrt{2\pi}} \sum_{k=-n}^{n} c_k e^{jkt}, \text{ for } c_k \text{ complex}$$

Such combinations are called *trigonometric polynomials*. According to Theorem 5.3, $c_k = \hat{p}(u_k)$, the Fourier coefficient of p with respect to u_k, which is given by

$$\hat{p}(u_k) = p \cdot \frac{e^{jkt}}{\sqrt{2\pi}} = \frac{1}{\sqrt{2\pi}} \int_0^{2\pi} p(t) \, \overline{e^{jkt}} \, dt$$

$$= \frac{1}{\sqrt{2\pi}} \int_0^{2\pi} p(t) \, e^{-jkt} \, dt$$

Moreover, according to part (ii) of the theorem,

$$\| p \|^2 = \sum_{k=-n}^{n} | \hat{p}(u_k) |^2$$

If we use the real trigonometric system, a trigonometric polynomial takes the form

$$p(t) = \frac{a_0}{\sqrt{2\pi}} + \sum_{k=1}^{n} a_k \frac{\cos(kt)}{\sqrt{\pi}} + b_k \frac{\sin(kt)}{\sqrt{\pi}}$$

where

$$a_0 = \frac{1}{\sqrt{2\pi}} \int_0^{2\pi} p(t) \, dt$$

$$a_k = \hat{p}[\cos(kt)/\sqrt{\pi}] = \frac{1}{\sqrt{\pi}} \int_0^{2\pi} p(t) \cos(kt) \, dt$$

and

$$b_k = \hat{p}[\sin(kt)/\sqrt{\pi}] = \frac{1}{\sqrt{\pi}} \int_0^{2\pi} p(t) \sin(kt) \, dt$$

These trigonometric polynomials make up the subspace $S\langle U \rangle$, where $U = \{1/\sqrt{2\pi}, \cos(kt)/\sqrt{\pi}, \sin(kt)/\sqrt{\pi}\}$, for $k = 1, 2, \ldots, n$. Henceforth, this system will be denoted by $\mathcal{T}_n^R$, and the corresponding subspace will be $S\langle \mathcal{T}_n^R \rangle$. The full real trigonometric system will be denoted by $\mathcal{T}^R$.

Example 5.19

Suppose $p(t)$ in $S\langle \mathcal{T}_n^C \rangle$ is defined by

$$p(t) = \frac{1}{\sqrt{2\pi}} \sum_{\substack{k=-n \\ k \neq 0}}^{n} \frac{1}{k} e^{jkt}$$

Then part (i) of Theorem 5.3 implies that

$$\frac{1}{k} = \hat{p}(u_k) = \frac{1}{\sqrt{2\pi}} \int_0^{2\pi} p(t) \, e^{-jkt} \, dt$$

Let us verify this last equality. The integral on the right-hand side is equal to

$$\frac{1}{\sqrt{2\pi}} \int_0^{2\pi} \left[\frac{1}{\sqrt{2\pi}} \sum_{\substack{r=-n \\ r \neq 0}}^{n} \frac{1}{r} e^{jrt} \right] e^{-jkt} \, dt$$

$$= \sum_{\substack{r=-n \\ r \neq 0}}^{n} \frac{1}{r} \left[\frac{1}{2\pi} \int_0^{2\pi} e^{jrt} e^{-jkt} \, dt \right]$$

But the $e^{jrt}/\sqrt{2\pi}$ form an orthonormal system over T, and hence the last integral equals unity if $r = k$ and zero otherwise. Hence, the entire expression reduces to $1/k$, as it must.

5.4 PROJECTIONS IN HILBERT SPACE

Given a finite orthonormal system $U = \{u_1, u_2, \ldots, u_m\}$ in a Hilbert space X and an arbitrary element x in X, the question naturally arises as to the type of approximation one can get by projecting x onto the subspace generated by U, $S\langle U \rangle$. We have seen that in R^n the projection of x onto $S\langle U \rangle$ gives the best approximation, where "best" is measured by proximity in the norm. An exact analogue to Theorem 5.1 holds in a Hilbert space. We denote the projection of x onto $S\langle U \rangle$, $\sum_{k=1}^{m} \hat{x}(u_k) u_k$, by x_U.

Theorem 5.4.[5] Suppose $U = \{u_1, u_2, \ldots, u_m\}$ is an orthonormal system in a Hilbert space X, and let $x \in X$. If z is an element in the subspace $S\langle U \rangle$, then

(i) $\| x - x_U \| \leq \| x - z \|$.
(ii) $\| x - x_U \|^2 = \| x \|^2 - \sum_{k=1}^{m} | \hat{x}(u_k) |^2$.

The consequences of Theorem 5.4 are enormous. Suppose $f(t)$ is an L_2 signal defined on $[0, 2\pi)$ and we wish to approximate this signal by a collection of sine and cosine waves from the trigonometric orthonormal system $\mathcal{T}^R$. The theorem states that the best way to do this is to use the Fourier coefficients of $f(t)$ as the

[5] Ibid., 84.

coefficients in a trigonometric polynomial. Of course, according to Theorem 5.3, if the signal happens already to be a collection of sine and cosine waves, then it will be identical to the resulting trigonometric polynomial. But this is not the interesting case. The problem is to fit a collection of waves to a given signal $f(t)$. It is Theorem 5.4 that tells us how to accomplish the best fit. More than that, part (ii) quantifies the error as measured by the L_2 norm. Needless to say, similar comments apply to the utilization of the orthonormal system $\mathcal{T}^C$.

Example 5.20

Consider the signal

$$f(t) = \frac{\pi - t}{2} I_{[0,2\pi)}(t)$$

which is illustrated in Figure 5.3. We wish to find the best L_2 approximation from the subspace $S\langle \mathcal{T}_n^R \rangle$, that is, the best approximation using sine and cosine waves in the orthonormal system $\mathcal{T}_n^R$. Theorem 5.4 gives the solution: we must use the projection of $f(t)$ onto the given subspace, and this projection is given by the trigonometric polynomial $p_f(t)$ whose coefficients are the appropriate Fourier coefficients of the signal. In the case of the signal $f(t)$ of this example, this means that

$$a_0 = \frac{1}{\sqrt{2\pi}} \int_0^{2\pi} \frac{\pi - t}{2} \, dt = 0$$

$$a_k = \frac{1}{\sqrt{\pi}} \int_0^{2\pi} \frac{\pi - t}{2} \cos(kt) \, dt = 0$$

$$b_k = \frac{1}{\sqrt{\pi}} \int_0^{2\pi} \frac{\pi - t}{2} \sin(kt) \, dt = \frac{\sqrt{\pi}}{k}$$

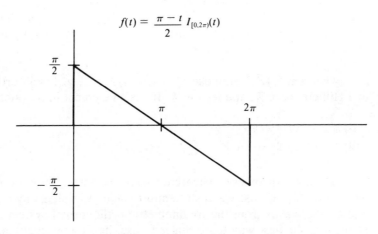

Figure 5.3 Triangular Pulse Waveform

Sec. 5.4   Projections in Hilbert Space

Consequently, the projection of $f(t)$ onto $S\langle \mathcal{T}_n^R \rangle$ is given by

$$p_f(t) = \sum_{k=1}^{n} \frac{\sqrt{\pi}}{k} \frac{\sin(kt)}{\sqrt{\pi}} = \sum_{k=1}^{n} \frac{\sin(kt)}{k}$$

But more is known from Theorem 5.4(ii); indeed, the square of the L_2 error is given by

$$\|f - p_f\|^2 = \|f\|^2 - \left[a_0^2 + \sum_{k=1}^{n} (a_k^2 + b_k^2) \right]$$

$$= \int_0^{2\pi} |f(t)|^2 \, dt - \sum_{k=1}^{n} \pi k^{-2}$$

$$= \int_0^{2\pi} \left[\frac{\pi - t}{2} \right]^2 dt - \pi \sum_{k=1}^{n} k^{-2}$$

$$= \frac{\pi^3}{6} - \pi \sum_{k=1}^{n} k^{-2}$$

An immediate corollary of Theorem 5.4(ii) is the Bessel inequality. It follows at once from the theorem by simply letting $m \to \infty$ and noting that $\|x - x_U\| \geq 0$.

Theorem 5.5[6] **(Bessel Inequality).**   Suppose $\{u_1, u_2, u_3, \ldots\}$ is an orthonormal system in a Hilbert space X and $x \in X$. Then

$$\sum_{k=1}^{\infty} |\hat{x}(u_k)|^2 \leq \|x\|^2$$

An immediate consequence of the Bessel inequality is that the sequence of Fourier coefficients is in the Hilbert space l_2.

Now suppose $U(m)$ denotes the subcollection consisting of u_1 through u_m from an infinite orthonormal system $\{u_1, u_2, \ldots\}$. Moreover, suppose strict inequality holds in the Bessel inequality. Then

$$\|x - x_{U(m)}\|^2 = \|x\|^2 - \sum_{k=1}^{m} |\hat{x}(u_k)|^2$$

$$\geq \|x\|^2 - \sum_{k=1}^{\infty} |\hat{x}(u_k)|^2 = \lambda > 0$$

In other words, the limit of the projections $x_{U(m)}$ as $m \to \infty$ does not equal x since x and $x_{U(m)}$ always differ in norm by more than $\sqrt{\lambda}$. The projections give the best subspace approximations, but in the limit they do not converge to the original element. In the case of the trigonometric system $\mathcal{T}^R$, this would mean that letting

[6] Ibid., 84.

$m \to \infty$ in the trigonometric series would not yield the signal $f(t)$ as the L_2 limit. Indeed, we would like to have, in the L_2 sense,

$$f = \lim_{m \to \infty} \sum_{k=1}^{m} \hat{f}(u_k) u_k = \sum_{k=1}^{\infty} \hat{f}(u_k) u_k$$

but this means precisely that equality is attained in the Bessel inequality. Fortunately for signal processing and image processing, as well as engineering in general, the Bessel inequality is actually an equality for the trigonometric system, in real and complex forms, as well as for other important orthonormal systems.

An orthonormal system is said to be *complete* if it is not a proper subsystem of any other orthonormal system. The trigonometric system is complete. The following important theorem sets out two equivalency conditions for a complete orthonormal system.

Theorem 5.6.[7] Suppose $\{u_1, u_2, \ldots\}$ is an orthonormal system in a Hilbert space X. Then the following conditions are equivalent:

(a) $\{u_1, u_2, \ldots\}$ is complete
(b) For every x in X, $\| x \|^2 = \sum_{k=1}^{\infty} | \hat{x}(u_k) |^2$
(c) For every x and y in X, $x \cdot y = \sum_{k=1}^{\infty} \hat{x}(u_k) \overline{\hat{y}(u_k)}$

Part (b) of Theorem 5.6 states that equality holds in the Bessel inequality, and therefore $\lim_{m \to \infty} \| x - x_{U(m)} \| = 0$. Hence, for any complete orthonormal system in L_2, including the trigonometric system on T, we can write $f = \sum_{k=1}^{\infty} \hat{f}(u_k) u_k$ (equality holding in L_2). Part (c), known as *Parseval's identity*, states that the Fourier coefficients actually determine the inner product. In the next section, Theorem 5.6, as well as other theorems, will be restated in terms of the trigonometric system on T.

Example 5.21

Let $f(t)$ be the signal given in Example 5.20. Since the trigonometric system is complete, we have

$$f(t) = \sum_{k=1}^{\infty} \frac{\sin(kt)}{k}$$

in the $L_2(T)$ norm. Be careful, however; nothing is claimed at this point concerning pointwise convergence. That is, Theorem 5.6 does not assure us that

$$f(t) = \lim_{n \to \infty} \sum_{k=1}^{n} \frac{\sin(kt)}{k}$$

[7] Ibid., 85.

Sec. 5.5　Trigonometric Fourier Series of Square Integrable Signals

for each t, which is the usual pointwise interpretation of infinite series convergence in terms of partial sums. Rather, it assures us that there is L_2 convergence, i.e.,

$$\lim_{n \to \infty} \int_0^{2\pi} \left[f(t) - \sum_{k=1}^{n} \frac{\sin(kt)}{k} \right]^2 dt = 0$$

Earlier, mention was made of the fact that an immediate corollary of the Bessel inequality was that the sequence $\{\hat{x}(u_k)\}$ was in l_2. In fact, much more can be said. Since the sequence of Fourier coefficients is in l_2, the mapping $\Phi: x \to \{\hat{x}(u_k)\}$ is a mapping from the Hilbert space X into l_2, i.e., $\Phi: X \to l_2$. It turns out that this mapping is onto, as is stated in the next theorem.

Theorem 5.7[8] **(Riesz-Fischer).** If $\{u_1, u_2, \ldots\}$ is an orthonormal system in a Hilbert space X and $\{c_k\}$ is a sequence in l_2, then there exists some element x in X such that $\Phi(x) = \{c_k\}$; in other words, $\hat{x}(u_k) = c_k$ for all k.

5.5 TRIGONOMETRIC FOURIER SERIES OF SQUARE INTEGRABLE SIGNALS

The major theorems for orthonormal projections in a Hilbert space will now be applied to the trigonometric systems $\mathcal{T}^C$ and $\mathcal{T}^R$, both of which are complete.[9] Although we have mentioned that $\mathcal{T}^C$ is simply the complex form of $\mathcal{T}^R$, we have yet to demonstrate the relationship. In fact, the equations

$$c_k = \begin{cases} a_0 & \text{for } k = 0 \\ \dfrac{a_k - jb_k}{\sqrt{2}} & \text{for } k > 0 \\ \dfrac{a_{-k} + jb_{-k}}{\sqrt{2}} & \text{for } k < 0 \end{cases}$$

hold, where a_k and b_k represent the Fourier coefficients with respect to $\cos(kt)/\sqrt{\pi}$ and $\sin(kt)/\sqrt{\pi}$, respectively, and c_k represents the Fourier coefficient with respect to $e^{jkt}/\sqrt{2\pi}$. For instance, for $k > 0$,

$$\frac{a_k - jb_k}{\sqrt{2}} = \frac{1}{\sqrt{2}} \frac{1}{\sqrt{\pi}} \int_0^{2\pi} f(t) [\cos(kt) - j\sin(kt)] \, dt$$

$$= \frac{1}{\sqrt{2\pi}} \int_0^{2\pi} f(t) [\cos(-kt) + j\sin(-kt)] \, dt = c_k$$

---

[8] Ibid., 85.
[9] Ibid., 89.

where the Euler formula, $e^{j\theta} = \cos\theta + j\sin\theta$, has been applied. As a consequence of the relationship between a_k, b_k, and c_k, the results for trigonometric series need be stated relative to only one of the systems; nevertheless, for the sake of completeness, we shall often include both forms.

We first state Theorem 5.4 in the form appropriate to trigonometric Fourier series. This new form employs the L_2 norm, and we state both parts of it using the square of the norm.

Theorem 5.8. If $f \in L_2(T)$, then

(i) For any trigonometric polynomial $p(t) = \frac{1}{\sqrt{2\pi}} \sum_{k=-n}^{n} \lambda_k e^{jkt}$,

$$\int_0^{2\pi} \left| f(t) - \frac{1}{\sqrt{2\pi}} \sum_{k=-n}^{n} c_k e^{jkt} \right|^2 dt \leq \int_0^{2\pi} |f(t) - p(t)|^2 dt$$

(ii)

$$\int_0^{2\pi} \left| f(t) - \frac{1}{\sqrt{2\pi}} \sum_{k=-n}^{n} c_k e^{jkt} \right|^2 dt = \int_0^{2\pi} |f(t)|^2 dt - \sum_{k=-n}^{n} |c_k|^2$$

In terms of the real trigonometric system, Theorem 5.8(i) takes the form

(i′)

$$\int_0^{2\pi} \left| f(t) - \left(\frac{a_0}{\sqrt{2\pi}} + \frac{1}{\sqrt{\pi}} \sum_{k=1}^{n} a_k \cos(kt) + b_k \sin(kt) \right) \right|^2 dt$$

$$\leq \int_0^{2\pi} \left| f(t) - \left(\frac{\lambda_0}{\sqrt{2\pi}} + \frac{1}{\sqrt{\pi}} \sum_{k=1}^{n} \lambda_k \cos(kt) + \mu_k \sin(kt) \right) \right|^2 dt$$

for any set of complex constants $\lambda_0, \lambda_1, \ldots, \lambda_n, \mu_1, \ldots, \mu_n$. It should be evident at once that it is far better to remember the theorems in the Hilbert space form: not only are they then more general, but, put simply, they are easier to remember! Example 5.20 provides an illustration of the result (i′). Insofar as the complex case is concerned, the next example is typical.

Example 5.22

Consider the signal $f(t) = I_{[0,\pi)}(t)$, which is equal to unity on the first half of T and zero on the second half. For this function,

$$c_0 = \frac{1}{\sqrt{2\pi}} \int_0^{2\pi} f(t)\, dt = \frac{1}{\sqrt{2\pi}} \int_0^{\pi} dt = \frac{\pi}{\sqrt{2\pi}}$$

Sec. 5.5 Trigonometric Fourier Series of Square Integrable Signals   259

while for $k \neq 0$,

$$c_k = \frac{1}{\sqrt{2\pi}} \int_0^{2\pi} f(t) e^{-jkt} dt = \frac{1}{\sqrt{2\pi}} \int_0^{\pi} e^{-jkt} dt$$

$$= \frac{-1}{\sqrt{2\pi} jk} e^{-jkt} \bigg|_0^{\pi} = \frac{j}{\sqrt{2\pi} k} [\cos(\pi k) - j \sin(\pi k) - 1]$$

$$= \frac{j}{\sqrt{2\pi} k} [(-1)^k - 1]$$

Hence, the projection of $f(t)$ onto $S\langle \mathcal{T}_n^C \rangle$ is given by

$$p_f(t) = \frac{1}{\sqrt{2\pi}} \left(\frac{\pi}{\sqrt{2\pi}} + \sum_{\substack{k=-n \\ k \neq 0}}^{n} \frac{j[(-1)^k - 1]}{\sqrt{2\pi} k} e^{jkt} \right)$$

Thus, according to Theorem 5.8, $p_f(t)$ is the best L_2 approximation to $f(t)$ in the subspace; i.e., it gives the best approximation by a trigonometric polynomial of exponentials $e^{jkt}/\sqrt{2\pi}$. Notice that except for $k = 0$, all even coefficients are zero. Applying part (ii) of the theorem, and noting that the square of the L_2 norm of $f(t)$ is π, we have

$$\int_0^{2\pi} |f(t) - p_f(t)|^2 dt = \pi - \frac{\pi}{2} - \sum_{\substack{k=-n \\ k \text{ odd}}}^{n} \frac{2k^{-2}}{\pi} = \frac{\pi}{2} - \frac{2}{\pi} \sum_{\substack{k=-n \\ k \text{ odd}}}^{n} k^{-2}$$

Since the trigonometric system is complete, Theorem 5.6 applies, and we consequently have the next theorem.

Theorem 5.9 (Parseval). For f in $L_2(T)$,

(i)
$$\int_0^{2\pi} |f(t)|^2 dt = \sum_{k=-\infty}^{\infty} |c_k|^2$$

(ii) If g is also in $L_2(T)$ and d_k is the kth Fourier coefficient relative to the complex trigonometric system, then

$$\int_0^{2\pi} f(t) \overline{g(t)} \, dt = \sum_{k=-\infty}^{\infty} c_k \overline{d}_k$$

As discussed earlier in regard to Theorem 5.6, an immediate corollary of Theorem 5.9 is that a square integrable signal on T must be the limit of its projections, where the limit is taken in the norm of the Hilbert space. Here that norm is the L_2 norm; consequently, for f in $L_2(T)$, we have

$$f(t) \stackrel{2}{=} \frac{1}{\sqrt{2\pi}} \sum_{k=-\infty}^{\infty} c_k e^{jkt}$$

where the symbol "$\stackrel{2}{=}$" is used to denote equality relative to the L_2 norm. In precise terms, this means that

$$\lim_{n \to \infty} \int_0^{2\pi} \left| f(t) - \frac{1}{\sqrt{2\pi}} \sum_{k=-n}^{n} c_k e^{jkt} \right|^2 dt = 0$$

This last point was emphasized in Example 5.21. As noted there, no claim is made with regard to pointwise convergence. Stated in terms of the real trigonometric system, we have

$$\lim_{n \to \infty} \int_0^{2\pi} \left| f(t) - \left(\frac{a_0}{\sqrt{2\pi}} + \frac{1}{\sqrt{\pi}} \sum_{k=1}^{n} a_k \cos(kt) + b_k \sin(kt) \right) \right|^2 dt = 0$$

or

$$f(t) \stackrel{2}{=} \frac{a_0}{\sqrt{2\pi}} + \frac{1}{\sqrt{\pi}} \sum_{k=1}^{\infty} a_k \cos(kt) + b_k \sin(kt)$$

Moreover, for the real trigonometric system, Theorem 5.9 takes the form

(i')

$$\int_0^{2\pi} |f(t)|^2 dt = |a_0|^2 + \sum_{k=1}^{\infty} (|a_k|^2 + |b_k|^2)$$

(ii')

$$\int_0^{2\pi} f(t) \overline{g(t)} \, dt = a_0 \overline{\alpha}_0 + \sum_{k=1}^{\infty} (a_k \overline{\alpha}_k + b_k \overline{\beta}_k)$$

where α_k and β_k represent the appropriate Fourier coefficients for $g(t)$.

Example 5.23

Referring to Example 5.20, part (i') yields

$$\frac{\pi^2}{6} = \sum_{k=1}^{\infty} k^{-2}$$

5.6 TRIGONOMETRIC FOURIER SERIES OF INTEGRABLE SIGNALS

The results of the previous section are dependent upon the use of signals in $L_2(T)$. Yet the Fourier coefficients are defined, in terms of integration, for all L_1 functions on T. In other words, the mapping $f \to \{c_k\}$, where

$$c_k = \frac{1}{\sqrt{2\pi}} \int_0^{2\pi} f(t) e^{-jkt} \, dt$$

Sec. 5.6 Trigonometric Fourier Series of Integrable Signals 261

is defined for all f in $L_1(T)$, and $L_1(T)$ properly contains $L_2(T)$. What, then, can be said concerning "Fourier series" of $L_1(T)$ signals?

In the first place, for $L_1(T)$ in general, one cannot speak of convergence in the L_2 norm. And second, anything that is said regarding the more general $L_1(T)$ case applies also to $L_2(T)$, but the converse is certainly not true. For instance, for signals known only to be integrable, the Hilbert space theory thus far introduced is inapplicable. However, the results of that same intuitive projection theory can certainly serve as a model for an extension to $L_1(T)$.

Accordingly, we define a mapping $\mathscr{F}_d$ on $L_1(T)$ by $\mathscr{F}_d(f) = \{c_k\}$. We then formally define the *Fourier series* of f to be $(1/\sqrt{2\pi}) \sum_{k=-\infty}^{\infty} c_k e^{jkt}$ and write

$$f \sim \frac{1}{\sqrt{2\pi}} \sum_{k=-\infty}^{\infty} c_k e^{jkt}$$

where the symbol "$\sim$" presupposes no direct relationship between an L_1 signal and its Fourier series. In fact, analogously to the inverse Fourier transform $\mathscr{F}^{-1}$, the series represents the inverse transform of $\mathscr{F}_d$ in a formal sense. Thus, just as Theorem 4.7 gave inversion conditions for $\mathscr{F}^{-1}(F)$ to equal f in $L_1(-\infty, \infty)$, inversion conditions are needed to see when

$$\mathscr{F}_d^{-1}[\{c_k\}] = \frac{1}{\sqrt{2\pi}} \sum_{k=-\infty}^{\infty} c_k e^{jkt}$$

equals f in $L_1(T)$, and in what sense the two are equal.

Before proceeding, it is worthwhile to point out that the relationship between the Fourier transform mapping $\mathscr{F}$ and the Fourier series mapping $\mathscr{F}_d$ is more than mere algebraic formality. Suppose $f(t)$ is an L_1 signal on T. Define f on the entire real line by periodic extension, i.e., $f(t) = f(t + 2\pi m)$, $m = \pm 1, 2, \ldots$. For instance, when extended periodically, the signal $f(t)$ of Example 5.20 becomes the "sawtooth" signal shown in Figure 5.4, whereas the signal defined in Example 5.22 becomes the square wave depicted in Figure 5.5. Since every trigonometric

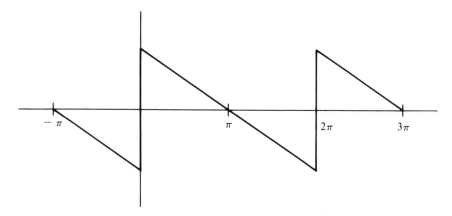

Figure 5.4 Sawtooth Pulse Train

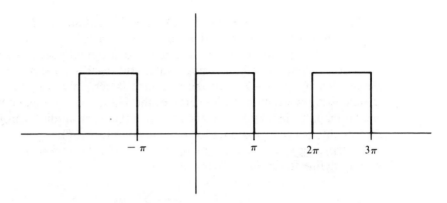

Figure 5.5 Square Waveform

polynomial, and hence every trigonometric Fourier series, consists of waves of periodicity 2π, every such series is automatically defined, wherever it converges, according to that periodicity. Consequently, although the previous mapping $\mathscr{F}_d$ is defined only in terms of integrable functions on $[0, 2\pi)$, it is actually a mapping on functions of periodicity 2π. Moreover, if the mapping can be inverted pointwise—that is, in the usual partial sums sense—then the inversion holds periodically throughout the real axis. This is precisely why trigonometic Fourier series have played such a central role in harmonic analysis and the study of wave propagation.

Suppose now that $f(t)$ is a signal of periodicity 2π on $(-\infty, \infty)$. Then, according to the preceding conventions,

$$[\mathscr{F}_d(f)](k) = \frac{1}{\sqrt{2\pi}} \int_0^{2\pi} f(t) e^{-jkt} \, dt$$

$$= \frac{1}{\sqrt{2\pi}} \int_{-\infty}^{\infty} f(t) I_{[0,2\pi)}(t) e^{-jkt} \, dt$$

$$= \frac{1}{\sqrt{2\pi}} [\mathscr{F}(f I_{[0,2\pi)})](k)$$

In other words, by restricting the periodic wave to a single period in the interval in which it is defined, $c_k = (1/\sqrt{2\pi}) F(k)$, where $f I_{[0,2\pi)} \leftrightarrow F$; that is, the spectral content of a periodic signal is contained within its Fourier coefficients. This latter point motivates the definition of the *discrete spectrum* by the formula $F(k) = \sqrt{2\pi} \, c_k$, which is a complex-valued function on the integers that is precisely equal to the Fourier transform of the restricted periodic signal at the integers. Moreover, in a manner corresponding to the Fourier transform, $|F(k)|$ is called the *discrete amplitude spectrum*.

Example 5.24

Let $f(t)$ be the periodic extension of the signal $[(\pi - t)/2]I_{[0,2\pi)}(t)$ which was given in Example 5.20. The signal is shown in Figure 5.4. Then

$$F(k) = \sqrt{2\pi}\, c_k = \int_0^{2\pi} \frac{\pi - t}{2} e^{-jkt}\, dt$$

$$= \begin{cases} -\dfrac{j\pi}{k} & \text{for } k \neq 0 \\ 0 & \text{for } k = 0 \end{cases}$$

Consequently,

$$f \sim -\frac{j}{2} \sum_{\substack{k=-\infty \\ k \neq 0}}^{\infty} \frac{e^{jkt}}{k}$$

and the amplitude spectrum is given by $|F(k)| = \pi/|k|$ for $k \neq 0$ and $|F(0)| = 0$. (See Figure 5.6.)

One immediate consequence of the relationship between c_k and the Fourier transform $F(\omega)$ is that the Riemann-Lebesgue lemma, Theorem 4.1(e), holds for the Fourier coefficients of an $L_1(T)$ extended signal; that is, $\lim_{k \to \infty} c_k = 0$. But, of course, this implies that the same holds for the real trigonometric system coefficients, i.e., $a_k, b_k \to 0$ as $k \to \infty$.

As mentioned earlier, conditions under which an L_1 signal of periodicity 2π equals (in some pointwise sense) its Fourier series are actually inversion conditions for the operator $\mathcal{F}_d$ on $L_1(T)$. We now state three well-known results concerning pointwise inversion. Remember that when f equals its Fourier series at some point x, we write, with the usual limit of the partial sums interpretation,

$$f(x) = \frac{1}{\sqrt{2\pi}} \sum_{k=-\infty}^{\infty} c_k e^{jkx}$$

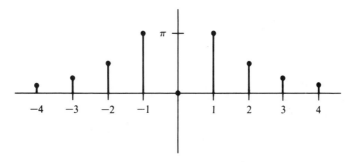

Figure 5.6 Amplitude Spectrum

Moreover, note that the theorems apply to the extensions of signals of periodicity 2π. Hence, it makes sense to talk about derivatives at zero and 2π.

Theorem 5.10[10] **(Dini's Test).** Suppose $f(t) \in L_1(T)$ and f satisfies a Hölder condition at x in T—i.e., there exist A and α, with $A > 0$ and $0 < \alpha \leq 1$, such that

$$|f(t) - f(x)| \leq A |t - x|^\alpha$$

whenever t is sufficiently close to x. Then the Fourier series of f converges to $f(x)$ at the point x. As a special case, if f is differentiable at x, then f equals its Fourier series at x.

Example 5.25

Since the signal $f(t)$ in Example 5.24 is differentiable at every point in the open inverval $(0, 2\pi)$, it follows that for any x such that $0 < x < 2\pi$,

$$f(x) = \frac{-j}{2} \sum_{\substack{k=-\infty \\ k \neq 0}}^{\infty} \frac{e^{jkx}}{k}$$

where the equality is the usual pointwise infinite series convergence. Using the real trigonometric system and referring to Example 5.20, where the Fourier coefficients were found, we see that for $0 < x < 2\pi$,

$$f(x) = \sum_{k=1}^{\infty} \frac{\sin(kx)}{k}$$

in the pointwise sense.

Theorem 5.11[11] **(Jordan's Test).** Suppose $f \in L_1(T)$ and f is of bounded variation in a neighborhood of x. Then the Fourier series of f converges to the average of the left- and right-hand limits at x, i.e.,

$$\frac{1}{2}[f(x^-) + f(x^+)] = \frac{1}{\sqrt{2\pi}} \sum_{k=-\infty}^{\infty} c_k e^{jkx}$$

In particular, if f also happens to be continuous at x, then the Fourier series converges to $f(x)$ itself.

Example 5.26

Consider the square-wave signal of Example 5.22. Its periodic extension is continuous except at integer multiples of π. Nevertheless, it is of bounded variation throughout. Consequently, its Fourier series converges to it at every point except the integer multiples of π. At those points, the series coverges to $\frac{1}{2}$, the average of

---

[10] E. C. Titchmarsh, *The Theory of Functions, 2nd Edition* (London: Clarendon Press, 1964), 406.

[11] Ibid., 406.

Sec. 5.6   Trigonometric Fourier Series of Integrable Signals

the left- and right-hand limits. In fact, Dini's test applies everywhere but at those points, where we have used Jordan's test instead.

Theorem 5.12.[12]   If $f(t)$ is periodic with period 2π, continuous, and of bounded variation on $[0, 2\pi]$, then its Fourier series converges uniformly to it.

Theorem 5.12 is rather strict, requiring that the 2π-periodic extension be continuous at the endpoints of T.

We close this section with an important point about the interval $T = [0, 2\pi)$. We have chosen to stay within this interval in order not to obscure the presentation with extraneous calculations. The fact of the matter is that one could proceed with both the L_1 and the L_2 developments on any interval $[a, b)$ with only a rescaling of the arguments in the Fourier coefficients occurring. Indeed, if we were to consider periodic functions with period λ, the entire theory would go through unaltered, except that we would be considering $L_1[0, \lambda)$ and $L_2[0, \lambda)$, and the Fourier series would have the form

$$\frac{1}{\sqrt{2\pi}} \sum_{k=-\infty}^{\infty} c_k e^{jk2\pi t/\lambda}$$

and the coefficients c_k would be given by

$$c_k = \frac{1}{\sqrt{\lambda}} \int_0^\lambda f(t)\, e^{-jk2\pi t/\lambda}\, dt$$

Using the sine and cosine expansion, we would then have the series

$$\frac{a_0}{\sqrt{\lambda}} + \sqrt{\frac{2}{\lambda}} \sum_{k=1}^{\infty} a_k \cos(2k\pi t/\lambda) + b_k \sin(2k\pi t/\lambda)$$

where

$$a_0 = \frac{1}{\sqrt{\lambda}} \int_0^\lambda f(t)\, dt$$

and for $k \geq 1$,

$$a_k = \sqrt{\frac{2}{\lambda}} \int_0^\lambda f(t) \cos(2k\pi t/\lambda)\, dt$$

and

$$b_k = \sqrt{\frac{2}{\lambda}} \int_0^\lambda f(t) \sin(2k\pi t/\lambda)\, dt$$

---

[12] Ibid., 410.

If we keep in mind that these new coefficients are still Fourier coefficients, in the inner-product sense, then it is a trivial exercise to apply the results of the chapter to trigonometric Fourier series on an interval $[0, \lambda)$.

5.7 WALSH EXPANSIONS

The *Rademacher functions* are square-type waves defined recursively on the interval $[0, 1)$ (and periodically extended to all of R) by

$$r_0(t) = 1$$

$$r_1(t) = \begin{cases} 1 & \text{for } 0 \leq t < \tfrac{1}{2} \\ -1 & \text{for } \tfrac{1}{2} \leq t < 1 \end{cases}$$

$$\vdots$$

$$r_{n+1}(t) = r_1(2^n t) \text{ for } n = 1, 2, \ldots$$

Each function $r_n(t)$ thus alternates between 1 and -1 on equal lengths of 2^{-n}. (See Figure 5.7.)

Although the Rademacher functions form an orthonormal system in $L_2[0, 1)$, the system is not complete. A larger system containing the Rademacher functions is needed. Such a system is provided by the Walsh functions.

The *Walsh functions* are given by the set of all finite products of the Rademacher functions. The 0th Walsh function is defined by $w_0(t) = 1$, and for $n \geq 1$,

$$w_n(t) = r_{n_1+1}(t) \times r_{n_2+1}(t) \times \cdots \times r_{n_p+1}(t)$$

where

$$n = 2^{n_1} + 2^{n_2} + \cdots + 2^{n_p}$$

and

$$n_1 < n_2 < \cdots < n_p$$

Though this notation is cumbersome, its only purpose is to specify some ordering to the manner in which all products of the Rademacher functions are to be assembled. The first eight Walsh functions are shown in Figure 5.8, together with a vector notation that is useful for labeling purposes. Rigorously, if $\mathbf{w}_k$ denotes the kth *Walsh column vector*, then when we are utilizing the first 2^n Walsh functions, the kth Walsh function is given by the product

$$w_k(t) = \mathbf{w}_k' \begin{pmatrix} I_{[0, 2^{-n})}(t) \\ I_{[2^{-n}, 2 \times 2^{-n})}(t) \\ \vdots \\ I_{[(2^n - 1) \times 2^{-n}, 1)}(t) \end{pmatrix}$$

Sec. 5.7 Walsh Expansions

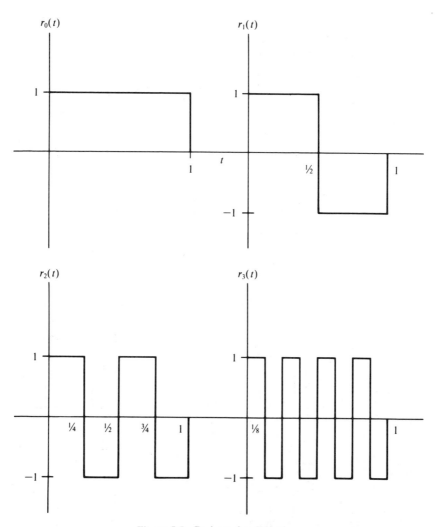

Figure 5.7 Rademacher Functions

where the prime on w'_k denotes the transpose of $\mathbf{w}_k$, $\mathbf{w}'_k$ being the kth *Walsh row vector*. The Walsh functions form a complete orthonormal system in $L_2[0, 1)$.[13]

Let $f(t)$ be an L_2 signal on $[0, 1)$, and let α_k denote the Fourier coefficient $\hat{f}(w_k)$ with respect to the kth Walsh function; that is,

$$\alpha_k = \int_0^1 f(t) w_k(t) \, dt$$

---

[13] Casper Goffman and George Pedrick, *First Course in Functional Analysis* (Englewood Cliffs, N.J.: Prentice-Hall, Inc., 1965), 200.

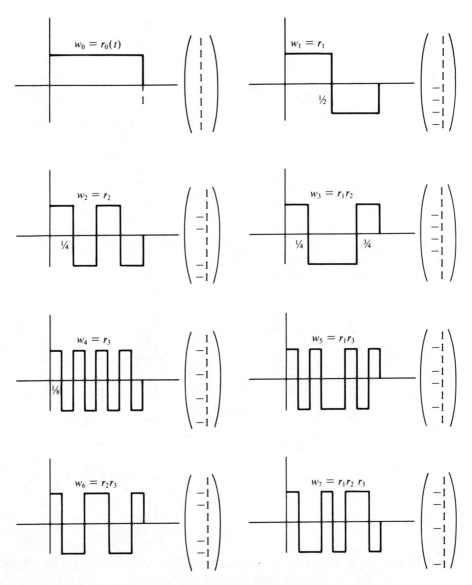

Figure 5.8 Walsh Functions

Since the Walsh functions form a complete orthonormal system, the partial sums of the series

$$\sum_{k=0}^{\infty} \alpha_k w_k(t)$$

converge to $f(t)$ in the L_2 norm. Moreover, if the signal $f(t)$ is of bounded varia-

Sec. 5.7 Walsh Expansions

tion, an assumption reasonable in practice, the series converges to $f(t)$ at all points of continuity.[14] Since $f(t)$ is of bounded variation, it is continuous everywhere except at a countable collection of points in [0, 1). Hence, the representation

$$f(t) = \sum_{k=0}^{\infty} \alpha_k w_k(t)$$

holds pointwise at all but a countable number of points. The constants α_k in the Walsh expansion are called *sequency* constants.

Suppose we wish to approximate $f(t)$ in $L_2[0, 1)$ by

$$f_{2^n}(t) = \sum_{k=0}^{2^n-1} \alpha_k w_k(t)$$

In order to facilitate the computation of the sequency constants (Fourier coefficients), we can divide [0, 1] into 2^n equal parts and compute the integrals of f over the subintervals. We obtain, for $i = 0, 1, 2, \ldots, 2^n - 1$,

$$A_i = \int_{i/2^n}^{(i+1)/2^n} f(t)\, dt$$

We use the A_i as components in the area vector

$$\mathbf{A}_{2^n} = \begin{pmatrix} A_0 \\ A_1 \\ \vdots \\ A_{2^n-1} \end{pmatrix}$$

The sequency terms α_k are linear combinations of the A_i and are found by

$$\boldsymbol{\alpha}_{2^n} = H_{2^n} \mathbf{A}_{2^n}$$

where

$$\boldsymbol{\alpha}_{2^n} = \begin{pmatrix} \alpha_0 \\ \alpha_1 \\ \vdots \\ \alpha_{2^n-1} \end{pmatrix}$$

and H_{2^n} is the Hadamard martrix, whose rows consist of the Walsh vectors $\mathbf{w}'_0$, $\mathbf{w}'_1, \ldots, \mathbf{w}'_{2^n-1}$.

Example 5.27

Let $f(t) = t$ on [0, 1). The Walsh expansion for $n = 3$ is of the form

$$f_8(t) = \alpha_0 w_0(t) + \cdots + \alpha_7 w_7(t)$$

---

[14] N. J. Fine, "On the Walsh Functions," *American Math Society Trans.*, 65, (1949):372.

The area vector, which is readily obtained, is given by

$$\mathbf{A}_8 = \frac{1}{128}\begin{pmatrix} 1 \\ 3 \\ 5 \\ 7 \\ 9 \\ 11 \\ 13 \\ 15 \end{pmatrix}$$

The Hadamard matrix is

$$H_8 = \begin{pmatrix} 1 & 1 & 1 & 1 & 1 & 1 & 1 & 1 \\ 1 & 1 & 1 & 1 & -1 & -1 & -1 & -1 \\ 1 & 1 & -1 & -1 & 1 & 1 & -1 & -1 \\ 1 & 1 & -1 & -1 & -1 & -1 & 1 & 1 \\ 1 & -1 & 1 & -1 & 1 & -1 & 1 & -1 \\ 1 & -1 & 1 & -1 & -1 & 1 & -1 & 1 \\ 1 & -1 & -1 & 1 & 1 & -1 & -1 & 1 \\ 1 & -1 & -1 & 1 & -1 & 1 & 1 & -1 \end{pmatrix}$$

Notice that H_8 is symmetric. Finally, the sequency vector is

$$\boldsymbol{\alpha}_8 = \frac{1}{128}\begin{pmatrix} 64 \\ -32 \\ -16 \\ 0 \\ -8 \\ 0 \\ 0 \\ 0 \end{pmatrix}$$

In general, when working with Walsh functions, we can always find the function $f_{2^n}(t)$ utilizing matrix notation. We simply define the vector $\mathbf{f}_{2^n}$ by the product relation

$$f_{2^n}(t) = \mathbf{f}'_{2^n}\begin{pmatrix} I_{[0,2^{-n})}(t) \\ \vdots \\ I_{[(2^n-1)2^{-n},1)}(t) \end{pmatrix}$$

Then we have

$$\mathbf{f}_{2^n} = H'_{2^n}\boldsymbol{\alpha}_{2^n} = H'_{2^n}H_{2^n}\mathbf{A}_{2^n} = 2^n\mathbf{A}_{2^n}$$

where H'_{2^n} is the transpose of the Hadamard matrix, and where the chain of equalities follows directly from the definitions and the symmetry of the Hadamard matrix. Consequently, $f_{2^n}(t)$ can be quickly found using the so-called area map technique:

$$f_{2^n}(t) = 2^n(I_{[0,2^{-n})}(t) \ldots I_{[(2^n-1)2^{-n},1)}(t))\mathbf{A}_{2^n}$$

Sec. 5.7   Walsh Expansions   271

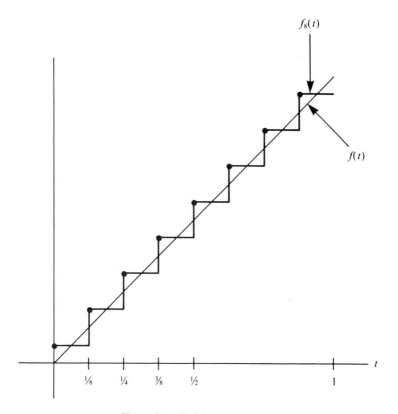

Figure 5.9   Walsh Approximation

Example 5.28

Let $f(t)$ be the signal of Example 5.27. Then, using the work of that example,

$$f_8(t) = 8(I_{[0,1/8)}(t)I_{[1/8,2/8)}(t) \ldots I_{[7/8,1)}(t))\mathbf{A}_8$$

$$= \frac{1}{16} I_{[0,1/8)}(t) + \frac{3}{16} I_{[1/8,1/4)}(t) + \cdots + \frac{15}{16} I_{[7/8,1)}(t)$$

The original signal $f(t)$ and its Walsh approximation are given in Figure 5.9.

The following theorem gives both L_∞ and L_2 bounds on the Walsh approximation.

Theorem 5.13.[15]   (i) Suppose $f(t)$ satisfies a Holder condition

$$|f(t + h) - f(t)| \leq B |h|^a, \text{ for } 0 < a \leq 1$$

---

[15] C. R. Giardina, "Bounds on the Truncation Error for Walsh Expansions," *Notices American Math Society,* Vol. 25, No. 2, (1978):A-311.

on [0, 1]. Then the maximum pointwise error is given by

$$\| f - f_{2^n} \|_\infty \leq \frac{B 2^{a-1}}{(2^a - 1)(2^a)^{n+1}}$$

The L_2 error, $\| f - f_{2^n} \|_2$, is bounded by the same quantity. (ii) Suppose $f(t)$ is of bounded variation with variation M. Then the L_2 error is bounded by $M/\sqrt{2^{n+1}}$.

Example 5.29

Using the signal $f(t)$ of Example 5.27, we see that the actual L_∞ error is given by $\| f - f_8 \|_\infty = \frac{1}{16}$. This agrees exactly with the bound that is obtained from Theorem 5.13, part (i). Indeed, the absolute value of the derivative $f'(t)$ is bounded by $B = 1$, and this guarantees that $f(t)$ satisfies a Holder condition with $B = 1$ and $a = 1$. (Recall the mean-value theorem from differential calculus.) Moreover, the same Holder condition gives

$$\| f - f_8 \|_2 \leq \frac{1}{16}$$

Notice that the bound on the L_2 error given by part (ii) of Theorem 5.13 is not as tight since the total variation is $M = 1$ and hence the bound is $\frac{1}{4}$. Note that the actual L_2 error for $f_8(t)$ is $1/\sqrt{48}$.

5.8 COMPLETE ORTHONORMAL SYSTEMS FOR EXPANSION OF IMAGES

Complete orthonormal systems can be obtained in two dimensions just as they are in the one-dimensional case. One of the simplest ways of constructing a complete orthonormal system of functions in two dimensions utilizes two complete orthonormal systems in one dimension. The procedure is to form the products of functions belonging to one system with functions belonging to the other system. Specifically, if $\varphi_1(t), \varphi_2(t), \ldots$ is a complete orthonormal system of functions on the interval $[a, b]$, and if $\psi_1(s), \psi_2(s), \ldots$ is a complete orthonormal system of functions on the interval $[c, d]$, then the functions

$$\Phi_{11}, \Phi_{12}, \Phi_{21}, \Phi_{13}, \Phi_{22}, \Phi_{31}, \ldots$$

where

$$\Phi_{ij}(t, s) = \varphi_i(t) \cdot \psi_j(s)$$

form a complete orthonormal system on the rectangle $[a, b] \times [c, d]$.[16] An important special case is $\varphi_i = \psi_i$. Then the functions

$$\Phi_{ij}(t, s) = \varphi_i(t) \cdot \varphi_j(s)$$

[16] R. Courant and D. Hilbert, *Methods of Mathematical Physics* (New York: Interscience Publishers, 1965), 56.

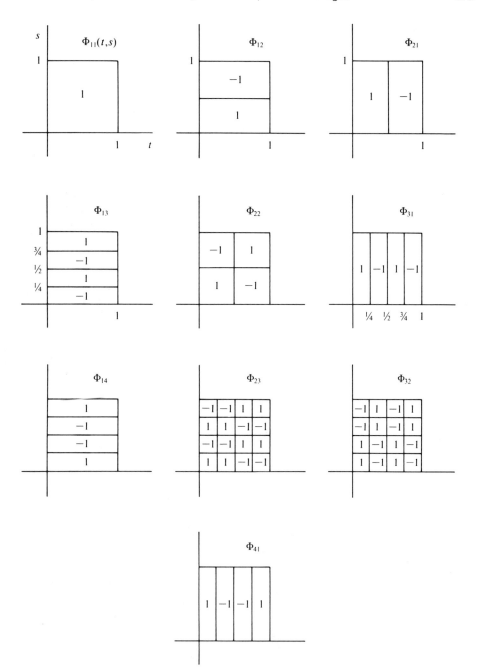

Figure 5.10 Two-Dimensional Walsh Functions

will form a complete orthonormal sequence of functions on the square $[a, b] \times [a, b]$. Analogous properties to those in the one-dimensional case hold and will not be listed here.

The Fourier coefficients are found in the usual manner. If $f(x, y)$ is an image defined on $[a, b] \times [c, d]$, then

$$\hat{f}(\Phi_{ij}) = \int_c^d \int_a^b f(x, y) \Phi_{ij}(x, y) \, dx \, dy$$

Example 5.30

If we let $\{\varphi_i(t)\}$ be the complete orthonormal system of Walsh functions on $[0, 1)$, and if $\psi_i(s) = \varphi_i(s)$, then $\Phi_{ij}(t, s)$ is the two-dimensional complete orthonormal system of Walsh functions. The first ten functions in the two-dimensional Walsh sequence are illustrated in Figure 5.10. Now consider the continuous image $f(x, y) = xy$ on $[0, 1) \times [0, 1)$. The Fourier (sequency) coefficient α_{22} associated with the two-dimensional Walsh function $\Phi_{22}(x, y)$, specified in Figure 5.10, is given by

$$\alpha_{22} = \int_0^{1/2} \int_0^{1/2} xy \, dx \, dy + \int_{1/2}^1 \int_{1/2}^1 xy \, dx \, dy$$

$$- \int_{1/2}^1 \int_0^{1/2} xy \, dx \, dy - \int_0^{1/2} \int_{1/2}^1 xy \, dx \, dy = \frac{1}{16}$$

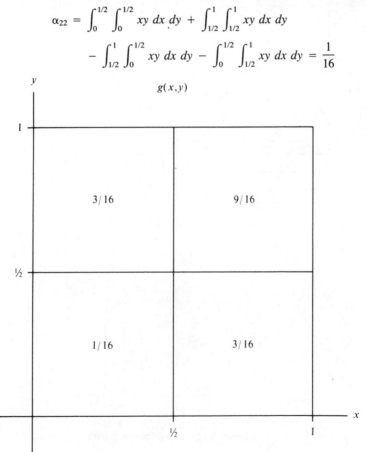

Figure 5.11 Walsh Image Approximation

If the sequency coefficients α_{11}, α_{12}, and α_{21} are also found, then an approximation to f is given by

$$g(x, y) = \alpha_{11}\Phi_{11}(x, y) + \alpha_{12}\Phi_{12}(x, y) + \alpha_{21}\Phi_{21}(x, y) + \alpha_{22}\Phi_{22}(x, y)$$

$$= \frac{1}{4}\Phi_{11}(x, y) - \frac{1}{8}\Phi_{12}(x, y) - \frac{1}{8}\Phi_{21}(x, y) + \frac{1}{16}\Phi_{22}(x, y)$$

Figure 5.11 shows the values of the step-type function $g(x,y)$ on its domain.

5.9 MATRIX IMAGE TRANSFORMS

Thus far in Chapters 4 and 5, we have considered the transforms $\mathcal{F}$ and $\mathcal{F}_d$, which are defined on continuous (analog) signals and images. In this section, we will digress and briefly consider a Fourier-type transform that is defined on the space $\mathcal{B}_{r,t}^{m,n}$ of saturated m by n images with upper left entry at pixel (r, t). Because of our desire to employ complex-valued entries, we shall make the assumption that the elements of $\mathcal{B}_{r,t}^{m,n}$ have complex-valued gray levels.

Given an image $f = (a_{pq})_{rt} \in \mathcal{B}_{r,t}^{m,n}$, we define the *discrete Fourier transform* (DFT) of f to be the image $F = (b_{kl})_{rt} \in \mathcal{B}_{r,t}^{m,n}$, defined by

$$b_{k,l} = (mn)^{-1} \sum_{p=1}^{m} \sum_{q=1}^{n} a_{pq} e^{-2\pi j[(k-1)(p-1)/m + (l-1)(q-1)/n]}$$

Letting

$$z(k, l)_{pq} = (mn)^{-1} e^{-2\pi j[(k-1)(p-1)/m + (l-1)(q-1)/n]}$$

we see that

$$z(k, l)_{pq} = v_{kp} w_{ql}$$

where

$$v_{kp} = m^{-1} e^{-2\pi j(k-1)(p-1)/m}$$

and

$$w_{ql} = n^{-1} e^{-2\pi j(l-1)(q-1)/n}$$

Define the bound matrix V in $\mathcal{B}_{r,t}^{m,m}$ and the bound matrix W in $\mathcal{B}_{r,t}^{n,n}$ by

$$V = (v_{kp})_{r,t}$$

and

$$W = (w_{ql})_{r,t}$$

respectively. Then let $X(A, B)$ denote "regular" matrix multiplication in $\mathcal{B}_{r,t}^{m,m}$—that is, multiply A by B as if they were regular matrices, and leave the result at

(r, t). Do the same for elements of $\mathcal{B}^{n;n}_{r,t}$. Straightforward matrix multiplication will verify that

$$F = X[X(V, f), W]$$

In general, if $f \in \mathcal{B}^{m;n}_{r,t}$, $A \in \mathcal{B}^{m;m}_{r,t}$, and $B \in \mathcal{B}^{n;n}_{r,t}$, then $X[X(A, f), B]$ is called a *matrix image transform* of f.[17] If A and B are fixed, then $f \to \text{MATRAN}(f) = X[X(A, f), B]$ maps $\mathcal{B}^{m;n}_{r,t}$ to $\mathcal{B}^{m;n}_{r,t}$. If A and B, treated as regular matrices, are invertible, then postmultiplication by B^{-1} and premultiplication by A^{-1} give

$$f = X[X(A^{-1}, \text{MATRAN}(f)), B^{-1}]$$

Put simply, if A and B are invertible, then so is the operation MATRAN.

In the case of the DFT, both V and W are invertible. Consequently, the inverse matrix image transform exists. This transform is called the *inverse discrete Fourier transform*. In the case of the DFT, a simple calculation shows that the inverses of V and W satisfy $V^{-1} = m\overline{V}$ and $W^{-1} = n\overline{W}$, where $\overline{V}$ and $\overline{W}$ are the respective conjugates of V and W.

Example 5.31

We now illustrate the matrix image transform method for the DFT. Let

$$f = \begin{pmatrix} 1 & 0 \\ 2 & 0 \\ 0 & 1 \\ -1 & 0 \end{pmatrix}_{4,5}$$

Since f is 4 by 2, the matrix image transform involves a 4 by 4 bound matrix V and a 2 by 2 bound matrix W. Using the Euler identity for $e^{j\theta}$ in both V and W gives

$$V = \frac{1}{4}\begin{pmatrix} 1 & 1 & 1 & 1 \\ 1 & -j & -1 & j \\ 1 & -1 & 1 & -1 \\ 1 & j & -1 & -j \end{pmatrix}_{4,5}$$

$$W = \frac{1}{2}\begin{pmatrix} 1 & 1 \\ 1 & -1 \end{pmatrix}_{4,5}$$

where, for notational simplicity, we have indicated scalar multiplication by putting the scalar in front of the bound matrix. Direct computation yields

$$F = X[X(V, f), W] = \frac{1}{8}\begin{pmatrix} 3 & 1 \\ -3j & 2 - 3j \\ 1 & -1 \\ 3j & 2 + 3j \end{pmatrix}_{4,5}$$

[17] E. R. Dougherty and C. R. Giardina, *Matrix-Structured Image Processing* (Englewood Cliffs, N.J.: Prentice-Hall, Inc., 1986), 185.

Sec. 5.10   Chain Codes for Boundary Representation of Images

Inversion utilizes

$$V^{-1} = \begin{pmatrix} 1 & 1 & 1 & 1 \\ 1 & j & -1 & -j \\ 1 & -1 & 1 & -1 \\ 1 & -j & -1 & j \end{pmatrix}_{4,5}$$

and

$$W^{-1} = \begin{pmatrix} 1 & 1 \\ 1 & -1 \end{pmatrix}_{4,5}$$

A direct calculation shows that $f = X[X(V^{-1}, F), W^{-1}]$.

5.10 CHAIN CODES FOR BOUNDARY REPRESENTATION OF IMAGES

Let A be a finite, nonempty set of symbols. Using terminology from automata theory, A is called an *alphabet* and its elements are called *letters*. The collection of all finite strings involving letters from A is denoted by A^*. Under the binary concatenation of strings in A^*, the set A^* forms a monoid. Note that concatenation is associative and that the empty string ϵ is a neutral element for concatenation. The latter condition means that for any string w in A^*, $\epsilon w = w\epsilon = w$.

In the chain encoding of linear-boundary-type images, the alphabet $A = \{0, 1, \ldots, 7\}$ is employed. A *chain code* c is a string from A^*. Each letter (integer) corresponds to a directed line segment of length 1 or $\sqrt{2}$ in the x-y plane, the directions of the segments being multiples of $45°$. Figure 5.12 gives the numeration scheme for the chain code. For each letter a_i in A, the directed line segment v_i

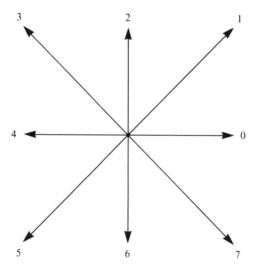

Figure 5.12 Chain-Code Line Segments

associated with a_i, as shown in the figure, can be written in phasor notation as

$$v_i = \left[1 + \frac{\sqrt{2} - 1}{2}(1 - (-1)^{a_i})\right] \measuredangle \left(\frac{\pi}{4}\right) a_i$$

If $c = a_1 a_2 \ldots a_m$ is a string of length m from A^*, then a line-type image can be associated with c by joining the segments v_i in a head-to-tail fashion while reading c from left to right. Specifically, if the string $a_i a_{i+1}$ appears within the string c, then the head of v_i should touch the tail of v_{i+1}. The process is begun by positioning the tail of v_1 at the origin. Using this procedure, for every c in A^* there exists a unique piecewise linear line-type image v associated with c. Should c be the empty string, we let v be the null image. Referring to the image characterizations of Section 4.1, the actual image associated with c is given by

$$f(x, y) = \begin{cases} 1 & \text{if } (x, y) \in B \\ \text{undefined} & \text{otherwise} \end{cases}$$

where B is the set of all points in R^2 which lie in one of the v_i. To avoid cumbersome terminology, we shall simply refer to the directed line segments as *piecewise-linear line images*. Note that the correspondence between strings in A^* and piecewise-linear line images is neither one-to-one nor onto. (See Example 5.32.)

Example 5.32

Consider the linear line image illustrated in Figure 5.13. This image is induced by the strings $c = 1\ 6\ 4\ 7\ 2$ and $d = 7\ 2\ 4\ 1\ 6\ 4$. Hence, the inducement of linear line images from A^* is not one-to-one. Moreover, if we were to choose a piecewise-linear

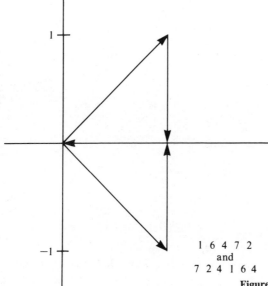

Figure 5.13 Non 1–1 Nature of Chain Encoding

line image that does not have a directed line segment with tail at the origin, then that image could not have been induced by a string from A^*. Hence, neither is the inducement onto.

In what follows, we shall assume that the piecewise-linear line image v has been induced by a fixed element c in A^*. Projections will be obtained by traversing the image $v_1v_2 \ldots v_k$ from the tail of v_1 to the head of v_k using arc length t as a

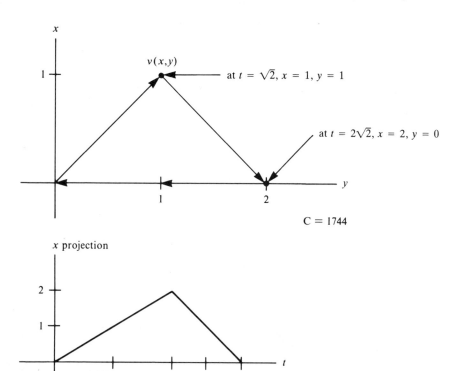

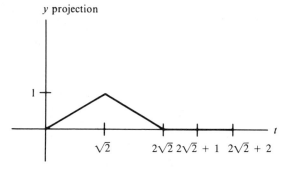

Figure 5.14 Projections of Chain-Encoded Images

parameter. If $c = a_1 a_2 \ldots a_k$, then the total arc length is given by

$$T = \sum_{i=1}^{K} \left[1 + \frac{(\sqrt{2} - 1)}{2} (1 - (-1)^{a_i}) \right]$$

As an arc in the plane, $v = v(t) = (x(t), y(t))$, where $x(t)$ and $y(t)$ specify the x and y coordinates, respectively, of v, and where $0 \le t \le T$. $x(t)$ and $y(t)$ are called the x and y *projections,* respectively. The time (arc length t being the parameter) required to traverse a particular link a_i is

$$\Delta t_i = 1 + \frac{(\sqrt{2} - 1)}{2} (1 - (-1)^{a_i})$$

where it should be noted that the apparently complicated expression yields either 1 or $\sqrt{2}$. The time required to traverse the first p links in the chain is

$$t_p = \sum_{i=1}^{p} \Delta t_i$$

The changes in the x-y projections of the image induced by the chain as the link a_i is traversed (see Figure 5.14) are

$$\Delta x_i = \text{sgn}(6 - a_i) \, \text{sgn}(2 - a_i)$$

$$\Delta y_i = \text{sgn}(4 - a_i) \, \text{sgn}(a_i)$$

where sgn is the *sign function,* given by

$$\text{sgn}(z) = \begin{cases} 1 & \text{if } z > 0 \\ 0 & \text{if } z = 0 \\ -1 & \text{if } z < 0 \end{cases}$$

5.11 FOURIER SERIES REPRESENTATION OF CHAIN-ENCODED, LINE-DRAWING-TYPE IMAGES

We now consider a direct procedure for obtaining the Fourier coefficients of a chain-encoded contour. Since the starting point of the chain code corresponds to the origin, the projections on the x and y axes for the first p links of the chain are given by

$$x_p = \sum_{i=1}^{p} \Delta x_i$$

and

$$y_p = \sum_{i=1}^{p} \Delta y_i$$

Sec. 5.11   Fourier Series Representation of Chain-Encoded Images

The Fourier series expansion for the x projection, as a function of the parameter t, of the entire chain code consisting of K symbols is given by

$$x(t) = A_0 + \sum_{n=1}^{\infty} A_n \cos \frac{2n\pi t}{T} + B_n \sin \frac{2n\pi t}{T}$$

where A_0, A_n, and B_n are the *nonnormalized* Fourier coefficients

$$A_0 = \frac{1}{T} \int_0^T x(t)\, dt$$

$$A_n = \frac{2}{T} \int_0^T x(t) \cos \frac{2n\pi t}{T}\, dt$$

$$B_n = \frac{2}{T} \int_0^T x(t) \sin \frac{2n\pi t}{T}\, dt$$

The Fourier coefficients A_n and B_n are easily found because the x projection $x(t)$ is piecewise linear and continuous for all time. A simple calculation shows that

$$A_n = \frac{T}{2n^2\pi^2} \sum_{p=1}^{K} \frac{\Delta x_p}{\Delta t_p} \left[\cos \frac{2n\pi t_p}{T} - \cos \frac{2n\pi t_{p-1}}{T} \right]$$

$$B_n = \frac{T}{2n^2\pi^2} \sum_{p=1}^{K} \frac{\Delta x_p}{\Delta t_p} \left[\sin \frac{2n\pi t_p}{T} - \sin \frac{2n\pi t_{p-1}}{T} \right]$$

The Fourier series expansion for the y projection of the chain code of the complete contour is similarly found to be

$$y(t) = C_0 + \sum_{n=1}^{\infty} C_n \cos \frac{2n\pi t}{T} + D_n \sin \frac{2n\pi t}{T}$$

where

$$C_n = \frac{T}{2n^2\pi^2} \sum_{p=1}^{K} \frac{\Delta y_p}{\Delta t_p} \left[\cos \frac{2n\pi t_p}{T} - \cos \frac{2n\pi t_{p-1}}{T} \right]$$

$$D_n = \frac{T}{2n^2\pi^2} \sum_{p=1}^{K} \frac{\Delta y_p}{\Delta t_p} \left[\sin \frac{2n\pi t_p}{T} - \sin \frac{2n\pi t_{p-1}}{T} \right]$$

The DC components A_0 and C_0 in these Fourier series representations are given by

$$A_0 = \frac{1}{T} \sum_{p=1}^{K} \frac{\Delta x_p}{2\Delta t_p} (t_p^2 - t_{p-1}^2) + \xi_p(t_p - t_{p-1})$$

$$C_0 = \frac{1}{T} \sum_{p=1}^{K} \frac{\Delta y_p}{2\Delta t_p} (t_p^2 - t_{p-1}^2) + \delta_p(t_p - t_{p-1})$$

where

$$\xi_p = \sum_{j=1}^{p-1} \Delta x_j - \frac{\Delta x_p}{\Delta t_p} \sum_{j=1}^{p-1} \Delta t_j$$

$$\delta_p = \sum_{j=1}^{p-1} \Delta y_j - \frac{\Delta y_p}{\Delta t_p} \sum_{j=1}^{p-1} \Delta t_j$$

and $\xi_1 = \delta_1 = 0$.

Example 5.33

The chain code image for $V_1 = 0005676644422123$ is illustrated in Figure 5.15. Since the induced contour is closed, both the x and y projections begin and end at $t = 0$. As a result, Theorem 5.12 applies; hence, $x(t)$ and $y(t)$ are uniform limits of their respective Fourier series.

Let $\tilde{X}_N(t)$, $N = 0, 1, 2, \ldots$, denote the function that is the Nth truncation of the Fourier series for x; that is, $\tilde{X}_N(t)$ is the Nth partial sum of the Fourier series for x. We often say that $\tilde{X}_N(t)$ is the Nth harmonic approximation, the number of harmonics being the number of Fourier coefficients, starting from 0, being employed in the truncation. In a similar manner, $\tilde{Y}_N(t)$ denotes the Nth harmonic approximation to the y projection.

If we let $\tilde{v}_N(t) = (\tilde{X}_N(t), \tilde{Y}_N(t))$, then, as the parameter t runs from 0 to T, $\tilde{v}_N(t)$ traces out an approximation to the piecewise-linear line image $v(t)$ that has been induced from the chain code V_1. Since $\tilde{X}_N(t)$ and $\tilde{Y}_N(t)$ converge uniformly to $x(t)$ and $y(t)$, respectively, $\tilde{v}_N(t)$ converges uniformly to $v(t)$. Figure 5.15 gives the first four harmonic approximations, $\tilde{v}_1, \tilde{v}_2, \tilde{v}_3$, and $\tilde{v}_4$, superimposed on the original chain code image v.

For a given image, it is useful to be able to specify the number of harmonics required in order that a truncated Fourier approximation to a contour be in error no greater than some prespecified amount in the x or y dimension. Let

$$\tilde{X}_N = A_0 + \sum_{n=1}^{N} A_n \cos \frac{2n\pi t}{T} + B_n \sin \frac{2n\pi t}{T}$$

$$\tilde{Y}_N = C_0 + \sum_{n=1}^{N} C_n \cos \frac{2n\pi t}{T} + D_n \sin \frac{2n\pi t}{T}$$

be the Fourier series truncated after N harmonics for the $x(t)$ and $y(t)$ projections, respectively, and define the error ϵ as

$$\epsilon = \max[\sup_t |x(t) - \tilde{X}_N(t)|, \sup_t |y(t) - \tilde{Y}_N(t)|]$$

It can be shown that, so long as $V(0) = V(T)$ (the induced image contour is closed),

$$\epsilon \leq \frac{T}{2\pi^2 N} \max[V_0^T(x'(t)), V_0^T(y'(t))]$$

where the total variation of the derivative $x'(t)$ has been symbolized as

Sec. 5.11 Fourier Series Representation of Chain-Encoded Images 283

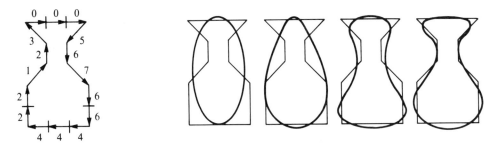

Figure 5.15 Chain Code of Image and First Four Harmonic Approximations

$V_0^T(x'(t))$ and the total variation of the derivative $y'(t)$ as $V_0^T(y'(t))$. The derivatives for the link a_i, everywhere except at the end points, are

$$x'_i = \frac{\Delta x_i}{\Delta t_i}$$

$$y'_i = \frac{\Delta y_i}{\Delta t_i}$$

These can be tabulated according to the value a_i and are given in Table 5.1. The total variations of $x'(t)$ and $y'(t)$ are, respectively,

$$V_0^T(x'(t)) = \sum_2^K |x'_i - x'_{i-1}|$$

$$V_0^T(y'(t)) = \sum_2^K |y'_i - y'_{i-1}|$$

A graph of the actual error ϵ versus the number N of harmonics is shown in Figure 5.16 for the chain code V_1 of Example 5.33. The predicted bounds on the error according to the maximum total variation estimate are shown superimposed on the graph.

TABLE 5.1 x AND y DERIVATIVES OF LINK a_i

a_i	x'_i	y'_i
0	1.0	0.0
1	0.707	0.707
2	0.0	1.0
3	−0.707	0.707
4	−1.0	0.0
5	−0.707	−0.707
6	0.0	−1.0
7	0.707	−0.707

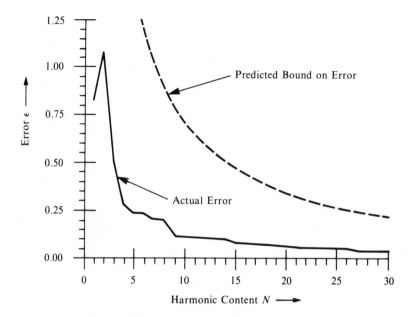

Figure 5.16 Predicted and Actual Error Versus Harmonic Content

Example 5.34

Consider the following four chain codes:

$V_2 = 11172206667666644444422$

$V_3 = 5412343001010007711075454450654134446$

$V_4 = 00310704547644571534504133142 0600$

$V_5 = 2333446766544433267700012232545433221566770101 4443221104$
$5566700032110773345567100077623344450 07776$

As shown in Figures 5.17 through 5.20, for each of these codes there are image contours of increasing complexity (wiggliness). Each figure displays the chain-code representation of the image contour together with superimposed Fourier approximations that incorporate increasingly higher harmonic content. The actual error and *one-half* the predicted bound on the error, in terms of the harmonic content N, are shown at the bottom of each figure. It is apparent from inspection of the figures that, as the contour becomes more complex, the predicted bound becomes more conservative. Consequently, to make a realistic comparison of the accuracy of harmonic reconstruction as it applies to different contours, both plots of ϵ should be normalized against $T = t_K$, i.e., use $\epsilon' = \epsilon/T$. In any event, the predicted bound on ϵ offers the advantage of quick and easy computation as compared to the calculation of the actual error.

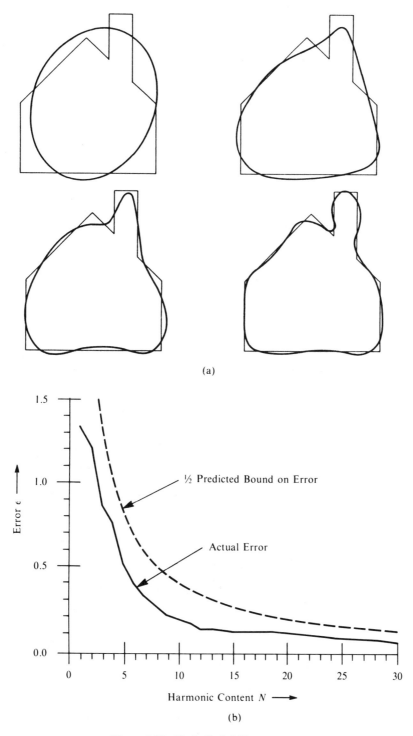

Figure 5.17 Chain-Coded House

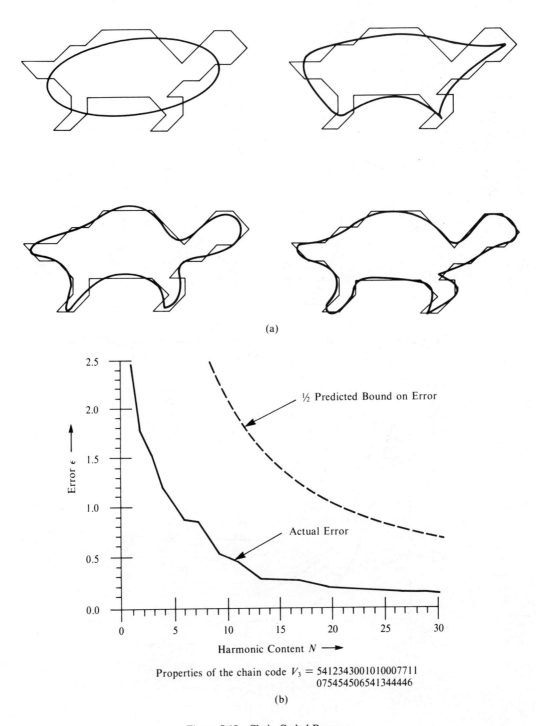

Figure 5.18 Chain-Coded Raccoon

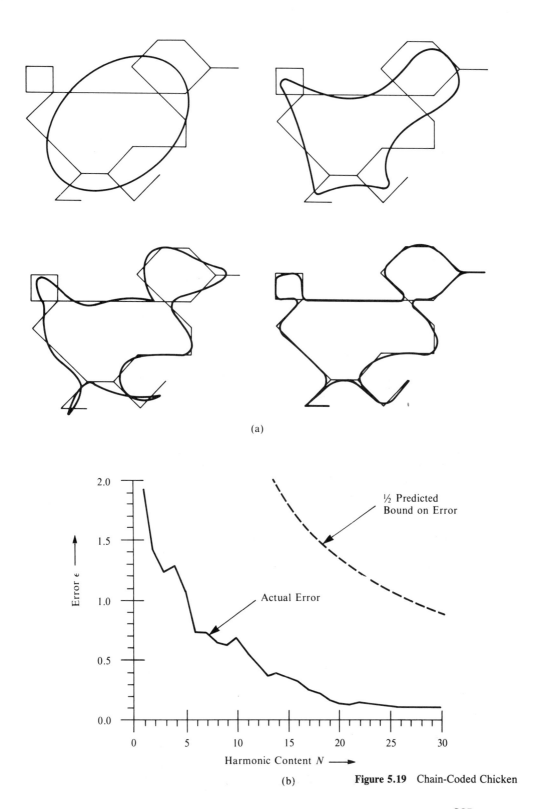

Figure 5.19 Chain-Coded Chicken

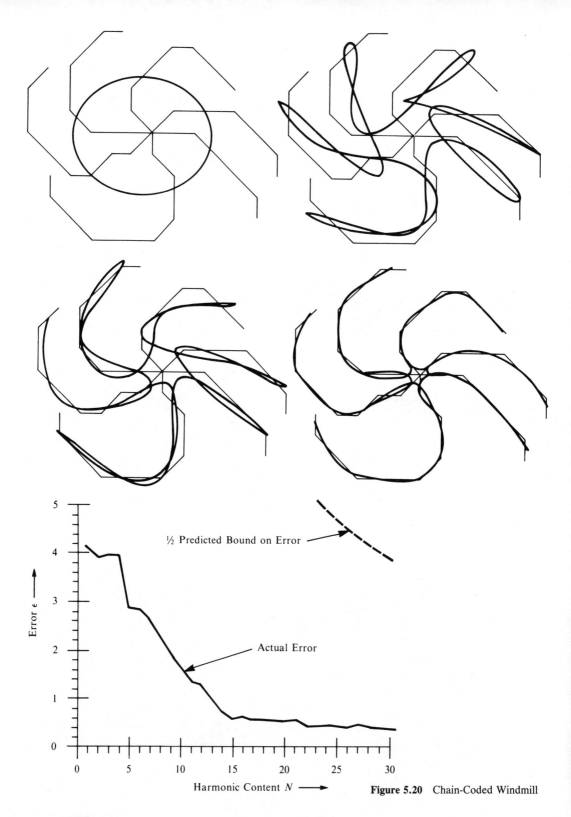

Figure 5.20 Chain-Coded Windmill

5.12 NORMALIZATION OF FOURIER SERIES REPRESENTATIONS

The Fourier series representation of the projections of a piecewise-linear line image is a compression method that can be used for pattern identification. The Fourier coefficients of an observed image may be compared to the corresponding coefficients of an archetypal pattern, the latter coefficients being stored in memory. Often this needs to be accomplished independently of the size, location, and orientation of the observed image. Due to these requirements, a self-consistent *normalization* based on intrinsic properties of the observed image must be employed in order to bring it into a standard form that allows comparison with the archetypes in the database.

In the present compression methodology, normalization must be accomplished with respect to (1) the starting points for obtaining the Fourier expansion and (2) rotation, translation, and magnification. This can be done through the utilization of the *harmonic ellipsi*.[18]

Using the results of Section 5.11, the truncated Fourier approximation to a closed piecewise linear line image can be written as

$$\tilde{X}_N(t) = A_0 + \sum_{n=1}^{N} X_n(t)$$

and

$$\tilde{Y}_N(t) = C_0 + \sum_{n=1}^{N} Y_n(t)$$

where, for $1 \leq n \leq N$,

$$X_n(t) = A_n \cos(2\pi nt/T) + B_n \sin(2\pi nt/T)$$

and

$$Y_n(t) = C_n \cos(2\pi nt/T) + D_n \sin(2\pi nt/T)$$

Let $v_n(t) = (X_n(t), Y_n(t))$ for any n. Then, as the parameter t runs from 0 to T, $v_n(t)$ describes a locus of points in the plane. This locus is elliptical. Indeed, suppose $A_n D_n - B_n C_n$ is nonzero. The elimination of the sine and cosine terms in the defining expressions for X_n and Y_n yields

$$\frac{(C_n^2 + D_n^2)X_n^2 + (A_n^2 + B_n^2)Y_n^2 - 2(A_n C_n + B_n D_n)X_n Y_n}{A_n D_n - B_n C_n} = 1$$

which is an ellipse. For the case where $A_n D_n - B_n C_n = 0$, we have an ellipse in a degenerate form—a straight line or a point. Finally, it should be noted that

---

[18] The use of harmonic ellipsi was suggested by F. Kuhl in C. R. Giardina and F. P. Kuhl, "Accuracy of Curve Approximations by Harmonically Related Vectors with Elliptical Loci," *Computer Graphics and Image Processing*, (1977):277–85.

the same elliptical loci will be obtained for the points $v_n = (X_n, Y_n)$ regardless of the starting point for taking projections of the contour.

We now proceed to the normalization problem outlined at the beginning of the section. To begin with, a difference in the starting points is displayed in the projected space as a phase shift. Hence, a starting point displaced λ units in the direction of rotation around the contour from the original starting point will have projections, for $n \geq 1$,

$$X_n(t^* + \lambda) = A_n \cos \frac{2\pi n}{T}(t^* + \lambda) + B_n \sin \frac{2\pi n}{T}(t^* + \lambda)$$

$$Y_n(t^* + \lambda) = C_n \cos \frac{2\pi n}{T}(t^* + \lambda) + D_n \sin \frac{2\pi n}{T}(t^* + \lambda)$$

where

$$t^* + \lambda = t$$

Expanding X_n and Y_n and collecting terms yields

$$X_n^*(t^*) = A_n^* \cos \frac{2\pi n t^*}{T} + B_n^* \sin \frac{2\pi n t^*}{T}$$

$$Y_n^*(t^*) = C_n^* \cos \frac{2\pi n t^*}{T} + D_n^* \sin \frac{2\pi n t^*}{T},$$

where

$$\begin{pmatrix} A_n^* & C_n^* \\ B_n^* & D_n^* \end{pmatrix} = \begin{pmatrix} \cos \frac{2\pi n \lambda}{T} & \sin \frac{2\pi n \lambda}{T} \\ -\sin \frac{2\pi n \lambda}{T} & \cos \frac{2\pi n \lambda}{T} \end{pmatrix} \begin{pmatrix} A_n & C_n \\ B_n & D_n \end{pmatrix}$$

To obtain a normalization with respect to rotation, note that a counterclockwise rotation of the X-Y coordinate axes through ψ degrees into the U-V axes is accomplished by

$$\begin{pmatrix} U \\ V \end{pmatrix} = \begin{pmatrix} \cos \psi & \sin \psi \\ -\sin \psi & \cos \psi \end{pmatrix} \begin{pmatrix} X \\ Y \end{pmatrix}$$

The effect of this rotation on the Fourier coefficients A_n^*, B_n^*, C_n^* and D_n^* is readily apparent when the projections X_n^*, Y_n^* are expressed in matrix form:

$$\begin{pmatrix} X_n^* \\ Y_n^* \end{pmatrix} = \begin{pmatrix} A_n^* & B_n^* \\ C_n^* & D_n^* \end{pmatrix} \begin{pmatrix} \cos \frac{2\pi n t^*}{T} \\ \sin \frac{2\pi n t^*}{T} \end{pmatrix}$$

Sec. 5.12   Normalization of Fourier Series Representations

Then the projections μ_n, ν_n on the U-V axes are

$$\begin{pmatrix} \mu_n \\ \nu_n \end{pmatrix} = \begin{pmatrix} \cos\psi & \sin\psi \\ -\sin\psi & \cos\psi \end{pmatrix} \begin{pmatrix} X_n^* \\ Y_n^* \end{pmatrix} = \begin{pmatrix} \cos\psi & \sin\psi \\ -\sin\psi & \cos\psi \end{pmatrix} \begin{pmatrix} A_n^* & B_n^* \\ C_n^* & D_n^* \end{pmatrix} \begin{pmatrix} \cos\dfrac{2\pi nt^*}{T} \\ \sin\dfrac{2\pi nt^*}{T} \end{pmatrix}$$

A rotated set of Fourier coefficients A_n^{**}, B_n^{**}, C_n^{**}, and D_n^{**} may be defined by

$$\begin{pmatrix} A_n^{**} & B_n^{**} \\ C_n^{**} & D_n^{**} \end{pmatrix} = \begin{pmatrix} \cos\psi & \sin\psi \\ -\sin\psi & \cos\psi \end{pmatrix} \begin{pmatrix} A_n^* & B_n^* \\ C_n^* & D_n^* \end{pmatrix}$$

The combined effects of the rotation, together with a displacement of the starting point, on the coefficients A_n, B_n, C_n, and D_n of the original expression are readily described in matrix notation as follows:

$$\begin{pmatrix} A_n^{**} & B_n^{**} \\ C_n^{**} & D_n^{**} \end{pmatrix} = \begin{pmatrix} \cos\psi & \sin\psi \\ -\sin\psi & \cos\psi \end{pmatrix} \begin{pmatrix} A_n & B_n \\ C_n & D_n \end{pmatrix} \begin{pmatrix} \cos\dfrac{2\pi n\lambda}{T} & -\sin\dfrac{2\pi n\lambda}{T} \\ \sin\dfrac{2\pi n\lambda}{T} & \cos\dfrac{2\pi n\lambda}{T} \end{pmatrix}$$

The preceding matrix equation gives us a transformation

$$\begin{pmatrix} A_a \\ B_n \\ C_n \\ D_n \end{pmatrix} \rightarrow \begin{pmatrix} A_n^{**} \\ B_n^{**} \\ C_n^{**} \\ D_n^{**} \end{pmatrix}$$

call it Λ, which takes the original Fourier coefficients and outputs a collection of coefficients normalized with respect to both starting point and orientation. For a given contour v, these new coefficients can be compared to sets stored in the database in order to decide upon a classification for v.

In order to utilize the mapping Λ, some canonical methodology must be set for deciding on the starting point and the final spatial orientation of the Fourier approximations of the original image. In other words, for a given observed image v, we need to know the inputs ψ and λ in order to compute A_n^{**}, B_n^{**}, C_n^{**}, and D_n^{**}. These inputs are derived from the first harmonic ellipse. Intuitively, the transformation is made so that the starting point is at one of the semi-major axes and the ellipse v_1 is rotated according to the position of the chosen axis. Because of the two possible choices of the semi-major axis, two possible canonical classifications result. These are simply 180° rotations of each other.

Though we will not go through the geometric details, the values of ψ and λ, based on the first harmonic ellipse, are given by

$$\lambda_1 = \frac{T}{4\pi} \arctan\left(\frac{2(A_1 B_1 + C_1 D_1)}{A_1^2 + C_1^2 - B_1^2 - D_1^2}\right)$$

and

$$\psi_1 = \arctan\left(\frac{C_1^*}{A_1^*}\right)$$

The normalization of the Fourier coefficients relative to starting point and orientation is achieved by using Λ, with the output coefficients based upon the first harmonic ellipse of the observed image.[19]

If we desire a classification independent of size, this normalization can be accomplished by dividing each coefficient by the magnitude of the semi-major axis, which is given by

$$E^* = [(A_1^*)^2 + (C_1^*)^2]^{1/2}$$

Finally, classification independent of translation can be achieved by simply ignoring the terms A_0 and B_0.

Example 5.35

The elliptic classification procedure is shown for a tank on an incline in Figure 5.21. The image and the first harmonic ellipse are shown in Figure 5.21(a), and the 30th harmonic normalized approximation of the tank is given in Figure 5.21(b). Two possible classifications involving the 30th harmonic approximation, as generated by the normalized coefficients, are shown by solid and dashed lines in Figure 5.21(c).

5.13 WALSH FUNCTION FEATURE REPRESENTATIONS

In this section we develop a procedure for replacing a line-drawing-type image by an image whose domain is finite. The resulting *Walsh image* differs from a bound-matrix image in that (1) the Walsh-type output is a constant image, whereas images given by bound matrices need not be constant; and (2) more importantly, the Walsh image involves dots (or points in the domain) other than the lattice points, i.e., those at (i, j), for i, j integers. Walsh images are also called *dot images*.

The strategy in locating a dot image is rigorously described using ϵ-*nets*. For an image $f: A \to R$, bounded $A \subset R^2$, an ϵ-net is an image W with finite domain (that is, W consists of a finite number of dots) such that every point in the domain of f must be within $\epsilon > 0$ of some point in the domain of W in both the vertical and horizontal directions. In other words, the points in the domain of W are scattered in such a way that each point in the domain of f is not more than ϵ away, vertically or horizontally, from some point in the domain of W. Quanti-

---

[19] F. P. Kuhl and C. R. Giardina, "Elliptic Fourier Features of a Closed Contour," *Computer Graphics and Image Processing*, 18, (1982):236–58.

Sec. 5.13 Walsh Function Feature Representations

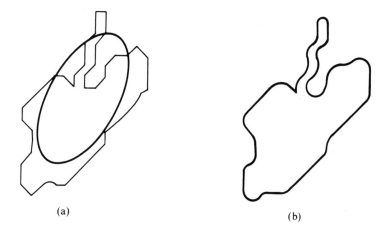

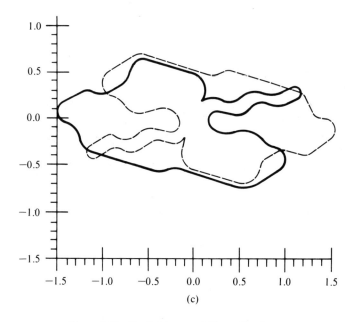

Figure 5.21 Tank Image and Normalization

tatively, given any point in the domain of f, say (x_α, y_α), there must exist at least one point in the domain of W, call it (x_i, y_i), such that

$$\| (x_\alpha, y_\alpha) - (x_i, y_i) \|_{\max} = \max\{| x_\alpha - x_i |, | y_\alpha - y_i |\} \le \epsilon$$

Since only constant images of value unity are dealt with in this section, we proceed as in Section 5.11 and elect not to distinguish between an image and its domain.

The task before us is as follows: given a chain-encoded image f and an $\epsilon > 0$, find a Walsh image that serves as an ϵ-net for f. To begin with, we proceed in a manner similar to that in Section 5.11 and find the x and y projections that result from traversing f at a uniform rate so that, in this instance, the traversal time is one unit. These projections are once again denoted by $x(t)$ and $y(t)$, and these must be expressed as truncated Walsh expansions. It can be shown that, for a given ϵ, the number of terms needed in the expansions is $N = \max\{n, k\}$, where n is the smallest integer greater than $\log_2(M/\epsilon) - 1$ and k is the smallest integer greater than $\log_2(L/\epsilon) - 1$, where, in turn, M and L are the maximum absolute slopes of the piecewise linear segments comprising $x(t)$ and $y(t)$, respectively. Note that M has been picked so that the absolute value of the derivative of $x(t)$ is bounded by M. Hence, according to Theorem 5.13, since $x(t)$ satisfies a Holder condition with $B = M$ and $a = 1$,

$$| x(t) - x_{2^n}(t) | \le M \times 2^{-n-1}$$

Similar remarks apply to $y(t)$ and L. (Note that whenever a function satisfies a Holder condition with B and $a = 1$, it is said to satisfy a *Lipshitz condition* with Lipshitz constant B.)

Once N has been found, we partition the domain interval $[0, 1]$ into 2^{-N} equally spaced subintervals, in each of which we find the 2^N-by-1 area vectors (see Section 5.7) for both $x(t)$ and $y(t)$. These are, respectively,

$$\mathbf{A}_x = \begin{pmatrix} A_{0x} \\ A_{1x} \\ \vdots \\ A_{(2^N-1)x} \end{pmatrix} \qquad \mathbf{A}_y = \begin{pmatrix} A_{0y} \\ A_{1y} \\ \vdots \\ A_{(2^N-1)y} \end{pmatrix}$$

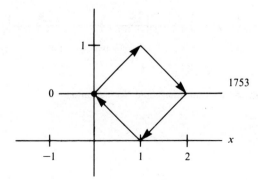

Figure 5.22 Image of Diamond

Sec. 5.13   Walsh Function Feature Representations   295

where each tuple in $\mathbf{A}_x$ is given by

$$A_{ix} = \int_{i/2^N}^{(i+1)/2^N} x(t)\, dt, \text{ for } i = 0, 1, \ldots 2^N - 1$$

and a similar integral gives A_{iy}.

The Walsh feature set may then be found. In vector form, and using matrix partitioning, we let

$$W_{2^N} = 2^N (\mathbf{A}_x \mid \mathbf{A}_y) = \begin{pmatrix} 2^N A_{0x} & 2^N A_{0y} \\ 2^N A_{1x} & 2^N A_{1y} \\ \vdots & \vdots \\ 2^N A_{(2^N-1)x} & 2^N A_{(2^N-1)y} \end{pmatrix}$$

The rows of W_{2^N}, treated as ordered pairs, specify the dots of the Walsh image.

Example 5.36

Consider the chain-encoded image in Figure 5.22. Traversing the figure at a constant clockwise rate such that it takes one unit of time to arrive back to the starting point, we obtain the x and y projections illustrated in Figure 5.23.

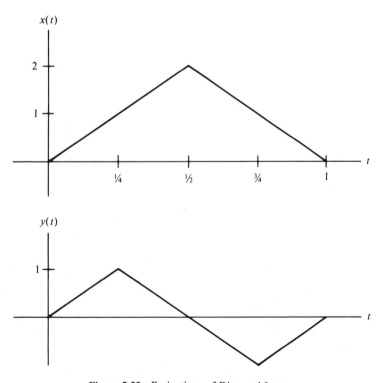

Figure 5.23   Projections of Diamond Image

We can see that

$$|x(t+h) - x(t)| \leq 4|h|$$

and

$$|y(t+h) - y(t)| \leq 4|h|$$

Thus, both the x and y projections satisfy a Lipshitz condition with constant 4. Hence, $M = L = 4$. Therefore, using the bound given in Theorem 5.13, we have

$$\|x(t) - x_{2^N}(t)\| \leq \frac{4}{2^{N+1}}$$

and a similar inequality holds for $y(t)$.

For purposes of illustration, if a maximum pointwise error of $\epsilon = \frac{1}{2}$ is desired in each direction, then $N = 2$. We need the first four terms of the Walsh expansion. We first obtain the 4×1 area vectors:

$$A_x = \frac{1}{8}\begin{pmatrix} 1 \\ 3 \\ 3 \\ 1 \end{pmatrix} \qquad A_y = \frac{1}{8}\begin{pmatrix} 1 \\ 1 \\ -1 \\ -1 \end{pmatrix}$$

The two-dimensional Walsh feature representation is given by

$$W_4 = 2^2(A_x | A_y) = \begin{pmatrix} \frac{1}{2} & \frac{1}{2} \\ \frac{3}{2} & \frac{1}{2} \\ \frac{3}{2} & -\frac{1}{2} \\ \frac{1}{2} & -\frac{1}{2} \end{pmatrix}$$

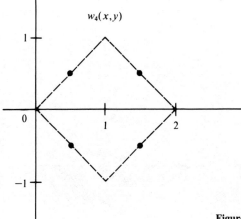

Figure 5.24 ϵ-Net Approximation to Diamond Image

Sec. 5.13   Walsh Function Feature Representations

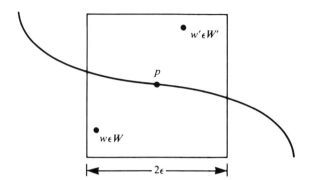

Figure 5.25 ε-Net Approximation to Curve

The ε-net W_4 is illustrated in Figure 5.24. Observe in the figure that each point in the original chain-encoded image is at most half a unit away in both the horizontal and vertical directions from a point in the Walsh dot image.

A key property of the Walsh image W is that it forms an ε-net for the original image f. Let us discuss this further with the help of Figure 5.25.

Select any point p on the original image f and draw a square of sides 2ϵ centered at p. Since W is an ε-net for f, there must exist a point $w \in W$ in this square. However, if W' is also an ε-net for the same image f, then there must also be a point $w' \in W'$ in the same square. The conclusion one comes to is the key to the recognition procedure: the farthest that w can be from w' is $2\sqrt{2}\epsilon$. The recognition technique, which is applied after normalization of the observed Walsh set, is to construct spheres of radius $2\sqrt{2}\epsilon$ about each point of this set and then find the knowledge-base pattern which has a point in each of these spheres. To implement the latter matching procedure, each of the stored patterns (archetypal images) which is to be input into the matching process must be represented in terms of a Walsh feature set of cardinality 2^N. Before checking any particular archetypal Walsh feature set, that set must be normalized in a manner analogous to the normalization of the observation set. If the collection of stored patterns is denoted by $P_1, P_2, \ldots, P_r$, then the observed image is declared to be of type P_j if for each sphere S centered at a point of the normalized observed Walsh set, there exists a point p in the normalized Walsh set for P_j such that p is in S. Ambiguous results will not be obtained if ϵ is chosen judiciously and the archetypal patterns are sufficiently distinct.

Normalization techniques for Walsh dot images involve techniques similar to those for Fourier-type images and will not be discussed here. Figures 5.26 through 5.30 illustrate Walsh-type images together with squares of sides 2ϵ centered about each Walsh dot. These diagrams are for an F104 computer-generated image utilizing 16, 32, 64, 128, and 256 dots, respectively.

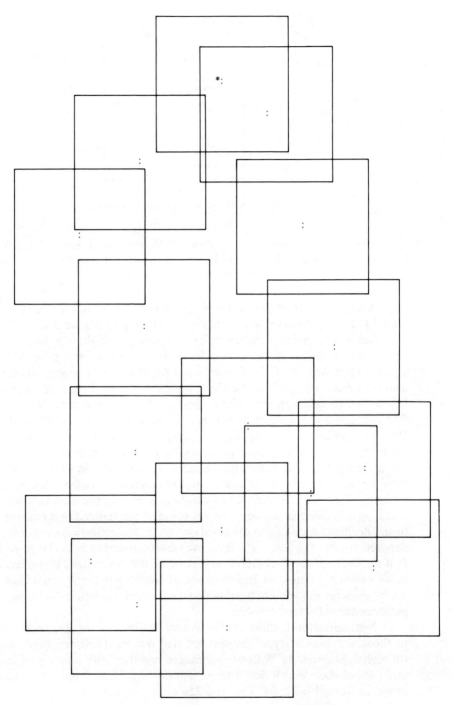

Figure 5.26 16-Walsh-Dot Approximation to a Plane

Sec. 5.13 Walsh Function Feature Representations

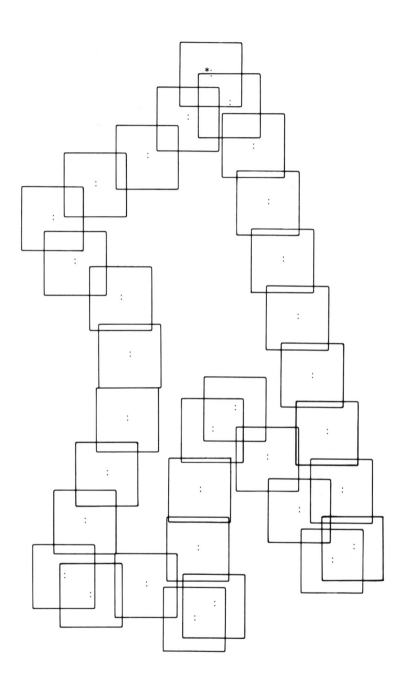

Figure 5.27 32-Walsh-Dot Approximation to a Plane

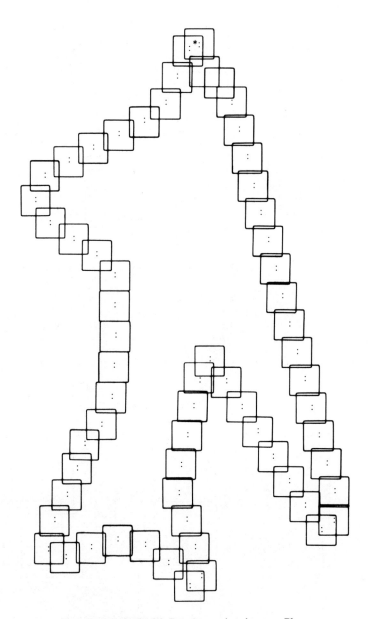

Figure 5.28 64-Walsh-Dot Approximation to a Plane

Sec. 5.13　Walsh Function Feature Representations

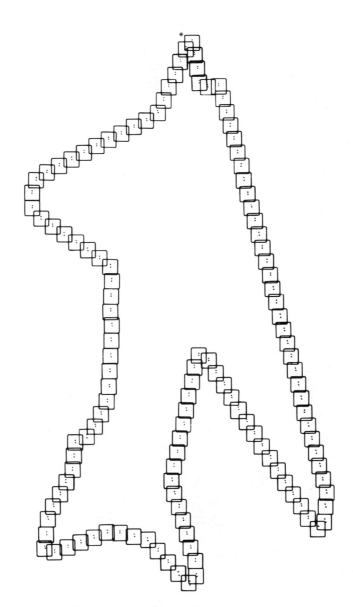

Figure 5.29　128-Walsh-Dot Approximation to a Plane

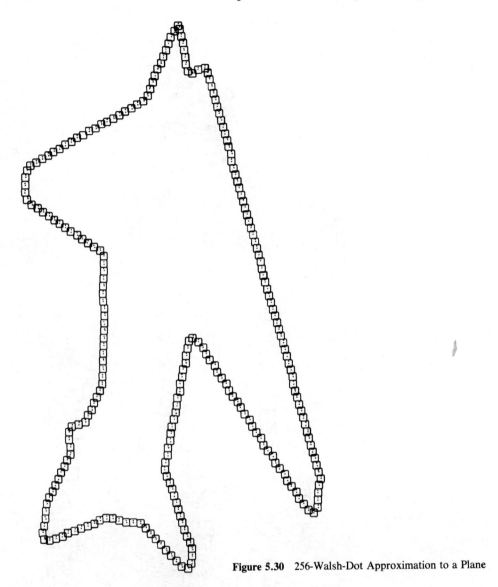

Figure 5.30 256-Walsh-Dot Approximation to a Plane

EXERCISES

5.1. As in Example 5.3, determine the best approximation x_P to x by a vector in the subspace determined by u_2 and u_3.

5.2. Show geometrically that $x_P = z$ determined in Exercise 5.1 minimizes the Euclidean norm $\|x - z\|$ among all vectors z lying in the plane determined by u_2 and u_3. What is $\|x - x_P\|$?

5.3. Give an example showing why absolute convergence is necessary for the definition of a vector space to hold in Example 5.6.

5.4. Consider the set S of all m by n bound matrices $(a_{pq})_{rs}$ with $m \geq 1$, $n \geq 1$ and fixed r, s. Assume that the information density is nonzero. Is S a vector space? Show that any subset of bound matrices in S having a common domain is an inner-product space.

5.5. Prove that the images u_1, u_2, u_3, and u_4 given in Example 5.16 form an orthonormal system for all saturated images in $\mathcal{B}_{0;1}^{2;2}$.

5.6. Consider the signal

$$f(t) = \begin{cases} \dfrac{\pi - t}{2} & \text{if } 0 \leq t < \pi \\ 0 & \text{if } \pi \leq t < 2\pi \end{cases}$$

Find the projection of $f(t)$ onto $S\langle \mathcal{T}_n^R \rangle$.

5.7. For the signal given in Exercise 5.6, find the square of the L_2 error.

5.8. Apply Parseval's theorem to $f(t)$ of Exercise 5.6.

5.9. Find the amplitude spectrum associated with the Fourier series of $f(t)$ given in Exercise 5.6. Is Theorem 5.12 applicable in this problem?

5.10. Let

$$f(t) = \begin{cases} 2t & \text{if } 0 \leq t < \frac{1}{2} \\ 1 & \text{if } \frac{1}{2} \leq t < 1 \end{cases}$$

Find the first eight terms $f_8(t)$ of the Walsh series for $f(t)$.

5.11. Using $f(t)$ and $f_8(t)$ of Exercise 5.10, compare the actual errors $\| f - f_8 \|$ and $\| f - f_8 \|_2$ with the upper bounds on the error explained in Theorem 5.13.

5.12. Consider the image $f(x, y)$ given in Example 5.30 and find the two-dimensional Walsh representation involving all ten functions Φ_{ij} in Figure 5.10.

5.13. Find the DFT of the image

$$f = \begin{pmatrix} 1 & 0 & -1 & 1 & 2 & 1 & 0 & 1 \\ 0 & 0 & 1 & 2 & 1 & 0 & 1 & 0 \end{pmatrix}_{2,3}$$

5.14. Consider the chain code 000543, which largely corresponds to the upper part of the image shown in Figure 5.15. Find the projections of this image and the corresponding Fourier approximation using four and eight harmonics.

5.15. Find the first harmonic ellipse associated with the image of Exercise 5.14. Then find the angle between the major axis of this ellipse and the abscissa.

5.16. For the image f given in Exercise 5.14, find the first eight terms in the Walsh approximations and discuss the ϵ-net associated with f.

6

Probability for Image Processing

6.1 PROBABILITY SPACE

A *probability space* is an ordered triple $(S, \mathcal{F}, P)$ consisting of a set S of outcomes (from some experiment), a collection $\mathcal{F}$ of subsets of S, and a probability measure P. S is called the *sample space*, and the subsets of S which are contained in $\mathcal{F}$ are known as *events*. The choice of which subsets of S to include in $\mathcal{F}$ is not purely arbitrary: $\mathcal{F}$ must satisfy the criteria of a sigma field to be subsequently discussed.

In image processing applications, the set S may be finite or infinite. Sometimes S may be a subset of the real line R or a subset of the plane R^2. Other times S may be a collection of images or feature vectors associated with specific images.

A *sigma field* (σ-field) $\mathcal{F}$ over S is a collection of subsets of S satisfying the following three constraints:

F1. The sample space S must be in $\mathcal{F}$.
F2. If A and B are in $\mathcal{F}$, then $A - B$ must also be in $\mathcal{F}$.
F3. If $A_1, A_2, \ldots$ are in $\mathcal{F}$, then $A_1 \cup A_2 \cup A_3 \cup \ldots$ is in $\mathcal{F}$.

An immediate corollary of F1 and F2 is that the empty set $\emptyset$ is in $\mathcal{F}$, since $\emptyset = S - S \in \mathcal{F}$. Moreover, if $A_1, A_2, \ldots, A_m \in \mathcal{F}$, then $A_1 \cup A_2 \cup \cdots \cup A_m \in \mathcal{F}$. This follows from F3 by letting $A_{m+1} = A_{m+2} = \cdots = \emptyset$. Notice that the power set 2^S is always a σ-field over S. Whenever S is finite, the set of all subsets of S will often be employed as the σ-field in question, and hence every subset of S is an event.

A probability measure must be defined on the class of events $\mathcal{F}$. This is done by assigning to each event a number between zero and one. In practice, these probabilities are usually determined by statistical procedures with the aid of controlled experiments. These empirically ascertained probabilities only approximate the true underlying probabilities. However, being more interested for now in the mathematical structure than the physical situation, we shall ignore this approximation. Mathematically, a probability measure P is a function

$$P: \mathcal{F} \to [0, 1]$$

such that

P1. $P(S) = 1$.
P2. If $A_1, A_2, \ldots$ are mutually exclusive events—i.e., $A_i \cap A_j = \emptyset$ for $i \neq j$—then

$$P(A_1 \cup A_2 \cup \cdots) = P(A_1) + P(A_2) + \cdots \quad \text{(countable additivity)}$$

Condition P1 states that the probability of the whole sample space is 1. Two immediate consequences of P1 and P2 are:

1. $P(\emptyset) = 0$.
2. $P(A \cup B) = P(A) + P(B)$ if $A \cap B = \emptyset$.

The latter property is simply called *additivity*. When S is finite and $\mathcal{F} = 2^S$, we often define P by $P(A) = \text{card}(A)/\text{card}(S)$. We call this probability the *equally likely* probability since it means that

$$P(\{x\}) = \frac{1}{\text{card}(S)}$$

for any singleton event $\{x\}, x \in S$.

Example 6.1

Let the sample space S be a four-point subset of $Z \times Z$ consisting of the points $(0, 0), (0, 1), (1, 0),$ and $(1, 1)$. As previously mentioned, we could use the set 2^S of all sixteen subsets of S as a sigma field over S. However, to illustrate the definition, we will instead let the sigma field be

$$\mathcal{F} = \{\emptyset, S, \{(0, 0), (0, 1)\}, \{(1, 0), (1, 1)\}\}$$

It is straightforward to verify that $\mathcal{F}$ satisfies the three conditions for a sigma field. To complete the definition of a probability space $(S, \mathcal{F}, P)$, a probability function P must be defined on $\mathcal{F}$. For $\emptyset$ and S, there is no option: we must have $P(\emptyset) = 0$ and $P(S) = 1$. Moreover, if we let $P(\{(0, 0), (0, 1)\}) = \frac{1}{3}$, then, by countable additivity, we must set $P(\{(1, 0), (1, 1)\}) = \frac{2}{3}$. In terms of image processing, the events in sigma field $\mathcal{F}$ might be interpreted as constant images (with value unity). This is similar to the identification of subsets of $Z \times Z$ with constant images that was made in the case of digital Minkowski algebra in Section 3.7. Under this identification, it follows that the image

$$\mathcal{E}_1 = \begin{pmatrix} 1 & * \\ \textcircled{1} & * \end{pmatrix}$$

Sec. 6.1   Probability Space   307

which corresponds to the event $\{(0, 0), (0, 1)\}$, occurs with probability $\frac{1}{3}$, whereas the image

$$\mathscr{E}_2 = \begin{pmatrix} * & 1 \\ \circledast & 1 \end{pmatrix}$$

which corresponds to the event $\{(1, 0), (1, 1)\}$, occurs with probability $\frac{2}{3}$.

The images $\mathscr{E}_1$ and $\mathscr{E}_2$ of the preceding example might be regarded as noise added onto a "deterministic image." Instrumental errors might cause this addition. For instance,

$$W_1 = \begin{pmatrix} 5 & 3 \\ \circled{7} & 2 \end{pmatrix}$$

might be observed $33\frac{1}{3}$ percent of the time and

$$W_2 = \begin{pmatrix} 4 & 4 \\ \circled{6} & 3 \end{pmatrix}$$

might be observed the other $66\frac{2}{3}$ percent of the time. Using the additivity assumption concerning the noise yields

$$W = \begin{pmatrix} 4 & 3 \\ \circled{6} & 2 \end{pmatrix}$$

as the deterministic image, since

$$W_1 = \text{EXTADD}(W, \mathscr{E}_1)$$

and

$$W_2 = \text{EXTADD}(W, \mathscr{E}_2)$$

More useful image processing probability spaces arise through the use of random fields. A *random field* on a finite subset X of $Z \times Z$ is a probability space $(S, \mathscr{F}, P)$ where the sample space S consists of all functions w from X into a finite subset Y of the real line R. Symbolically,

$$w: X \to Y$$

and w is therefore an image. An extended version of a random field, to be employed shortly, will allow a star ($*$) to be included in the codomain Y. For convenience, the sigma field $\mathscr{F}$ is taken to be the set of all subsets of S. Consequently, events consisting of sets of images and probabilities of those events are defined in accordance with all the axioms above. Extended definitions of a random field exist for the case where X or Y has infinite cardinality.

For a random field, as just defined, if card $(X) = n$ and card $(Y) = m$, then there are m^n images comprising the sample space S. Therefore, the probabilities of $2^{(m^n)}$ events must be defined.

Example 6.2

Let $X = \{(i, j): i, j = 0, 1, 2\}$ and $Y = \{-2, -1, 0, 1, 2\}$. Then the sample space S has 5^9 images as outcomes. One of these is

$$w = \begin{pmatrix} -2 & -1 & 0 \\ 2 & 1 & 1 \\ ② & -1 & 2 \end{pmatrix}$$

and five others are constant images

$$w_a = \begin{pmatrix} a & a & a \\ a & a & a \\ ⓐ & a & a \end{pmatrix}$$

where $a = -2, -1, 0, 1, 2$.

If we define the probability P to be equally likely, then for any event A, $P(A)$ = card (A)/card (S). For instance, $P(\{w_{-2}, w_1, w_0\}) = 3/5^9$. In this example, the probability of any singleton event is very small, namely, $1/5^9$, while the probability of obtaining an image that has no zero-valued pixels is $(4/5)^9$. Combination and permutation formulas can be utilized in calculating numerous other probabilities; however, we shall not proceed in that direction.

The following theorem presents some of the most fundamental properties of a probability space.

Theorem[1] 6.1. Let $(S, \mathcal{F}, P)$ be an arbitrary probability space, and let A and B be events. Then

(a) If $A_1, A_2, \ldots, A_n$ are mutually exclusive events, then

$$P\left(\bigcup_{i=1}^{n} A_i\right) = \sum_{i=1}^{n} P(A_i).$$

(b) $P(A - B) = P(A) - P(A \cap B)$.
(c) If $B \subset A$, then $P(A - B) = P(A) - P(B)$.
(d) If $B \subset A$, then $P(A) \geq P(B)$.
(e) $P(A^c) = 1 - P(A)$.
(f) $P(A \cup B) = P(A) + P(B) - P(A \cap B)$.
(g) If $B_1, B_2, \ldots, B_N$ are events that *partition* S, i.e., if $B_1, B_2 \ldots, B_N$ are mutually exclusive and their union is S, then

$$P(A) = \sum_{n=1}^{N} (A \cap B_n)$$

---

[1] Marek Fisz, *Probability Theory and Mathematical Statistics* (New York: John Wiley & Sons, Inc., 1963), pp. 12–16.

Sec. 6.2   Conditional Probabilities   309

(h) If $A_1, A_2, \ldots$ are events such that $A_{i+1} \subset A_i$, then $P(A_1 \cap A_2 \cap \ldots) = \lim_{n \to \infty} P(A_n)$.

(i) If $A_1, A_2, \ldots$ are events such that $A_i \subset A_{i+1}$, then $P(A_1 \cup A_2 \cup \ldots) = \lim_{n \to \infty} P(A_n)$.

6.2 CONDITIONAL PROBABILITIES

A problem that often arises is to determine the probability of some event A given that another event B is known to occur. This probability is called the *conditional probability of A given B* and is denoted by $P(A \mid B)$. It is given by

$$P(A \mid B) = \begin{cases} \dfrac{P(A \cap B)}{P(B)} & \text{if } P(B) \neq 0 \\ \text{undefined} & \text{if } P(B) = 0 \end{cases}$$

When $P(B) \neq 0$, cross-multiplication yields

$$P(A \mid B) \cdot P(B) = P(A \cap B)$$

Example 6.3

Consider an extended random field with $X = \{(0, 0), (0, 1), (1, 1)\}$ and $Y = \{-1, 0, 1, *\}$. Then a total of 64 different images, including the empty image, constitute the sample space S. We shall again use the equally likely model for the probability function P. Let A be the event *the image has at least one gray value zero*; let B be the event *the image has two activated pixels*; let C be the event *the image contains only one activated pixel*; and let D be the event *the image contains one or two activated pixels*. Since $B \cap C = \emptyset$, $P(D) = P(B \cup C) = P(B) + P(C)$. But $P(B) = \frac{27}{64}$ and $P(C) = \frac{9}{64}$; therefore, $P(D) = \frac{36}{64}$. Notice that $P(B \mid C) = 0$. More interestingly, using $P(A) = \frac{37}{64}$,

$$P(B \mid A) = \frac{P(B \cap A)}{P(A)} = \frac{15}{37}$$

This is so because there are only 15 images that contain two activated pixels, at least one of which is a zero. (See Figure 6.1.)

Since

$$P(A \mid B)P(B) = P(A \cap B) = P(B \cap A) = P(B \mid A)P(A)$$

it follows that

$$P(B \mid A) = \frac{P(A \mid B)P(B)}{P(A)}$$

whenever $P(A) \neq 0$. This is often called *Bayes' Theorem*. By using part (g) of Theorem 6.1, we obtain a more practical form of Bayes' Theorem: if B_1, B_2,

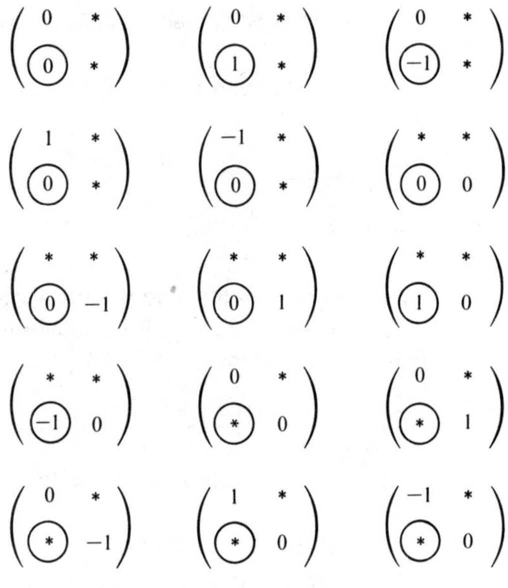

$$\begin{pmatrix} 0 & * \\ 0 & * \end{pmatrix} \quad \begin{pmatrix} 0 & * \\ 1 & * \end{pmatrix} \quad \begin{pmatrix} 0 & * \\ -1 & * \end{pmatrix}$$

$$\begin{pmatrix} 1 & * \\ 0 & * \end{pmatrix} \quad \begin{pmatrix} -1 & * \\ 0 & * \end{pmatrix} \quad \begin{pmatrix} * & * \\ 0 & 0 \end{pmatrix}$$

$$\begin{pmatrix} * & * \\ 0 & -1 \end{pmatrix} \quad \begin{pmatrix} * & * \\ 0 & 1 \end{pmatrix} \quad \begin{pmatrix} * & * \\ 1 & 0 \end{pmatrix}$$

$$\begin{pmatrix} * & * \\ -1 & 0 \end{pmatrix} \quad \begin{pmatrix} 0 & * \\ * & 0 \end{pmatrix} \quad \begin{pmatrix} 0 & * \\ * & 1 \end{pmatrix}$$

$$\begin{pmatrix} 0 & * \\ * & -1 \end{pmatrix} \quad \begin{pmatrix} 1 & * \\ * & 0 \end{pmatrix} \quad \begin{pmatrix} -1 & * \\ * & 0 \end{pmatrix}$$

Figure 6.1 Event $A \cap B$ of Example 6.3

..., B_n are events that partition the sample space, then

$$P(B_i \mid A) = \frac{P(A \mid B_i)P(B_i)}{\sum_{j=1}^{n} P(A \mid B_j)P(B_j)}$$

Example 6.4

Referring to Example 6.3, where $P(B \mid A) = \frac{15}{37}$, $P(A) = \frac{37}{64}$, and $P(B) = \frac{27}{64}$, it follows from Bayes' Theorem that $P(A \mid B) = \frac{15}{27}$.

In image-classification-type applications of Bayes' Theorem, $B_1, B_2, \ldots, B_n$ are often feature vectors or pattern classes representing images that are to be identified. In this context, $P(B_i)$, $i = 1, 2, \ldots, n$, is known as an *a priori probability*. However, as mentioned in the previous section, in practice these probabilities are almost always approximated using experimental statistical procedures. For now, we shall assume that such probabilities, as well as the conditional probabilities $P(A \mid B_i)$, $i = 1, 2, \ldots, n$, are known. The objective is to find the conditional probability $P(B_i \mid A)$, which is called the *a posteriori probability* of B_i. In this context the event A is often referred to as the *observation* event.

A very important concept in probability is the independence of two events A and B in $\mathcal{F}$ for a fixed probability space $(S, \mathcal{F}, P)$. Whenever

$$P(A \cap B) = P(A) \cdot P(B)$$

we say that A and B are *independent*; if A and B are not independent, then they are *dependent*. From the conditional probability formula $P(A \mid B)P(B) = P(A \cap B)$ and the assumption that A and B are independent, it follows that $P(A \mid B) = P(A)$ whenever $P(B) \neq 0$. Conversely, if $P(A \mid B) = P(A)$, then A and B are independent. From an intuitive point of view, the independence of A and B means that knowledge concerning the occurrence of event B does not affect our perception regarding the likelihood of occurrence of event A.

Example 6.5

Referring to Example 6.3, since $P(B \mid A) = \frac{15}{37}$ and $P(B) = \frac{27}{64}$, it follows that events A and B are dependent. In that same example, let F be the event that the image has gray value one at pixel (0, 0). Then $P(F) = \frac{1}{4}$. If G is the event that the image has gray value zero at pixel (1, 0), then $P(F \mid G) = \frac{1}{4}$. Hence, F and G are independent events in this probability space.

6.3 RANDOM VARIABLES AND PROBABILITY DISTRIBUTION FUNCTIONS

Intuitively, a random variable is a real-valued function on the sample space S; however, for theoretical reasons this very general notion will have to be somewhat restricted. Random variables are defined on the probability space $(S, \mathcal{F}, P)$ for numerous reasons. Among the most important of them is the fact that they provide both a numerical and a common basis for treating very diverse and, possibly, nonnumerical sample spaces. Often, once a random variable is defined, the original sample space is forgotten and the values of the random variable are used as the "new sample space." Subsequent manipulations then involve subsets of R and are therefore amenable to conventional computational techniques.

Rigorously, ξ is a random variable on the probability space $(S, \mathcal{F}, P)$ if

RV1. $\xi: S \to R$.
RV2. For every real number r, $\xi^{-1}((-\infty, r])$ is an event (is in $\mathcal{F}$).

RV1 says that ξ is a real-valued function on S. RV2 requires that the inverse image of any half-infinite ray $(-\infty, r]$ be a member of the given sigma field. Notice that ξ is a point function: it maps points in S into real numbers. On the other hand, P is a set function, taking "special" subsets of S, events, into real numbers (between zero and one).

Example 6.6

Consider the probability space $(S, \mathcal{F}, P)$ of Example 6.1, where the sample space $S = \{(0, 0), (0, 1), (1, 0), (1, 1)\}$ and the sigma field $\mathcal{F} = \{\emptyset, S, \{(0, 0), (0, 1)\}, \{(1, 0), (1, 1)\}\}$. The function ξ defined by $\xi(0, 0) = \xi(0, 1) = 2$, and $\xi(1, 0) = \xi(1, 1) = -3$ is a random variable on this probability space. (See Figure 6.2.)

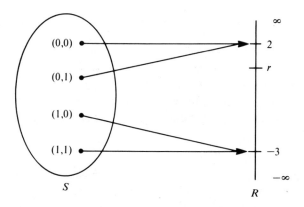

Figure 6.2 A Random Variable on a Probability Space

Notice that if $r < -3$, then

$$\xi^{-1}((-\infty, r]) = \varnothing$$

which is an event. (There are no arrows from points in S arriving below -3 in the real line.) Furthermore, if $-3 \leq r < 2$, then

$$\xi^{-1}((-\infty, r]) = \{(1, 0), (1, 1)\}$$

which is also a member of the sigma field $\mathscr{F}$. (Only two arrows leave points $(1, 0)$ and $(1, 1)$ in S and arrive in R below any value r that is less than 2 and greater than or equal to -3.) Finally, if $2 \leq r$, then

$$\xi^{-1}((-\infty, r]) = S$$

which is always an event. In this example a random variable will be any function that maps points $(0, 0)$ and $(0, 1)$ onto the same real number and also maps $(1, 0)$ and $(1, 1)$ onto a common value. If either of these conditions is violated, then a random variable will not result. Note that, from a mathematical viewpoint, a random variable is arbitrary as long as it satisfies the defining conditions.

Whenever all points of a sample space are mapped onto a single real number, the resulting function will be a random variable. In this case it is called a *constant* random variable.

As mentioned earlier, once a random variable is given, the original sample space is usually forgotten. Instead of saying that an outcome s in S is observed, we say that $\xi(s)$ in R is observed. Moreover, the probability structure of the given probability space $(S, \mathscr{F}, P)$ induces a "somewhat identical" probability space, where, as has just been mentioned, the new sample space is $\xi(S)$, the range of the random variable. Often, all of R is used instead of $\xi(S)$, with events in $\xi(S)^c$ having probability zero.

The new σ-field for this induced space is called the *Borel* field and is denoted by $\mathscr{B}$. It is the smallest σ-field over R which contains all open, closed, and half-open–half-closed intervals. The subsets of R which are in $\mathscr{B}$ are known as *Borel*

Sec. 6.3   Random Variables and Probability Distribution Functions   313

sets. For any Borel set B, $\xi^{-1}(B)$, the set of all points in S such that $\xi(s)$ is in B, is an event in $\mathcal{F}$. $\xi^{-1}(B)$ is often denoted by $\xi \in B$. For instance, $\xi^{-1}((-\infty, r])$ is written $\xi \in (-\infty, r]$, or $\xi \leq r$. The induced probability function on R is given by

$$P_\xi(B) = P(\xi^{-1}(B)) = P(\xi \in B)$$

In sum, given any probability space $(S, \mathcal{F}, P)$ and a random variable ξ, a new and more experimentally suitable probability space $(R, \mathcal{B}, P_\xi)$ can be found which preserves the probabilistic model. Events in the new space have an inverse image event in the original space, and both events have the same probability of occurrence. Moreover, the probability function P_ξ is completely characterized by its values on half-rays $(-\infty, r]$. Consequently, in the rest of this text we shall usually begin with the probability space $(R, \mathcal{B}, P_\xi)$ and ignore the original setting which induced this space. We shall be more interested in studying and applying the values of the given random variable. Toward this end, we define the probability *distribution function*

$$F_\xi : R \to R$$

associated with ξ by

$$F_\xi(x) = P_\xi((-\infty, x]) = P(\xi \leq x)$$

Example 6.7

Referring to Examples 6.1 and 6.6, for $r < -3$,

$$F_\xi(r) = P_\xi((-\infty, r]) = P(\xi \leq r) = P(\emptyset) = 0$$

For $-3 \leq r < 2$

$$F_\xi(r) = P_\xi((-\infty, r]) = P(\xi \leq r)$$
$$= P(\{(1, 0), (1, 1)\}) = \frac{2}{3}$$

And for $2 \leq r$,

$$F_\xi(r) = P_\xi((-\infty, r]) = P(\xi \leq r) - P(S) = 1$$

The distribution function $F_\xi(r)$ is plotted in Figure 6.3.

The following four properties of a probability distribution function, illustrated in the preceding example, follow easily from the definition:

D1. $F_\xi(-\infty) = 0$, which means that $\lim_{x \to -\infty} F_\xi(x) = 0$.
D2. $F_\xi(\infty) = 1$, which means that $\lim_{x \to +\infty} F_\xi(x) = 1$.
D3. $F_\xi(x)$ is monotonically increasing; that is, if $x \leq r$, then $F_\xi(x) \leq F_\xi(r)$.
D4. $F_\xi(x)$ is continuous from the right; that is, $\lim_{x \to r^+} F_\xi(x) = F_\xi(r)$.

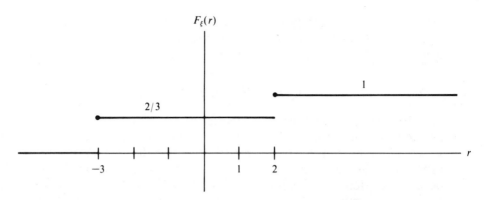

Figure 6.3 Probability Distribution Function

Among the most important facts in probability is that if a function $F(x)$ obeys properties D1–D4, then it will be a probability distribution function for some random variable ξ.[2] Consequently, the starting point for many investigations of probability in image processing will be the specification of a probability distribution function.

For instance, given a nonnull digital image $a = (a_{pq})_{rt}$, we define the associated *gray value distribution* $F_\xi(x)$ by

$$F_\xi(x) = P(\xi \leq x) = \frac{\text{card}\{a_{pq}: a_{pq} \leq x\}}{\text{card }(D_a)}$$

where D_a denotes the domain of a. We can then study $F_\xi(x)$ without regard to any underlying sample space.

Example 6.8

The associated gray value distribution $F_\xi(x)$ for the image

$$a = \begin{pmatrix} 1 & 2.2 & 2 \\ 3 & * & * \\ \textcircled{0} & * & 1 \end{pmatrix}$$

is given in Figure 6.4.

Example 6.9

A possible gray level distribution for a continuous image might be

$$F_\xi(x) = \frac{1}{2} + \frac{1}{\pi} \tan^{-1}(x)$$

since $F_\xi(-\infty) = 0$, $F_\xi(\infty) = 1$, $F_\xi(x)$ is monotonically increasing, and $F_\xi(x)$ is continuous (and therefore continuous from the right).

---

[2] William Feller, *An Introduction to Probability Theory and Its Applications* (New York: John Wiley & Sons, Inc., 1966), p. 127.

Sec. 6.4 The Probability Density Function

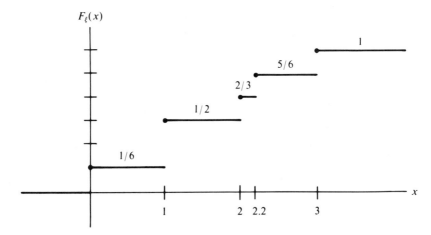

Figure 6.4 Gray Value Distribution Function

Example 6.10

If

$$F_\xi(x) = \begin{cases} 0 & \text{if } x < -1 \\ \tfrac{1}{2} & \text{if } -1 \leq x \leq 0 \\ \tfrac{1}{2} + x/2 & \text{if } 0 < x < 1 \\ 1 & \text{if } 1 \leq x \end{cases}$$

then $F_\xi(x)$ is a distribution function. Moreover, if $F_\xi(x)$ is the gray level distribution function of a continuous image, then "half the image" will have gray values equal to minus one, while the "other half" will have gray values equally distributed between zero and one. To see the first point, note that $P(\xi \leq s) = \tfrac{1}{2}$ for any s such that $-1 \leq s \leq 0$, while $P(\xi \leq s) = 0$ for $s < -1$.

6.4 THE PROBABILITY DENSITY FUNCTION

The probability distribution function $F_\xi(x)$ introduced in the previous section is extremely well behaved. Since it is bounded and monotonic, and therefore of bounded variation, it is continuous everywhere except at a countable set of points.[3] Moreover, it has a derivative $f_\xi(x)$ almost everywhere;[4] specifically, $F'_\xi(x) = f_\xi(x)$ everywhere except at a set of points on the real line whose "total length is zero."[5] Indeed, $F_\xi(x)$ cannot have a derivative wherever it is not con-

---

[3] Ralph P. Boas, Jr., *A Primer of Real Functions* (New York: John Wiley & Sons, Inc., 1963), p. 129.

[4] Ibid., p. 134.

[5] See footnote 5, Chapter 4.

tinuous, and even at points of continuity there may not be a derivative. Of more theoretical concern is the fact that even if $F_\xi(x)$ is continuous, in general,

$$\int_a^b f_\xi(x)\, dx \leq F_\xi(b) - F_\xi(a)$$

with equality not necessarily holding, and where the integral is understood in the Lebesgue sense.

In order to avoid pathological functions, which are not of practical significance, whenever $F_\xi(x)$ is continuous, we shall assume equality holds in the preceding integral. In such a case, ξ is called a *continuous* random variable and we have

$$P(a < \xi \leq b) = F_\xi(b) - F_\xi(a) = \int_a^b f_\xi(x)\, dx$$

Employing mathematical terminology, this amounts to the assumption that $F_\xi(x)$ is *absolutely continuous*.[6] Furthermore, in the general situation, we shall assume that $F_\xi(x)$ can be represented as the sum of an absolutely continuous component plus a discontinuous step-type component, even though in the most general case there might also be a *singular* part.[7] These components will temporarily be denoted by F_1 and F_2, respectively. Step-type functions are illustrated in Figures 6.3 and 6.4, and in these cases the absolutely continuous components are zero. In general, there may be an infinite number of discontinuities in the step-type function F_2; however, we assume that no real point is a limit point for the set of discontinuities.

The rest of this section will be presented in a formal, nonrigorous manner; however, it is rigorous when $F_\xi(x)$ contains no discontinuities or singular part. Under the restricted conditions mentioned in the preceding paragraph, the *probability density function*, also called a *probability frequency function*, is denoted by $f_\xi(x)$ and given by

$$f_\xi(x) = \frac{dF_1(x)}{dx} + \sum_{n=-\infty}^{\infty} \alpha_n \delta(x - a_n)$$

where a_n is a point of discontinuity of $F_2(x)$, $\delta(\cdot)$ is the delta "function" discussed in the appendix, and α_n is the probability of a_n occurring. (Equivalently, α_n is the jump value of F_2 at a_n; that is, $\alpha_n = F_2(a_n) - \lim_{x \to a_n^-} F_2(x)$.)

Whenever $F_\xi(x)$ is continuous, so that there is no step discontinuity component F_2,

$$f_\xi(x) = \frac{dF_\xi(x)}{dx} = \frac{dF_1(x)}{dx}$$

---

[6] Walter Rudin, *Real and Complex Analysis* (New York: McGraw-Hill, 1966), p. 121.

[7] I. P. Natanson, *Theory of Functions of a Real Variable*, rev. ed., trans. Leo F. Boron (New York: Frederick Ungar Publishing Co., 1961), p. 264.

Sec. 6.4 The Probability Density Function   317

On the other hand, whenever $F_1(x) = 0$, we have a pure "discrete" density

$$f_\xi(x) = \sum_{n=-\infty}^{\infty} a_n \delta(x - a_n)$$

and formally, using the properties of the delta function,

$$F_\xi(x) = \int_{-\infty}^{x} f_\xi(t)\, dt$$

Example 6.11

For $F_\xi(x)$ given in Example 6.7, we have the pure discrete density $f_\xi(x) = \tfrac{2}{3}\delta(x + 3) + \tfrac{1}{3}\delta(x - 2)$. (See Figure 6.5.)

Example 6.12

Referring to Example 6.9, let

$$F_\xi(x) = \frac{1}{2} + \frac{1}{\pi} \tan^{-1} x$$

Then

$$f_\xi(x) = \frac{dF_\xi(x)}{dx} = \frac{1}{\pi}\left(\frac{1}{1 + x^2}\right)$$

which is called the *Cauchy density function*. Notice that

$$F_\xi(x) = \frac{1}{\pi} \int_{-\infty}^{x} \frac{1}{1 + t^2}\, dt$$

and is known as the *Cauchy distribution function*.

Example 6.13

Consider the gray level distribution function $F_\xi(x) = F_1(x) + F_2(x)$ of Example 6.10. The continuous part of this function is given by

$$F_1(x) = \begin{cases} 0 & \text{if } x < 0 \\ x/2 & \text{if } 0 \leq x < 1 \\ \tfrac{1}{2} & \text{if } 1 \leq x \end{cases}$$

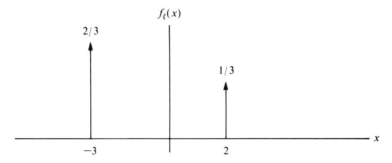

Figure 6.5 Density Function Involving Delta Functions

and the discontinuous, or discrete, component is given by

$$F_2(x) = \begin{cases} 0 & \text{if } x < -1 \\ \frac{1}{2} & \text{if } -1 \leq x \end{cases}$$

Hence,

$$f_\xi(x) = \frac{1}{2} I_{(0,1)}(x) + \frac{1}{2} \delta(x + 1)$$

since $(d/dx)(x/2) = \frac{1}{2}$ and, in a formal sense, $(d/dx)u(x) = \delta(x)$, where $u(x)$ is the unit step function at the origin. (See Figure 6.6.) As usual,

$$F_\xi(x) = \int_{-\infty}^{x} f_\xi(t)\, dt$$

The probability density function satisfies the following two basic properties:

d1. $f_\xi(t) \geq 0$.
d2. $\int_{-\infty}^{\infty} f_\xi(t)\, dt = 1$.

These properties characterize probability density functions in the sense that, given $f(t)$ satisfying d1 and d2, there exists a corresponding probability distribution function $F_\xi(x)$ given by

$$F_\xi(x) = \int_{-\infty}^{x} f_\xi(t)\, dt$$

Furthermore, as mentioned previously,

$$P(a < \xi \leq b) = \int_{a}^{b} f_\xi(t)\, dt$$

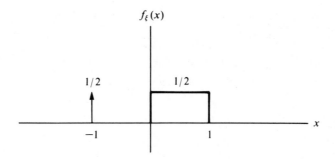

Figure 6.6 Density Function Corresponding to Mixed (Continuous and Discrete) Distribution

Sec. 6.4   The Probability Density Function

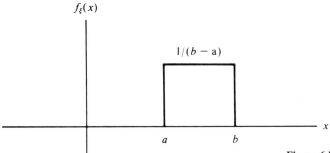

Figure 6.7   Uniform Density Function

Example 6.14

Consider the *uniform* density function

$$f_\xi(x) = \frac{1}{b-a} I_{(a,b)}(x)$$

for $b > a$. (See Figure 6.7.) The corresponding distribution function is

$$F_\xi(x) = \frac{x-a}{b-a} I_{(a,b)}(x) + I_{[b,\infty)}(x)$$

(See Figure 6.8.)

Example 6.15

The *Gaussian*, or *normal*, density function is given by

$$f_\xi(x) = \frac{1}{\sqrt{2\pi}\sigma} e^{-(x-\mu)^2/2\sigma^2}$$

where μ and $\sigma > 0$ are real-valued parameters. Notice that by changing μ and σ an entire family of densities, and therefore distribution functions, results (see Figure 6.9).

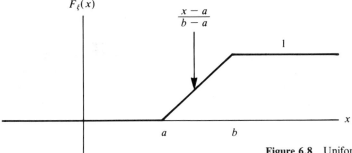

Figure 6.8   Uniform Distribution Function

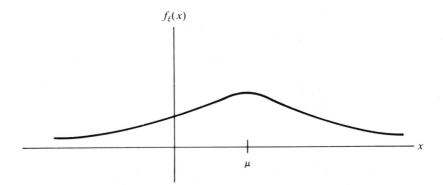

Figure 6.9 Gaussian Density Function

6.5 MATHEMATICAL EXPECTATION AND PARAMETERS

A principal purpose of this section is to describe functions that are useful in characterizing the main features of a probability density function. We begin with moments about the origin. The *nth moment* about the origin of the random variable ξ, $n = 0, 1, 2, \ldots$, is denoted by $E(\xi^n)$ and is given by the Lebesgue-Stieltjes integral

$$E(\xi^n) = \int_{-\infty}^{\infty} x^n dF_\xi(x)$$

provided that

$$\int_{-\infty}^{\infty} |x|^n \, dF_\xi(x) < \infty$$

In keeping with the formal presentation of the previous section, the *n*th moment can also be given in terms of the probability density function by

$$E(\xi^n) = \int_{-\infty}^{\infty} x^n f_\xi(x) \, dx$$

provided that

$$\int_{-\infty}^{\infty} |x|^n \, f_\xi(x) \, dx < \infty$$

Integrals must be absolutely convergent for the existence of moments. When $n = 1$, $E(\xi)$ is called the *mean* or *average*. In general, $E(\xi)$ corresponds to the center of gravity on the x axis of the density. For $n = 0$, we obtain $E(1) = 1$.

Example 6.16

Consider the Cauchy density

$$f_\xi(x) = \frac{1}{\pi} \frac{1}{1 + x^2}$$

introduced in Example 6.12. There is no mean value in this case since

$$\int_{-\infty}^{\infty} \frac{|x|}{1 + x^2} dx = \infty$$

Obviously, no other moments exist either.

Example 6.17

For the uniform density presented in Example 6.14,

$$E(\xi^n) = \frac{1}{b - a} \int_{a}^{b} x^n \, dx$$

$$= \frac{b^{n+1} - a^{n+1}}{(n + 1)(b - a)}$$

In general, for a real-valued, piecewise-continuous function g on R,[8] the *expected value* $E(g(\xi))$ can be defined by

$$E(g(\xi)) = \int_{-\infty}^{\infty} g(x) f_\xi(x) \, dx$$

provided that

$$\int_{-\infty}^{\infty} |g(x)| f_\xi(x) \, dx < \infty$$

Note that an immediate consequence of the definition is the linearity of expectation; that is,

$$E[\alpha g(\xi) + \beta h(\xi)] = \alpha E(g(\xi)) + \beta E(h(\xi))$$

for constants α and β.

Throughout the rest of the text, the absolute convergence property will be assumed to hold and will not be stated explicitly. With that in mind, some important expected values will now be presented. Many of these involve the *central nth moment* $E((\xi - \mu)^n)$, where μ denotes the mean $E(\xi)$. The *variance* is then given by $E((\xi - \mu)^2)$ and is denoted by σ^2. The *standard deviation* is the positive square root of the variance, i.e.,

$$\sigma = \sqrt{E((\xi - \mu)^2)}$$

Since

$$\sigma^2 = E(\xi^2) - 2\mu E(\xi) + \mu^2 E(1)$$

$E(\xi) = \mu$ and $E(1) = 1$,

$$\sigma^2 = E(\xi^2) - \mu^2$$

---

[8] Though, for the sake of simplicity, we have assumed that g is piecewise continuous, more generally g can be a *Baire* function. For a discussion of Baire functions, see M. E. Munroe, *Measure and Integration*, 2d ed. (Reading, MA: Addison-Wesley, 1971), p. 97.

Intuitively, the variance is indicative of the "width" of the density function. The *mean deviation* is given by $E(|\xi - \mu|)$ and, like the variance, is indicative of the "width" of the density function $f_\xi(x)$.

We now introduce several transform techniques associated with expected value. The *moment generating function* for ξ is

$$E(e^{-\xi s})$$

where s is a complex quantity. (We can refer to Section 4.6 for numerous properties; the moment generating function is the bilateral Laplace transform of $f_\xi(x)$.) Similarly, the *characteristic function* of ξ is given by $E(e^{-j\omega\xi})$, with $j = \sqrt{-1}$. This quantity is the Fourier transform of $f_\xi(x)$. Moments associated with this random variable can be found using Theorem 4.2. Finally, the *moment factorial generating function* is given by $E(z^\xi)$, which is also called the z transform of $f_\xi(x)$. These three transform techniques are employed in similar ways; however, the first two are usually used in the continuous case while the z transform is usually employed in the discrete case.

Example 6.18

Consider the *geometric density*

$$f_\xi(x) = p \sum_{n=1}^{\infty} (1-p)^{n-1} \delta(x-n), \text{ for } 0 < p < 1$$

which is discrete. $f_\xi(x)$ is a density since $f_\xi(x) \geq 0$ and

$$\int_{-\infty}^{\infty} f_\xi(x) \, dx = p \sum_{n=1}^{\infty} (1-p)^{n-1} = \frac{p}{1-(1-p)} = 1$$

Formally, the z transform of $f_\xi(x)$ is given by

$$E(z^\xi) = p \sum_{n=1}^{\infty} (1-p)^{n-1} \int_{-\infty}^{\infty} z^x \delta(x-n) \, dx = p \sum_{n=1}^{\infty} z^n (1-p)^{n-1}$$

Therefore,

$$E(z^\xi) = \frac{pz}{1 - z(1-p)}$$

provided that $|z(1-p)| < 1$. Taking the derivative of $E(z^\xi)$ with respect to z and setting $z = 1$ formally gives the mean:

$$\mu = E(\xi) = \frac{d}{dz}[E(z^\xi)]\bigg|_{z=1} = \frac{d}{dz}\left(\frac{pz}{1-z(1-p)}\right)\bigg|_{z=1} = \frac{1}{p}$$

Using

$$\frac{d^2 E(z^\xi)}{dz^2}\bigg|_{z=1} = E(\xi(\xi-1)) = E(\xi^2) - \mu$$

gives

$$E(\xi^2) = \frac{2 - p}{p^2}$$

so that

$$\sigma^2 = E(\xi^2) - \mu^2 = \frac{1 - p}{p^2}$$

Higher order moments can also be found using the moment factorial generating function. This name for $E(z^\xi)$ is based on the identity

$$\left.\frac{d^n E(z^\xi)}{dz^n}\right|_{z=1} = E(\xi(\xi - 1)(\xi - 2) \cdots (\xi - n + 1))$$

which is called a *factorial moment*.

One of the most celebrated results in probability theory is the Chebyshev inequality, which is given in the following theorem.

Theorem 6.2[9] **(Chebyshev Inequality).** Let ξ be a random variable with mean μ and variance σ^2. Then for any $\epsilon > 0$,

$$P(|\xi - \mu| \geq \epsilon) \leq \frac{\sigma^2}{\epsilon^2}$$

In particular, if $\epsilon = k\sigma$, with $k > 0$ and $\sigma > 0$, then

$$P(|\xi - \mu| \geq k\sigma) \leq \frac{1}{k^2}$$

The Chebyshev inequality gives meaning to the standard deviation; it shows that when σ is large, the spread (or width) of the density $f_\xi(x)$ about the mean μ is large, and when σ is small, the spread of $f_\xi(x)$ about the mean is small. Figure 6.10 depicts the case $k = 2$; there the probability that ξ is greater than or equal to 2σ from the mean is less than or equal to $\frac{1}{4}$.

Example 6.19

Consider the image

$$a = \begin{pmatrix} -2 & 0 & 0 \\ 0 & \circledast & 0 \\ 0 & 0 & 2 \end{pmatrix}$$

---

[9] Harald Cramer, *Mathematical Methods of Statistics* (Princeton: Princeton University Press, 1966), p. 182.

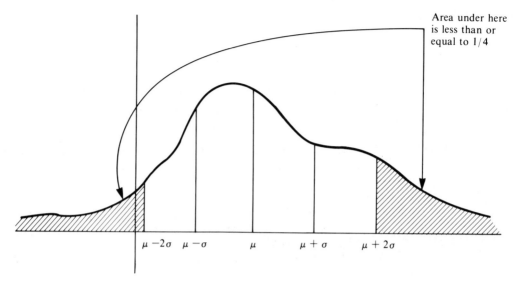

Figure 6.10 Chebyshev Inequality

and let ξ be the gray level random variable associated with this image. The gray level distribution function is

$$F_\xi(x) = \frac{1}{8} I_{[-2,\infty)}(x) + \frac{3}{4} I_{[0,\infty)}(x) + \frac{1}{8} I_{[2,\infty)}(x)$$

(See Figure 6.11(a).) The corresponding (discrete) density is given by

$$f_\xi(x) = \frac{1}{8} \delta(x + 2) + \frac{3}{4} \delta(x) + \frac{1}{8} \delta(x - 2)$$

(See Figure 6.11(b).) The mean is $\mu = 0$ and the variance is $\sigma^2 = 1$. An application of Chebyshev's inequality with $k = 2$ gives

$$P(|\xi - 0| \geq 2) \leq \frac{1}{4}$$

In the preceding example, $P(|\xi| \geq 2) = P(\xi = 2) + P(\xi = -2) = \frac{1}{4}$ precisely, and Chebyshev's inequality is actually equality. This shows that, in general, the inequality cannot be improved. On the other hand, most of the time the inequality will not be equality. Sharper, nonparametric (distribution-free) results will shortly be given involving similar parameters.

The *median* for the distribution function $F_\xi(x)$ is any point r on the abscissa for which $F_\xi(r) = \frac{1}{2}$. Intuitively, in terms of the density function, half the area under $f_\xi(x)$ is to the left of r, and the other half is to the right of r. There are three possibilities: (1) There is one point r such that $F_\xi(r) = \frac{1}{2}$. Then this point r is the median. (2) There are many points r such that $F_\xi(r) = \frac{1}{2}$. Then all these points are called medians. (3) When a horizontal line of height $\frac{1}{2}$ is superimposed on the graph of $F_\xi(x)$, it punctures a unique vertical line at $x = r$ which is indicative

Sec. 6.5 Mathematical Expectation and Parameters   325

of a jump value for $F_\xi(x)$. In this case, we agree to call r the median. Rigorously, $F_\xi(r) > \frac{1}{2}$ and $\lim_{x \to r^-} F_\xi(x) < \frac{1}{2}$. The three possibilities are illustrated in Figure 6.12.

Like the mean, the median is often used to measure central location; however, the median always exists while the mean sometimes does not. A more specialized measure of central tendency is given by the *mode m*. Here only the continuous-case situation is involved, and the mode is any point (if it exists) for which $f_\xi(x)$ has an absolute maximum. When there is a single maximum for the density function, the density is called *unimodal*. (See Figure 6.13.) When there are two modes, the density is called *bimodal*.

Similar to the median, the *pth quantile*, where $0 < p < 1$, is defined as a value r on the abscissa such that

$$F_\xi(r) = p$$

Quantiles are used for purposes similar to those for which the variance is used. For instance, the *w*th *interquantile range*, where $0 < w < 1$, is an interval $[r, r']$ which is found by letting

$$r = F_\xi\left(\frac{1-w}{2}\right)$$

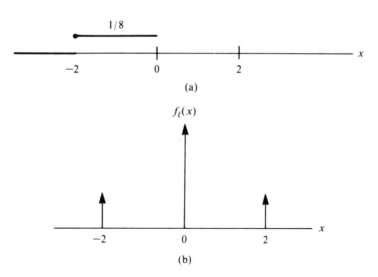

Figure 6.11 Grey Level Distribution Function and Density Function

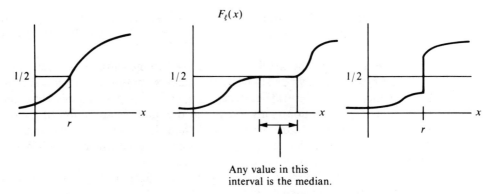

Figure 6.12 Three Cases for Finding the Median

and

$$r' = F_\xi\left(\frac{1+w}{2}\right)$$

A measure of nonsymmetry is given by the *Pearsonian skewness*, which for a unimodal density is defined by

$$s = \frac{\mu - m}{\sigma}$$

where μ is the mean, m is the mode, and $\sigma > 0$ is the standard deviation (when they exist). A result similar to Chebyshev's inequality makes use of some of the parameters just described.

Theorem 6.3[10] **(Gauss inequality).** Let ξ be a continuous random variable with mean μ and variance σ^2. Assume further that $f_\xi(x)$ is unimodal with mode m. Then for any $k > 0$,

$$P(|\xi - m| \geq k t) \leq \frac{4}{9k^2}$$

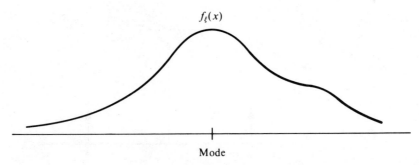

Figure 6.13 Mode: The Highest Point in a Density Function

[10] Ibid., p. 183.

where $t^2 = \sigma^2 + (\mu - m)^2$. Moreover, for $k > |s|$,

$$P(|\xi - \mu| \geq k\sigma) \leq \frac{4}{9} \frac{1 + s^2}{(k - |s|)^2}$$

where s is the Pearsonian skewness.

We conclude the section by mentioning several properties of some of the parameters thus far introduced. If ξ is a random variable with mean μ and variance σ^2, then for any real number c,

$$E((\xi - c)^2) = \sigma^2 + (\mu - c)^2$$

It follows from this that $E((\xi - c)^2)$ is minimized by letting $c = \mu$. Hence, the mean value provides a least-squares solution regarding the quantity $E((\xi - c)^2)$. Similar least-squares problems are discussed in Sections 7.3 and 7.4. Whereas the mean minimizes a two-norm-type criterion, the median provides the minimum for a one-norm criterion. Specifically, let ξ be a random variable for which the first moment exists, and let r denote the median value. Then, among all real numbers c, the value $c = r$ minimizes the quantity[11]

$$E(|\xi - c|) = \int_{-\infty}^{\infty} |x - c| f_\xi(x) \, dx$$

The mode m also satisfies a certain optimality property. For a unimodal density, since m is the value on the abscissa corresponding to the largest value of $f_\xi(x)$, m provides the value x which is "most likely to occur." Consequently, the mode is often said to be the *maximum-likelihood value*.

6.6 SOME IMPORTANT PROBABILITY DENSITY FUNCTIONS

In this short section, several important families of density functions are listed along with some basic properties. Some of these families, such as the geometric and uniform densities, have been introduced in previous sections. Table 6.1 gives discrete densities, while Table 6.2 gives continuous ones. Example 6.20 illustrates the gamma distribution.

Example 6.20

The gamma density function is given by

$$f_\xi(x) = \frac{x^{\alpha-1} e^{-x/\beta}}{\beta^\alpha \Gamma(\alpha)} I_{(0,\infty)}(x)$$

for $\alpha, \beta > 0$. Here, $\Gamma(\alpha)$ is the gamma function

$$\Gamma(\alpha) = \int_0^\infty t^{\alpha-1} e^{-t} \, dt$$

[11] Ibid., p. 179.

TABLE 6.1 DISCRETE PROBABILITY DENSITIES AND SOME RELATED PROPERTIES

Name	Density $f_\xi(x) =$	Constraints on Parameters	Mean $E(\xi) =$	Variance $E(\xi - E(\xi))^2 =$
Geometric	$p \sum_{n=1}^{\infty} (1 - p)^{n-1} \delta(x - n)$	$0 < p < 1$	$\dfrac{1}{p}$	$\dfrac{1-p}{p^2}$
Binomial	$\sum_{k=0}^{n} \binom{n}{k} p^k q^{n-k} \delta(x - k)$	$0 < p < 1$ $q = 1 - p$	np	npq
Negative Binomial	$\sum_{n=0}^{\infty} \binom{n+r-1}{n} p^r q^n \delta(x - n)$	$r = 1, 2, 3, \ldots$ $0 < p < 1$ $q = 1 - p$	$\dfrac{r(r+1)q^2}{p^2}$	$\dfrac{rq}{p^2}$
Poisson	$e^{-\lambda} \sum_{n=0}^{\infty} \dfrac{\lambda^n}{n!} \delta(x - n)$	$\lambda > 0$	λ	λ

An integration by parts shows that $\Gamma(\alpha) = (\alpha - 1)\Gamma(\alpha - 1)$ for $\alpha > 1$. Since $\Gamma(1) = 1$,

$$\Gamma(n) = (n - 1)! \text{ for } n = 1, 2, \ldots$$

The mean can be found by a change of variables $t = x/\beta$:

$$\mu = \frac{1}{\beta^\alpha \Gamma(\alpha)} \int_0^\infty x^\alpha e^{-x/\beta} \, dx = \frac{\beta}{\Gamma(\alpha)} \int_0^\infty t^\alpha e^{-t} \, dt = \frac{\beta \Gamma(\alpha + 1)}{\Gamma(\alpha)} = \alpha\beta$$

Similarly, $E(\xi^2) = \alpha\beta^2(\alpha + 1)$, and so the variance is $\sigma^2 = \alpha\beta^2$. It is important to note that when $\alpha = 1$, the gamma frequency family reduces to the exponential family (in Table 6.2, $\lambda = 1/\beta$), and consequently the mean and variance for the exponential case are β and β^2, respectively.

In the beta density in Table 6.2, the beta function

$$B(x, y) = \int_0^1 t^{x-1}(1 - t)^{y-1} \, dt, \text{ for } x, y > 0$$

implicitly appears since by a change of variables it can be seen that

$$B(x, y) = \frac{\Gamma(x)\Gamma(y)}{\Gamma(x + y)}$$

6.7 FUNCTIONS OF A SINGLE RANDOM VARIABLE

Given a random variable ξ and a (Baire) function $g: R \to R$, the composite $g(\xi)$ is also a random variable. In this section we shall examine the distribution function of several such composites, where the function g will always be piecewise continuous. $F_g(y)$ will denote the distribution function for $g(\xi)$.

Sec. 6.7 Functions of a Single Random Variable

TABLE 6.2 CONTINUOUS PROBABILITY DENSITIES AND SOME RELATED PROPERTIES

Name	Density $f_\xi(x) =$	Constraints on Parameters	Mean Value $E(\xi) =$	Variance $E(\xi - E(\xi))^2 =$
Uniform	$\dfrac{1}{b-a} I_{(a,b)}(x)$	$b > a$	$\dfrac{a+b}{2}$	$\dfrac{(a-b)^2}{12}$
Gaussian	$\dfrac{1}{\sqrt{2\pi}\sigma} e^{-1/2\left(\frac{x-\mu}{\sigma}\right)^2}$	$\sigma > 0$	μ	σ^2
Gamma	$\dfrac{x^{\alpha-1} e^{-x/\beta}}{\beta^\alpha \Gamma(\alpha)} I_{(0,\infty)}(x)$	$\alpha, \beta > 0$	$\alpha\beta$	$\alpha\beta^2$
Exponential	$\lambda e^{-\lambda x} I_{(0,\infty)}(x)$	$\lambda > 0$	$\dfrac{1}{\lambda}$	$\dfrac{1}{\lambda^2}$
Weiball	$\lambda\beta x^{\beta-1} e^{-\lambda t^\beta} I_{(0,\infty)}(x)$	$\lambda, \beta > 0$	$\lambda^{-1/\beta}\Gamma\left(\dfrac{1}{\beta}+1\right)$	$\lambda^{-2/\beta}\left(\Gamma\left(\dfrac{2}{\beta}+1\right) - \Gamma^2\left(\dfrac{1}{\beta}+1\right)\right)$
Beta	$\dfrac{\Gamma(\alpha+\beta)}{\Gamma(\alpha)\Gamma(\beta)} x^{\alpha-1} \cdot (1-x)^{\beta-1} I_{[0,1]}(x)$	$\alpha, \beta > 0$	$\dfrac{\alpha(\alpha+1)}{(\alpha+\beta)(\alpha+\beta+1)}$	$\dfrac{\alpha\beta}{(\alpha+\beta)^2(\alpha+\beta+1)}$
Lognormal	$\dfrac{1}{\sqrt{2\pi}x\sigma} e^{[-(\ln x - \mu)^2/2\sigma^2]} \cdot I_{(0,\infty)}(x)$	$\sigma > 0$	$e^{\mu + \frac{1}{2}\sigma^2}$	$e^{2\mu+2\sigma^2} - e^{2\mu+\sigma^2}$
Laplace	$\dfrac{1}{2c} e^{-|x-a|/c}$	$c > 0$	a	$2c^2$
Pareto	$\dfrac{\lambda a^\lambda}{x^{\lambda+1}} I_{(a,\infty)}(x)$	$a, \lambda > 0$	$\dfrac{\lambda a}{\lambda - 1}$ when $\lambda > 1$; nonexistent otherwise	$\dfrac{\lambda a^2}{(\lambda-1)^2(\lambda-2)}$ when $\lambda > 2$; nonexistent otherwise
Cauchy	$\dfrac{1}{\pi\beta\left[1 + \left(\dfrac{x-a}{\beta}\right)^2\right]}$	$\beta > 0$	nonexistent	nonexistent

Consider the affine function $g(x) = ax + b$, where a and b are real and $a \neq 0$. For $a > 0$, we have

$$F_g(y) = P(g(\xi) \leq y) = P(a\xi \leq y - b) = P\left(\xi \leq \frac{y-b}{a}\right) = F_\xi\left(\frac{y-b}{a}\right)$$

On the other hand, for $a < 0$, the distribution function for $a\xi + b$ is

$$F_g(y) = P(a\xi \leq y - b) = P\left(\xi \geq \frac{y-b}{a}\right) = 1 - P\left(\xi < \frac{y-b}{a}\right)$$

$$= 1 - P\left(\xi \leq \frac{y-b}{a}\right) + P\left(\xi = \frac{y-b}{a}\right)$$

$$= 1 - F_\xi\left(\frac{y-b}{a}\right) + P\left(\xi = \frac{y-b}{a}\right)$$

Putting both cases together, for $a \neq 0$, we have

$$F_g(y) = \begin{cases} F_\xi\left(\frac{y-b}{a}\right) & \text{if } a > 0 \\ 1 - F_\xi\left(\frac{y-b}{a}\right) + P\left(\xi = \frac{y-b}{a}\right) & \text{if } a < 0 \end{cases}$$

In the case of a continuous random variable ξ, $F_\xi(x)$ is differentiable almost everywhere and $P(\xi = (y - b)/a) = 0$. Hence, for $a \neq 0$,

$$f_g(y) = \frac{dF_g(y)}{dy} = \frac{1}{|a|} f_\xi\left(\frac{y-b}{a}\right)$$

Example 6.21

Let ξ be uniformly distributed on $(0, 1)$. Then
$$f_\xi(x) = I_{(0,1)}(x)$$

If $g(\xi) = -2\xi + 4$, then

$$f_g(y) = \frac{1}{2} f_\xi\left(\frac{y-4}{-2}\right) = \frac{1}{2} I_{(0,1)}\left(\frac{y-4}{-2}\right) = \begin{cases} \frac{1}{2} & \text{if } 0 < \frac{y-4}{-2} < 1 \\ 0 & \text{otherwise} \end{cases}$$

$$= \begin{cases} \frac{1}{2} & \text{if } 0 > y - 4 > -2 \\ 0 & \text{otherwise} \end{cases} = \frac{1}{2} I_{(2,4)}(y)$$

Now consider the absolute value function $g(\xi) = |\xi|$. If $y < 0$, then

Sec. 6.7   Functions of a Single Random Variable

$F_g(y) = 0$ since $F_g(y) = P(|\xi| \leq y)$, which is the probability of the empty set. For $y \geq 0$,

$$F_g(y) = P(|\xi| \leq y) = P(-y \leq \xi \leq y) = P(\xi \leq y) - P(\xi < -y)$$
$$= F_\xi(y) - F_\xi(-y) + P(\xi = -y)$$

Consequently,

$$F_g(y) = \begin{cases} 0 & \text{if } y < 0 \\ F_\xi(y) - F_\xi(-y) + P(\xi = -y) & \text{if } y \geq 0 \end{cases}$$

If it happens that ξ is a continuous random variable, then $P(\xi = -y) = 0$, and we have the probability density given by

$$f_g(y) = \frac{dF_g(y)}{dy}$$

$$= \begin{cases} 0 & \text{if } y < 0 \\ f_\xi(y) + f_\xi(-y) & \text{if } y \geq 0 \end{cases}$$

Let $g(\xi) = \xi^2$. Then it can be shown that

$$F_g(y) = \begin{cases} 0 & \text{if } y < 0 \\ F_\xi(\sqrt{y}) - F_\xi(-\sqrt{y}) + P(\xi = -\sqrt{y}) & \text{if } y \geq 0 \end{cases}$$

In the continuous case,

$$f_g(y) = \begin{cases} 0 & \text{if } y \leq 0 \\ \dfrac{f_\xi(\sqrt{y}) + f_\xi(-\sqrt{y})}{2\sqrt{y}} & \text{if } y > 0 \end{cases}$$

Example 6.22

Let $f_\xi(x) = \frac{1}{2} I_{(-1,1)}(x)$ and $g(\xi) = \xi^2$. For $y \leq 0$, $f_g(y) = 0$. For $y > 0$,

$$f_g(y) = \frac{1}{2\sqrt{y}} \left(\frac{1}{2} I_{(-1,1)}(\sqrt{y}) + \frac{1}{2} I_{(-1,1)}(-\sqrt{y}) \right)$$

But

$$I_{(-1,1)}(\sqrt{y}) = \begin{cases} 1 & \text{if } 0 < y < 1 \\ 0 & \text{if } y \geq 1 \end{cases}$$

and

$$I_{(-1,1)}(-\sqrt{y}) = \begin{cases} 1 & \text{if } 0 < y < 1 \\ 0 & \text{if } y \geq 1 \end{cases}$$

Consequently, for all y, we obtain

$$f_g(y) = \frac{1}{2\sqrt{y}} I_{(0,1)}(y)$$

Notice that

$$F_g(y) = \sqrt{y} I_{(0,1)}(y) + I_{[1,\infty)}(y)$$

An instance of a piecewise-continuous change of variables occurs if

$$g(\xi) = \begin{cases} 1 & \text{if } \xi \geq 0 \\ -1 & \text{if } \xi < 0 \end{cases}$$

which is similar, but not equal, to $\text{sgn}(\xi)$. For $y < -1$,

$$F_g(y) = P(g(\xi) \leq y) = 0$$

When $-1 \leq y < 1$,

$$F_g(y) = P(g(\xi) \leq y) = P(g(\xi) = -1) = P(\xi < 0) = F_\xi(0) - P(\xi = 0)$$

Finally, when $y \geq 1$, $F_g(y) = 1$. Thus,

$$F_g(y) = \begin{cases} 0 & \text{if } y < -1 \\ F_\xi(0) - P(\xi = 0) & \text{if } -1 \leq y < 1 \\ 1 & \text{if } 1 \leq y \end{cases}$$

More generally, if $c > b$ and

$$g(\xi) = \begin{cases} c & \text{if } \xi \geq a \\ b & \text{if } \xi < a \end{cases}$$

then

$$F_g(y) = \begin{cases} 0 & \text{if } y < b \\ F_\xi(a) - P(\xi = a) & \text{if } b \leq y < c \\ 1 & \text{if } c \leq y \end{cases}$$

The corresponding probability density function in this case is

$$f_g(y) = [F_\xi(a) - P(\xi = a)]\delta(y - b) + [1 - F_\xi(a) + P(\xi = a)]\delta(y - c)$$

A good example of this piecewise-continuous change of variables occurs in the process of thresholding (see Section 2.7), where

$$g(\xi) = \begin{cases} 1 & \text{if } \xi \geq a \\ 0 & \text{if } \xi < a \end{cases}$$

Consequently,

$$F_g(y) = \begin{cases} 0 & \text{if } y < 0 \\ F_\xi(a) - P(\xi = a) & \text{if } 0 \leq y < 1 \\ 1 & \text{if } 1 \leq y \end{cases}$$

and the corresponding density function is

$$f_g(y) = [F_\xi(a) - P(\xi = a)]\delta(y) + [1 - F_\xi(a) + P(\xi = a)]\delta(y - 1)$$

Example 6.23

Suppose the gray level density function of an image is

$$f_\xi(x) = \frac{1}{4}I_{(-1,3)}(x)$$

and we threshold the image with $a = 0$. In this case, $P(\xi = 0) = 0$ and $F_\xi(0) = \frac{1}{4}$. Therefore, the probability density function of the thresholded image is

$$f_g(y) = \frac{1}{4}\delta(y) + \frac{3}{4}\delta(y - 1)$$

which makes sense intuitively since

$$P(\xi < 0) = \int_{-\infty}^{0} f_\xi(x)\, dx = \frac{1}{4}$$

and

$$P(\xi \geq 0) = \int_{0}^{\infty} f_\xi(x)\, dx = \frac{3}{4}$$

Another important piecewise-continuous operation in image processing is the clipping (truncation) operation for gray values, where a form of clipping circuit is given by

$$g(\xi) = \begin{cases} \xi & \text{if } \xi \geq a \\ 0 & \text{if } \xi < a \end{cases}$$

where we assume $a \geq 0$. If $y < 0$, then $F_g(y) = 0$. If $0 \leq y < a$, then

$$F_g(y) = P(g(\xi) \leq y) = P(\xi < a) = F_\xi(a) - P(\xi = a)$$

Finally, if $y \geq a$,

$$F_g(y) = P(g(\xi) \leq y) = P(\xi \leq y) = F_\xi(y)$$

Therefore,

$$F_g(y) = \begin{cases} 0 & \text{if } y < 0 \\ F_\xi(a) - P(\xi = a) & \text{if } 0 \leq y < a \\ F_\xi(y) & \text{if } a \leq y \end{cases} \tag{1}$$

Example 6.24

Consider the bound matrix

$$c = \begin{pmatrix} 2 & 4 & 2 \\ 1 & 3 & 2 \\ 0 & 1 & * \end{pmatrix}_{0,0}$$

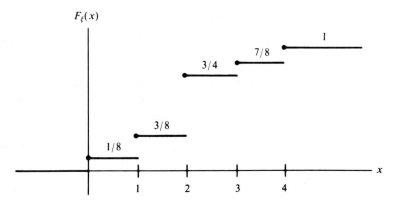

Figure 6.14 Gray Level Distribution

The gray level distribution function $F_\xi(x)$ associated with this bound matrix is plotted in Figure 6.14. Letting

$$g(\xi) = \begin{cases} \xi & \text{if } \xi \geq 3 \\ 0 & \text{if } \xi < 3 \end{cases}$$

and performing this operation on the gray values in question, we obtain the specific clipped image

$$d = \begin{pmatrix} 0 & 4 & 0 \\ 0 & 3 & 0 \\ 0 & 0 & * \end{pmatrix}_{0,0}$$

According to equation (1) above, the gray level distribution of the resulting image must be

$$F_g(y) = \begin{cases} 0 & \text{if } y < 0 \\ \frac{7}{8} - \frac{1}{8} & \text{if } 0 \leq y < 3 \\ F_\xi(y) & \text{if } 3 \leq y \end{cases}$$

with corresponding density given by

$$f_g(y) = \frac{3}{4}\delta(y) + \frac{1}{8}\delta(y - 3) + \frac{1}{8}\delta(y - 4)$$

But this is precisely what would be obtained by computing the gray level distribution function of the image d directly from its bound matrix.

The following result is sometimes useful for obtaining the density of a composite $g(\xi)$.

Theorem 6.4.[12] Let ξ be a continuous random variable with density $f_\xi(x)$. Let $\eta = g(\xi)$ and assume that the intervals $J_1, J_2, \ldots, J_n$ form a partition covering

[12] Emanuel Parzen, *Modern Probability Theory and its Applications* (New York: John Wiley & Sons, Inc., 1960), p. 312.

the support region of $f_\xi(x)$. Suppose that g is increasing or decreasing and is differentiable on the interior J_i^0 of each interval J_i. Then

$$f_\eta(y) = \sum_{j=1}^{n} f_\xi(h_j(y)) \left| \frac{dh_j(y)}{dy} \right| I_{g(J_j^0)}(y)$$

where h_j is the inverse function of g on J_j and $g(J_j^0)$ is the range of the interval J_j^0 under g.

Example 6.25

Let ξ be a continuous random variable with density $f_\xi(x)$. Consider $g(\xi) = \xi^2$. Here, $J_1 = (-\infty, 0) = J_1^0$, $J_2 = [0, \infty)$ and $J_2^0 = (0, \infty)$. The function g is everywhere differentiable, decreasing on J_1, and increasing on J_2. On J_1, $h_1(y) = -\sqrt{y}$, and on J_2, $h_2(y) = \sqrt{y}$. Thus, assuming the support of $f_\xi(x)$ to be $(-\infty, \infty)$,

$$f_\eta(y) = \frac{f_\xi(-\sqrt{y})}{2\sqrt{y}} I_{g(-\infty, 0)}(y) + \frac{f_\xi(\sqrt{y})}{2\sqrt{y}} I_{g(0, \infty)}(y)$$

Since $g(-\infty, 0) = (0, \infty) = g(0, \infty)$,

$$f_\eta(y) = \frac{f_\xi(-\sqrt{y}) + f_\xi(\sqrt{y})}{2\sqrt{y}} I_{(0, \infty)}(y)$$

as was previously mentioned.

6.8 RANDOM VECTORS AND JOINT DISTRIBUTIONS

Random vectors are defined similarly to random variables. For the probability space $(S, \mathcal{F}, P)$, $\mathbf{v}$ is an n-dimensional random vector if

(1) $\mathbf{v}: S \to R^n$.
(2) $\mathbf{v}^{-1}(B)$ is an event (in $\mathcal{F}$) for any n-dimensional Borel set B.

Condition 2 could be replaced with a condition analogous to the one-dimensional case. For

$$\mathbf{v} = \begin{pmatrix} \xi_1 \\ \xi_2 \\ \vdots \\ \xi_n \end{pmatrix}$$

all that is required is that, for any real number r, $\xi_i^{-1}((-\infty, r])$ be an element of $\mathcal{F}$. An example will be given shortly.

Just as in the one-dimensional case, the random vector $\mathbf{v}$ induces a probability measure on the class of all Borel sets in R^n. As a result, a new probability space $(R^n, \mathcal{B}, P_\mathbf{v})$ arises. The sigma field $\mathcal{B}$ consists of all Borel sets in R^n, and $P_\mathbf{v}$ is the set function $P_\mathbf{v}(B) = P(\mathbf{v} \in B)$ for every $B \in \mathcal{B}$. Most importantly, there is a *joint* probability distribution function

$$F_v: R^n \to R$$

defined by

$$F_v(x_1, x_2, \ldots, x_n) = P(\xi_1 \leq x_1, \xi_2 \leq x_2, \ldots, \xi_n \leq x_n)$$

Example 6.26

Consider the extended random field S with $X = \{(0, 0), (0, 1), (0, 2)\}$ and $Y = \{*, 1\}$. S has eight elements, viz.,

$$\begin{pmatrix} * \\ * \\ * \end{pmatrix} \begin{pmatrix} * \\ * \\ \textcircled{*} \end{pmatrix} \begin{pmatrix} 1 \\ * \\ * \end{pmatrix} \begin{pmatrix} 1 \\ 1 \\ * \end{pmatrix} \begin{pmatrix} * \\ * \\ \textcircled{1} \end{pmatrix} \begin{pmatrix} * \\ 1 \\ \textcircled{1} \end{pmatrix} \begin{pmatrix} 1 \\ * \\ \textcircled{1} \end{pmatrix} \begin{pmatrix} 1 \\ 1 \\ \textcircled{1} \end{pmatrix}$$

respectively denoted by $\mathscr{E}_1, \ldots, \mathscr{E}_8$. Let the probability P of each of these occurring be equally likely, and consider the random vector

$$\mathbf{v} = \begin{pmatrix} \xi \\ \eta \end{pmatrix}$$

where ξ gives the total number of activated pixels and η equals zero if $(0, 0)$ is not activated and unity if $(0, 0)$ is activated. For instance,

$$\mathbf{v}(\mathscr{E}_1) = \begin{pmatrix} \xi(\mathscr{E}_1) \\ \eta(\mathscr{E}_1) \end{pmatrix} = \begin{pmatrix} 0 \\ 0 \end{pmatrix}$$

and

$$\mathbf{v}(\mathscr{E}_6) = \begin{pmatrix} \xi(\mathscr{E}_6) \\ \eta(\mathscr{E}_6) \end{pmatrix} = \begin{pmatrix} 2 \\ 1 \end{pmatrix}$$

The joint distribution $F_v(x, y)$, which is sometimes written $F_{\xi\eta}(x, y)$, is "plotted" in Figure 6.15. In reading such a "plot," one must be careful on the vertical and horizontal lines because of the less than or equal to signs in the definition of $F_v(x, y)$. As a case in point, in this example, $F_v(\frac{3}{2}, 1) = \frac{1}{2}$, since $\mathscr{E}_1, \mathscr{E}_2, \mathscr{E}_3$, and $\mathscr{E}_5$ are in the event $(\xi \leq \frac{3}{2}, \eta \leq 1)$.

In Example 6.26, $\mathbf{v}$ is a feature vector in that its components consist of parameters that give data regarding pixel activation. The distribution function yields probabilistic information regarding the features ξ and η.

As in the one-dimensional case, probability distribution functions can be characterized by certain properties. In the two-dimensional case (generalizations to n dimensions are immediate), any function satisfying the following six conditions is the joint distribution function of some two-dimensional random vector:[13]

M1. $F(-\infty, y) = F(x, -\infty) = 0$.
M2. $F(\infty, \infty) = 1$.

---

[13] Cramer, *Mathematical Methods of Statistics*, p. 80.

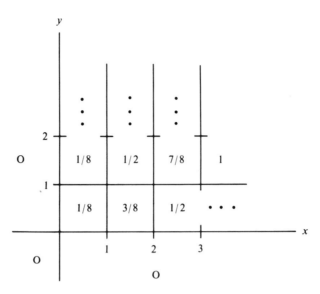

Figure 6.15 Joint Distribution of Number of Activated Pixels and Activation of (0, 0)

M3. If $x_1 \leq x_2$, then $F(x_1, y) \leq F(x_2, y)$ for all y. Similarly, if $y_1 \leq y_2$, then $F(x, y_1) \leq F(x, y_2)$ for all x.

M4. $\lim_{x \to r^+} F(x, y) = F(r, y)$ for all y.

M5. $\lim_{y \to s^+} F(x, y) = F(x, s)$ for all x.

M6. If $x_1 < x_2$ and $y_1 < y_2$, then
$$F(x_2, y_2) - F(x_1, y_2) - F(x_2, y_1) + F(x_1, y_1) \geq 0$$

Condition M6, illustrated in Figure 6.16, has no analog in the one-dimensional case. This condition also makes it a little more difficult to obtain and check joint probability distribution functions without first defining a random variable. In practice, one usually begins with probability density functions, which we shall discuss next.

As in the one-dimensional case, the joint distribution function is monotonic in each variable and therefore satisfies certain conditions on the type of discontinuities it may possess. Moreover, a representation for $F(x, y)$ holds in terms of an absolutely continuous part, a singular part, and a step-type discontinuous part. As in the case of one dimension, in our work the singular part will never be used and therefore will be omitted. An example of a step-type discontinuous distribution was given in Example 6.26. In this text, whenever there are no discontinuities in $F(x_1, x_2, \ldots, x_n)$, this will be called the continuous case. In such a situation, there exists an *n*-dimensional *joint* probability density function $f: R^n \to R$ such that

1. $f(x_1, x_2, \ldots, x_n) \geq 0$.
2. $\int_{-\infty}^{\infty} \cdots \int_{-\infty}^{\infty} f(x_1, x_2, \ldots, x_n) \, dx_1 \, dx_2 \cdots dx_n = 1$.

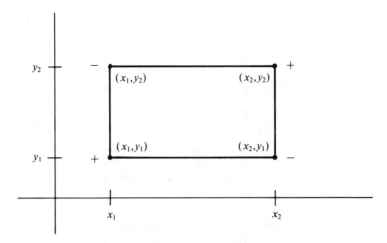

Figure 6.16 Criterion M6 for Joint Distribution Function

In addition,

$$F(x_1, x_2, \ldots, x_n) = \int_{-\infty}^{x_n} \int_{-\infty}^{x_{n-1}} \cdots \int_{-\infty}^{x_1} f(t_1, t_2, \ldots, t_n) \, dt_1 \, dt_2 \cdots dt_n$$

and

$$\frac{\partial^n F(x_1, x_2, \ldots, x_n)}{\partial x_1 \partial x_2 \cdots \partial x_n} = f(x_1, x_2, \ldots, x_n)$$

at all points for which f is continuous. In general, assuming there is no singular part, the distribution function F can be written as

$$F = F_1 + F_2$$

with F_1 absolutely continuous and F_2 a step-type function. Restricting ourselves to the two-dimensional case, a *density function* associated with F takes the form

$$f(x, y) = \frac{\partial^2 F_1(x, y)}{\partial x \, \partial y} + \sum_{i=-\infty}^{\infty} \sum_{j=-\infty}^{\infty} \alpha_{ij} \delta(x - a_i, y - b_j)$$

where each $\alpha_{ij} \geq 0$ arises from a jump discontinuity in F_2 and

$$\frac{\partial^2 F_1(x, y)}{\partial x \, \partial y} \geq 0$$

We will assume that the set of points (a_i, b_j) has no finite limit point. Moreover,

$$\int_{-\infty}^{\infty} \int_{-\infty}^{\infty} f(x, y) \, dx \, dy = 1$$

Sec. 6.8   Random Vectors and Joint Distributions   339

and, most importantly,

$$F(x, y) = \int_{-\infty}^{x} \int_{-\infty}^{y} f(r, s) \, ds \, dr$$

Whenever $F_1 \equiv 0$, the discrete case, the density consists exclusively of delta functions and the joint distribution function is made up of horizontal planes of various heights.

Example 6.27

Consider the extended random field given in Example 6.26. The distribution function is a step-type function that has the discrete density

$$f(x, y) = \frac{1}{8} \delta(x, y) + \frac{2}{8} \delta(x - 1, y) + \frac{1}{8} \delta(x - 2, y)$$
$$+ \frac{1}{8} \delta(x - 1, y - 1) + \frac{2}{8} \delta(x - 2, y - 1) + \frac{1}{8} \delta(x - 3, y - 1)$$

(See Figure 6.17.)

Example 6.28

An example of a (pure) continuous density function is

$$f(x, y) = I_{(0,1) \times (0,1)}(x, y)$$

which is "plotted" in Figure 6.18. The corresponding distribution function is

$$F(x, y) = \begin{cases} 0 & \text{if } x \leq 0 \text{ or } y \leq 0 \\ xy & \text{if } 0 \leq x \leq 1 \text{ and } 0 \leq y \leq 1 \\ x & \text{if } 0 \leq x \leq 1 \text{ and } 1 \leq y \\ y & \text{if } 0 \leq y \leq 1 \text{ and } 1 \leq x \\ 1 & \text{if } 1 \leq x \text{ and } 1 \leq y \end{cases}$$

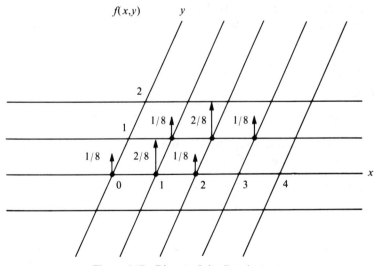

Figure 6.17   Discrete Joint Density

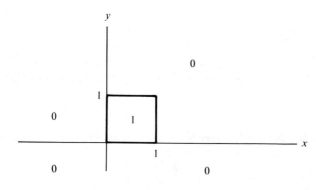

Figure 6.18 Uniform Joint Density

Notice that

$$\frac{\partial^2 F(x, y)}{\partial x \, \partial y} = f(x, y)$$

at points of continuity of f. This example is a generalization of the uniform density to two dimensions. Here the support region is a square, but obviously any shape can be used. In Section 6.13 an ellipse will be employed.

6.9 MARGINAL AND CONDITIONAL DISTRIBUTIONS

Whenever $F(x_1, x_2, \ldots, x_n)$ is a joint distribution function of an n-dimensional random vector, $\binom{n}{k} = n!/(n - k)! \, k!$ functions of only $k \geq 1$ of the n variables can be found by sending $n - k$ of the variables x_i to infinity. The resulting functions are themselves probability distribution functions and are called *marginal* distribution functions of the random vector. In the two-dimensional case ($n = 2$), there are two marginal distribution functions (corresponding to the case $k = 1$):

$$F_{\xi\eta}(x, \infty) = P(\xi \leq x, \eta < \infty)$$

and

$$F_{\xi\eta}(\infty, y) = P(\xi < \infty, \eta \leq y)$$

Whenever the joint density function $f(x, y)$ exists (which we will assume from here on), we have

$$F(x, y) = \int_{-\infty}^{x} \int_{-\infty}^{y} f(r, s) \, ds \, dr$$

and

$$F(x, \infty) = \int_{-\infty}^{x} \int_{-\infty}^{\infty} f(r, s) \, ds \, dr$$

Sec. 6.9   Marginal and Conditional Distributions   341

Formally differentiating with respect to x gives the marginal density function with respect to x:

$$f(x, \infty) = \int_{-\infty}^{\infty} f(x, s) \, ds$$

The marginal density $f(x, \infty)$ is often denoted by $f_\xi(x)$, as in the one-dimensional case. Similarly, the marginal density function with respect to y is given by

$$f_\eta(y) = f(\infty, y) = \int_{-\infty}^{\infty} f(r, y) \, dr$$

It should be noted that we write $f(x, y) = f_{\xi\eta}(x, y)$ and $F(x, y) = F_{\xi\eta}(x, y)$ whenever we wish to emphasize a random vector $\mathbf{v} = \binom{\xi}{\eta}$ associated with f and F. Recall that whenever $F(x, y)$ satisfies M1 through M6, there exist ξ and η such that $F = F_{\xi\eta}$.

Motivated by the definition of conditional probability given in Section 6.2, we introduce the concept of conditional distributions and conditional density functions. The following discussion is of the two-dimensional situation only; it is readily generalizable to any number n of dimensions.

The *conditional* density functions of ξ given $\eta = y$ and of η given $\xi = x$ are formally defined by

$$f(x \mid y) = \frac{f_{\xi\eta}(x, y)}{f_\eta(y)} = \frac{f(x, y)}{f(\infty, y)}$$

and

$$f(y \mid x) = \frac{f_{\xi\eta}(x, y)}{f_\xi(x)} = \frac{f(x, y)}{f(x, \infty)}$$

respectively. Thus, conditional densities are obtained by dividing the joint density by the appropriate marginal density whenever the latter is nonzero. Furthermore, it is assumed that these conditional density functions are zero elsewhere. If we wish to draw attention to the random variables ξ and η, we write $f_{\xi\mid\eta}(x \mid y)$ or $f_{\eta\mid\xi}(y \mid x)$.

The conditional distributions $F(x \mid y)$ and $F(y \mid x)$ are defined similarly as

$$F(x \mid y) = \int_{-\infty}^{x} f(r \mid y) \, dr = \frac{\int_{-\infty}^{x} f(r, y) \, dr}{\int_{-\infty}^{\infty} f(r, y) \, dr}$$

and

$$F(y \mid x) = \int_{-\infty}^{y} f(s \mid x) \, ds = \frac{\int_{-\infty}^{y} f(x, s) \, ds}{\int_{-\infty}^{\infty} f(x, s) \, ds}$$

These definitions apply directly for the case of continuous random variables having continuous joint and marginal densities.

Example 6.29

Consider the joint density function

$$f(x, y) = \left(\frac{1}{3} + \frac{xy}{3}\right) I_{(0,2) \times (0,1)}(x, y)$$

$$= \left(\frac{1}{3} + \frac{xy}{3}\right) I_{(0,2)}(x) I_{(0,1)}(y)$$

The marginal densities $f_\xi(x)$ and $f_\eta(y)$ are respectively given by

$$f(x, \infty) = I_{(0,2)}(x) \int_0^1 \left(\frac{1}{3} + \frac{xy}{3}\right) dy = \left(\frac{1}{3} + \frac{x}{6}\right) I_{(0,2)}(x)$$

and

$$f(\infty, y) = I_{(0,1)}(y) \int_0^2 \left(\frac{1}{3} + \frac{xy}{3}\right) dx = \left(\frac{2}{3} + \frac{2y}{3}\right) I_{(0,1)}(y)$$

The conditional density of ξ given $\eta = y$ is

$$f(x \mid y) = \left(\frac{1/3 + xy/3}{2/3 + 2y/3}\right) I_{(0,2) \times (0,1)}(x, y)$$

$$= \frac{1}{2} \left(\frac{1 + xy}{1 + y}\right) I_{(0,2) \times (0,1)}(x, y)$$

For fixed y,

$$\int_0^2 f(x \mid y) \, dx = \frac{(1 + xy)^2}{4(1 + y)y} \bigg|_{x=0}^{2} = \frac{4y(1 + y)}{4y(1 + y)} = 1$$

as it must. The conditional density of η given $\xi = x$ is

$$f(y \mid x) = \left(\frac{1 + xy}{1 + x/2}\right) I_{(0,2) \times (0,1)}(x, y)$$

The next example involves noncontinuous random variables; consequently, all manipulations are formal and require special interpretation.

Example 6.30

Let

$$f(x, y) = \frac{1}{8} \delta(x, y) + \frac{1}{4} \delta(x, y - 1) + \frac{1}{4} \delta(x - 1, y) + \frac{3}{8} \delta(x, y - 2)$$

This joint density might result from a co-occurrence-type relation. In $f(x, y)$, delta functions such as $\alpha_{ij} \delta(x - a_i, y - b_j)$ result from α_{ij}'s being the probability that gray value a_i appears in some relationship to gray value b_j. In the present example, the relationship might be that a_i is to the left of b_j. If this were the case, an image that has the specified density would be

$$c = \begin{pmatrix} 0 & 0 & 1 & 0 & 1 \\ 0 & 2 & * & 1 & 0 \\ 0 & 2 & * & * & * \\ \boxed{0} & 2 & * & * & * \end{pmatrix}$$

The marginal density with respect to x is given by

$$f(x, \infty) = \frac{3}{4} \delta(x) + \frac{1}{4} \delta(x - 1)$$

which says that the "left gray value" is zero three-fourths of the time and one one-fourth of the time. The other marginal density is

$$f(\infty, y) = \frac{3}{8} \delta(y) + \frac{1}{4} \delta(y - 1) + \frac{3}{8} \delta(y - 2)$$

and has a similar meaning with respect to the "right gray value" of the pair. By formally applying the conditional formula for continuous densities, we have

$$f(x \mid y) = \frac{\frac{1}{8} \delta(x, y) + \frac{1}{4} \delta(x, y - 1) + \frac{1}{4} \delta(x - 1, y) + \frac{3}{8} \delta(x, y - 2)}{\frac{3}{8} \delta(y) + \frac{1}{4} \delta(y - 1) + \frac{3}{8} \delta(y - 2)}$$

which is valid only at $y = 0$, 1, and 2 since $f(\infty, y) = 0$ elsewhere because $\delta(x - a, y - b)$ and $\delta(y - b)$ are zero for $y \neq b$. We emphasize the formality of these manipulations: rigorously, $f(\infty, y)$ is undefined at $y = 0$, 1, and 2; however, so is $f(x, y)$. In any event, letting $y = 0$, 1, and 2 in turn yields

$$f(x \mid 0) = \frac{\frac{1}{8} \delta(x, y) + \frac{1}{4} \delta(x - 1, y)}{\frac{3}{8} \delta(y)}$$

$$= \frac{\frac{1}{8} \delta(x) \delta(y) + \frac{1}{4} \delta(x - 1) \delta(y)}{\frac{3}{8} \delta(y)}$$

$$= \frac{1}{3} \delta(x) + \frac{2}{3} \delta(x - 1)$$

$$f(x \mid 1) = \frac{\frac{1}{4} \delta(x, y - 1)}{\frac{1}{4} \delta(y - 1)} = \delta(x)$$

$$f(x \mid 2) = \frac{\frac{3}{8} \delta(x, y - 2)}{\frac{3}{8} \delta(y - 2)} = \delta(x)$$

It should be noted that

$$\int_{-\infty}^{\infty} f(x \mid 0) \, dx = 1$$

and that the same is true for the other conditional densities. In terms of the application under consideration, $f(x \mid 0)$ has the following meaning: among all the times that the right pixel has the gray value zero (three times in the image c), one-third of the time the left pixel will also have gray value zero and two-thirds of the time the left pixel will have gray value one. This interpretation follows from the fact that in the discrete case $f(x \mid y) = P(\xi = x \mid \eta = y)$, i.e., $f(x \mid y)$ is the conditional probability that $\xi = x$ given that $\eta = y$. Analogously,

$$f(y|x) = \frac{\frac{1}{8}\delta(x, y) + \frac{1}{4}\delta(x, y - 1) + \frac{1}{4}\delta(x - 1, y) + \frac{3}{8}\delta(x, y - 2)}{\frac{3}{4}\delta(x) + \frac{1}{4}\delta(x - 1)}$$

so that

$$f(y \mid 0) = \frac{1}{6}\delta(y) + \frac{1}{3}\delta(y - 1) + \frac{1}{2}\delta(y - 2)$$

and

$$f(y \mid 1) = \delta(y)$$

As presented thus far in this section, marginal and conditional densities and distributions in the two-dimensional random vector situation are similar to densities and distributions described in the one-dimensional case. Consequently, all the parameters associated with the one-dimensional situation can be employed in the two-dimensional case.

When they exist, moments in the marginal case are found exactly as before, i.e.,

$$E(\xi^n) = \int_{-\infty}^{\infty} x^n f(x, \infty) \, dx$$

and

$$E(\eta^m) = \int_{-\infty}^{\infty} y^m f(\infty, y) \, dy$$

For the conditional situation, we speak of *conditional moments*. The conditional nth moment of ξ given y is

$$E(\xi^n \mid y) = \int_{-\infty}^{\infty} x^n f(x \mid y) \, dx$$

and similarly,

$$E(\eta^m \mid x) = \int_{-\infty}^{\infty} y^m f(y \mid x) \, dy$$

All integrals are assumed to be absolutely convergent.

Example 6.31

Consider the co-occurrence distribution given in Example 6.30. According to it, the average gray value of the left pixel is given by

$$E(\xi) = \int_{-\infty}^{\infty} x f(x, \infty) \, dx = \frac{1}{4}$$

The average gray value of the right pixel is given by

$$E(\eta) = \int_{-\infty}^{\infty} y f(\infty, y) \, dy = 1$$

Moreover, $E(\xi^n) = \frac{1}{4}$ for $n \geq 1$, and so $\sigma_\xi^2 = \frac{3}{16}$. Similarly, $E(\eta^m) = \frac{1}{4} + (\frac{3}{8})2^m$ for $m \geq 1$, and so $\sigma_\eta^2 = \frac{3}{4}$. As for conditional moments,

$$E(\xi^n \mid y) = \frac{\frac{1}{4}\delta(y)}{\frac{3}{8}\delta(y) + \frac{1}{4}\delta(y-1) + \frac{3}{8}\delta(y-2)}, \text{ for } n \geq 1$$

and so

$$E(\xi^n \mid 0) = \frac{2}{3}, \quad E(\xi^n \mid 1) = 0, \text{ and } E(\xi^n \mid 2) = 0$$

Similarly,

$$E(\eta^m \mid x) = \int_{-\infty}^{\infty} y^m f(y \mid x) \, dy = \frac{\frac{1}{4}\delta(x) + 2^m(\frac{3}{8})\delta(x)}{\frac{3}{4}\delta(x) + \frac{1}{4}\delta(x-1)}$$

and therefore,

$$E(\eta^m \mid 0) = \frac{1}{3} + \left(\frac{1}{2}\right)2^m, \text{ for } m \geq 1$$

and

$$E(\eta^m \mid 1) = 0, \text{ for } m \geq 1$$

The meanings of parameters associated with conditional and marginal densities are similar to those associated with the same in the one-dimensional case. Examples are given in the next section, where moments of vector-valued random variables are introduced.

To complete this section, we consider the crucial concept of independence for random variables. Rigorously, ξ and η are said to be *independent* whenever

$$F_{\xi\eta}(x, y) = F_\xi(x) F_\eta(y) = F_{\xi\eta}(x, \infty) F_{\xi\eta}(\infty, y)$$

holds for all pairs (x, y). Whenever ξ and η are not independent, they are said to be *dependent*. In practice, the test for independence involves probability density functions. Thus, ξ and η are independent whenever

$$f_{\xi\eta}(x, y) = f_\xi(x) f_\eta(y)$$

Observe that in this case $f(x \mid y) = f_\xi(x)$ and $f(y \mid x) = f_\eta(y)$.

In the more general case, the random variables $\xi_1, \xi_2, \ldots, \xi_n$ are said to be independent if for every n-tuple $(x_1, x_2, \ldots, x_n)$ in R^n,

$$F_{\xi_1 \xi_2 \cdots \xi_n}(x_1, x_2, \ldots, x_n) = F_{\xi_1}(x_1) F_{\xi_2}(x_2) \cdots F_{\xi_n}(x_n)$$

When this is not true, $\xi_1, \xi_2, \ldots, \xi_n$ are said to be dependent.

Example 6.32

Consider the random field of Example 6.26. For that field,

$$f_\xi(x) = \frac{1}{8}\delta(x) + \frac{3}{8}\delta(x-1) + \frac{3}{8}\delta(x-2) + \frac{1}{8}\delta(x-3)$$

and

$$f_\eta(y) = \frac{1}{2}\delta(y) + \frac{1}{2}\delta(y-1)$$

The interpretations of these equations are evident. Given that the sample space consists of eight images and the probability P is equally likely, $P(\xi = 0) = \frac{1}{8}$, $P(\xi = 1) = \frac{3}{8}$, $P(\xi = 2) = \frac{3}{8}$, and $P(\xi = 3) = \frac{1}{8}$. Moreover, $P(\eta = 0) = \frac{1}{2}$ and $P(\eta = 1) = \frac{1}{2}$. Note that

$$f_\xi(x)f_\eta(y) = \frac{1}{16}\delta(x, y) + \frac{3}{16}\delta(x-1, y) + \frac{3}{16}\delta(x-2, y)$$

$$+ \frac{1}{16}\delta(x-3, y) + \frac{1}{16}\delta(x, y-1) + \frac{3}{16}\delta(x-1, y-1)$$

$$+ \frac{3}{16}\delta(x-2, y-1) + \frac{1}{16}\delta(x-3, y-1)$$

$$\neq f_{\xi\eta}(x, y)$$

where $f_{\xi\eta}$ has been computed in Example 6.27. Consequently ξ and η are not independent. This should be intuitively clear from the manner in which ξ and η were described.

Independent random variables play a major role when functions of more than one random variable are needed. It is often assumed (hopefully after statistical testing is performed) that the random variables involved are independent. If they are not, the relationship between f_ξ and f_η is usually not easily describable in closed form.

6.10 EXPECTED VALUES FOR RANDOM VECTORS

Just as in the one-dimensional case, the expected value of a (Baire) function g of the random variables $\xi_1, \xi_2, \ldots, \xi_n$ can be found provided that the following integrals are absolutely convergent:

$$E(g(\xi_1, \xi_2, \ldots, \xi_n))$$

$$= \int_{-\infty}^{\infty} \cdots \int_{-\infty}^{\infty} g(x_1, x_2, \ldots, x_n)\, dF(x_1, x_2, \ldots, x_n)$$

$$= \int_{-\infty}^{\infty} \cdots \int_{-\infty}^{\infty} g(x_1, x_2, \ldots, x_n) f(x_1, x_2, \ldots, x_n)\, dx_1\, dx_2 \cdots dx_n$$

Sec. 6.10   Expected Values for Random Vectors   347

We have assumed, as usual, that the Lebesgue-Stieltjes integral reduces to a regular Lebesgue integral.

Of special importance are the $(p + q)$-*order moments* associated with the random variables ξ and η. For $p, q \geq 0$, these moments are given by

$$E(\xi^p \eta^q) = \int_{-\infty}^{\infty} \int_{-\infty}^{\infty} x^p y^q f_{\xi\eta}(x, y) \, dx \, dy$$

Moments similar to these are utilized as features for image recognition systems. (See Section 6.13.)

Example 6.33

Let

$$f_{\xi\eta}(x, y) = \frac{1}{9} \delta(x + 1, y - 1) + \frac{1}{9} \delta(x, y - 1) + \frac{2}{9} \delta(x, y)$$

$$+ \frac{2}{9} \delta(x, y + 1) + \frac{3}{9} \delta(x - 1, y - 1)$$

For $p, q > 0$,

$$E(\xi^p \eta^q) = \frac{1}{9}(-1)^p + \frac{3}{9}$$

$$E(\xi^p) = \frac{1}{9}(-1)^p + \frac{3}{9}, \text{ for } p > 0$$

$$E(\eta^q) = \frac{5}{9} + \frac{2}{9}(-1)^q, \text{ for } q > 0$$

The density $f_{\xi\eta}$ arises from the averaging mask (Section 2.9)

$$c = \begin{pmatrix} 1/9 & 1/9 & 3/9 \\ 0 & 2/9 & 0 \\ 0 & 2/9 & 0 \end{pmatrix}$$

by viewing each gray value as a probability at the respective pixel. Such a model is useful for feature generation by the method of moments (Section 6.13).

In all cases, $E(1) = 1$ and the mean values of ξ and η are $\bar{\xi} = E(\xi)$ and $\bar{\eta} = E(\eta)$, respectively. These agree with values obtained from the appropriate marginal densities. Second-order moments involve the two marginal second-order moments $E(\xi^2)$ and $E(\eta^2)$ and the *mixed* moment $E(\xi\eta)$. Central moments are defined as they are in one dimension, the n-plus-m-order central moment being

$$E((\xi - \bar{\xi})^n (\eta - \bar{\eta})^m)$$

Note that for the first-order central moments we have

$$E(\xi - \bar{\xi}) = E(\eta - \bar{\eta}) = 0$$

and for the variances,

$$\sigma_\xi^2 = E(\xi - \bar{\xi})^2$$

and

$$\sigma_\eta^2 = E(\eta - \bar{\eta})^2$$

As usual, σ_ξ and σ_η are called the standard deviations of ξ and η, respectively. The *covariance* of ξ and η is denoted by $\sigma_{\xi\eta}$ and is defined by

$$\sigma_{\xi\eta} = E((\xi - \bar{\xi})(\eta - \bar{\eta})) = E(\xi\eta) - \bar{\xi}\bar{\eta}$$

Theorem 6.5. Suppose that the mean values of ξ and η exist. Then

(a) $E(\xi + \eta) = E(\xi) + E(\eta)$; i.e., $\overline{\xi + \eta} = \bar{\xi} + \bar{\eta}$.
(b) If ξ and η are independent, then $E(\xi\eta) = E(\xi)E(\eta)$.
(c) If σ_ξ^2 and σ_η^2 exist, then

$$\sigma_{\xi+\eta}^2 = E(\xi + \eta - \overline{(\xi + \eta)})^2 = \sigma_\xi^2 + \sigma_\eta^2 + 2\sigma_{\xi\eta}$$

Theorem 6.5 is important in many areas of image processing—in particular, the covariance analysis of image sensitivity.

A quantity that measures the degree of linear dependence between two random variables is the *correlation coefficient* ρ, which is defined, when $\sigma_\xi \neq 0$ and $\sigma_\eta \neq 0$, by

$$\rho = \frac{\sigma_{\xi\eta}}{\sigma_\xi \sigma_\eta}$$

When $\rho = 0$, there is no linear dependence and the variables ξ and η are said to be *uncorrelated*. From part (b) of Theorem 6.5, it can be seen that whenever ξ and η are independent and the variances exist, then they are also uncorrelated; indeed, in such a case,

$$\sigma_{\xi\eta} = E(\xi\eta) - \bar{\xi}\bar{\eta} = 0$$

Whereas (Baire) functions of independent random variables are themselves independent random variables, (Baire) functions of uncorrelated random variables need not be uncorrelated.

Theorem 6.6[14]. If ξ and η are random variables for which there exist $E(\xi^2)$ and $E(\eta^2)$, then $-1 \leq \rho \leq 1$. Moreover, $\rho^2 = 1$ if and only if $P(\eta = a\xi + b) = 1$ holds true for some real numbers a and b.

Theorem 6.6 motivates the use of ρ as a measure of linear dependence between η and ξ. Indeed, the closer $|\rho|$ is to unity, the more "linearity" exists

[14] Cramer, p. 264.

Sec. 6.10   Expected Values for Random Vectors   349

between ξ and η. On the other hand, as mentioned previously, $\rho = 0$ does not imply that ξ and η are independent.

Example 6.34

Consider any three-by-three averaging mask with zero at the center, with all strong neighbors of (0, 0) having the same positive gray value, and all weak neighbors of (0, 0) having the same nonnegative gray value. For instance, the image

$$c = \begin{pmatrix} 0 & 1/4 & 0 \\ 1/4 & \circledcirc & 1/4 \\ 0 & 1/4 & 0 \end{pmatrix}$$

is a typical member of this class. As in Example 6.33, we shall interpret this image as a joint density. Then

$$f_{\xi\eta}(x, y) = \frac{1}{4} [\delta(x, y - 1) + \delta(x + 1, y) + \delta(x, y + 1) + \delta(x - 1, y)]$$

Notice that $\bar{\xi} = \bar{\eta} = 0$. Furthermore, $\sigma_\xi^2 = \sigma_\eta^2 = \frac{1}{2}$ and $\sigma_{\xi\eta} = 0$. Consequently, $\rho = 0$. However,

$$f_\xi(x) = f_{\xi\eta}(x, \infty) = \int_{-\infty}^{\infty} f_{\xi\eta}(x, y) \, dy$$

$$= \frac{1}{2} \delta(x) + \frac{1}{4} \delta(x + 1) + \frac{1}{4} \delta(x - 1)$$

and

$$f_\eta(y) = \frac{1}{2} \delta(y) + \frac{1}{4} \delta(y - 1) + \frac{1}{4} \delta(y + 1)$$

So

$$f_{\xi\eta}(x, y) \neq f_\xi(x) f_\eta(y)$$

Therefore, ξ and η are not independent.

We note in passing that manipulations similar to those performed in this example are sometimes useful in describing the texture of images.

We conclude this section with a discussion of matrix manipulation and a presentation of several definitions involving random vectors. We assume that all moments that are mentioned exist. Let

$$\mathbf{v} = E(\mathbf{v}) = \begin{pmatrix} \xi \\ \eta \end{pmatrix}$$

Then

$$\bar{\mathbf{v}} = \begin{pmatrix} \bar{\xi} \\ \bar{\eta} \end{pmatrix}$$

is called the *mean value of the random vector*. Note that $E(\mathbf{v} - \bar{\mathbf{v}}) = 0$. The *second moment matrix* is the symmetric, nonnegative definite matrix

$$E(\mathbf{v}\mathbf{v}') = \begin{pmatrix} E(\xi^2) & E(\xi\eta) \\ E(\xi\eta) & E(\eta^2) \end{pmatrix}$$

where dot product of **v** and its transpose **v**′ gives the two-by-two matrix

$$\begin{pmatrix} \xi^2 & \xi\eta \\ \xi\eta & \eta^2 \end{pmatrix}$$

and the expected value of a matrix is defined as the matrix of expected values. The *covariance matrix* associated with **v** is given by

$$\text{Cov}(\mathbf{v}) = E[(\mathbf{v} - \bar{\mathbf{v}})(\mathbf{v} - \bar{\mathbf{v}})']$$

$$= \begin{pmatrix} \sigma_\xi^2 & \sigma_{\xi\eta} \\ \sigma_{\xi\eta} & \sigma_\eta^2 \end{pmatrix}$$

This matrix is always symmetric and nonnegative definite. Furthermore, $\text{Cov}(\mathbf{v}) = E(\mathbf{v}\mathbf{v}') - \bar{\mathbf{v}}\bar{\mathbf{v}}'$. Covariance matrices in higher dimensions are defined in the same way and have the same properties as their two-dimensional counterpart. Among the applications involving covariance matrices is the Kalman filtering presented in Section 7.5.

Example 6.35

Let the random vector $\mathbf{v} = \binom{\xi}{\eta}$ be uniformly distributed with

$$f_{\xi\eta}(x, y) = \frac{1}{3} I_{(0,1)}(x) I_{(0,3)}(y - x)$$

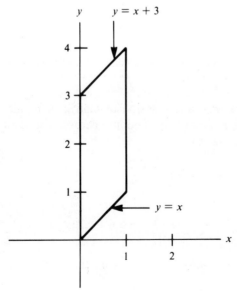

Figure 6.19 Support Region of Joint Density

(A diagram of the support region for f is shown in Figure 6.19.) Then

$$E(\xi^p \eta^q) = \int_{x=0}^{1} \int_{y=x}^{x+3} \frac{1}{3} x^p y^q \, dy \, dx$$

and hence

$$\bar{\mathbf{v}} = \begin{pmatrix} \frac{1}{2} \\ 2 \end{pmatrix}$$

Finally,

$$\text{Cov}(\mathbf{v}) = \begin{pmatrix} \frac{1}{12} & \frac{1}{12} \\ \frac{1}{12} & \frac{5}{6} \end{pmatrix}$$

6.11 FUNCTIONS OF TWO OR MORE RANDOM VARIABLES

In this section we find the distribution functions for certain functions g of two or more random variables. In the case where the random vector is two dimensional, we have

$$\mathbf{v}: S \to R^2$$

and

$$g: R^2 \to R$$

where S denotes the sample space. The more general case is given by

$$\mathbf{v}: S \to R^n$$

and

$$g: R^n \to R^m$$

Example 6.36

Given a nonnull digital image $a = (a_{pq})_{rt}$, we define the associated gray value density by

$$f_\zeta(x) = \frac{\text{card } \{a_{pq} : a_{pq} = x\}}{\text{CARD}(a)} = \frac{[\text{HIST}(a)](x)}{\text{CARD}(a)}$$

f_ζ is related to the gray value distribution by

$$F_\zeta(x) = \sum_{t \leq x} f_\zeta(t).$$

(See Example 6.8.)

Suppose a is composed of a foreground image b and a background image c in the sense that $a = \text{ADD}(b, c)$. Moreover, suppose that the gray level density of the background image is

$$f_\xi(x) = \frac{1}{2} \delta(x + 1) + \frac{1}{4} \delta(x) + \frac{1}{4} \delta(x - 1)$$

and that the gray level density of the foreground image is

$$f_\eta(y) = \frac{1}{5}\delta(y) + \frac{3}{5}\delta(y - 1) + \frac{1}{5}\delta(y - 3)$$

Suppose also that the respective background and foreground gray value random variables ξ and η are independent and that we desire the gray level density of a, whose gray level random variable is

$$\gamma = g(\xi, \eta) = \xi + \eta$$

Since ξ can only equal -1, 0, or 1 (with nonzero probability) and η can only equal 0, 1, or 3, γ can only equal -1, 0, 1, 2, 3, or 4. The density function $f_\gamma(z)$ is easily found by using the independence axiom and the fact that certain events are mutually exclusive. We obtain

$$P(\gamma = -1) = P(\xi = -1)P(\eta = 0) = \left(\frac{1}{2}\right)\left(\frac{1}{5}\right) = \frac{1}{10}$$

$$P(\gamma = 0) = P(\xi = -1)P(\eta = 1) + P(\xi = 0)P(\eta = 0) = \frac{7}{20}$$

$$P(\gamma = 1) = P(\xi = 1)P(\eta = 0) + P(\xi = 0)P(\eta = 1) = \frac{4}{20}$$

$$P(\gamma = 2) = P(\xi = 1)P(\eta = 1) + P(\xi = -1)P(\eta = 3) = \frac{5}{20}$$

$$P(\gamma = 3) = P(\xi = 0)P(\eta = 3) = \frac{1}{20}$$

$$P(\gamma = 4) = P(\xi = 1)P(\eta = 3) = \frac{1}{20}$$

Consequently,

$$f_\gamma(z) = \frac{2}{20}\delta(z + 1) + \frac{7}{20}\delta(z) + \frac{4}{20}\delta(z - 1)$$

$$+ \frac{5}{20}\delta(z - 2) + \frac{1}{20}\delta(z - 3) + \frac{1}{20}\delta(z - 4)$$

Example 6.37

In this example, we present a more organized and universal approach for finding $g(\xi, \eta) = \gamma$. We consider the discrete case involving only a finite number of delta functions where the random variables are assumed to be independent. The approach is akin to the truth tables used in propositional calculus. For instance, let $f_\xi(x)$ and $f_\eta(y)$ be as in Example 6.36; however, this time we want to find $\gamma = g(\xi, \eta) = $ maximum$(\xi, \eta) = \xi \vee \eta$. The first step is to make a table of all possibilities as in Table 6.3. The first column gives all possible values of ξ, the second gives all possible values of η, and the last gives the maximum of ξ and η.

Sec. 6.11   Functions of Two or More Random Variables

TABLE 6.3

ξ	η	$\gamma = \xi \vee \eta$
−1	0	0
0	0	0
1	0	1
−1	1	1
0	1	1
1	1	1
−1	3	3
0	3	3
1	3	3

since the larger of −1 and 0 is 0

From the first two rows of the table,

$$P(\gamma = 0) = P(\xi = -1)P(\eta = 0) + P(\xi = 0)P(\eta = 0) = 3/20$$

where the probabilities are read off from the densities f_ξ and f_η. Similarly, the next four rows of the table give

$$P(\gamma = 1) = \frac{1}{20} + \frac{3}{10} + \frac{3}{20} + \frac{3}{20} = \frac{13}{20}$$

and the last three rows yield

$$P(\gamma = 3) = \frac{1}{10} + \frac{1}{20} + \frac{1}{20} = \frac{4}{20}$$

Consequently,

$$f_\gamma(z) = \frac{3}{20}\delta(z) + \frac{13}{20}\delta(z-1) + \frac{4}{20}\delta(z-3)$$

Example 6.37 dealt with discrete random vectors. Now assume that $\binom{\xi}{\eta}$ is a continuous random vector and that

$$g: R^2 \to R^1$$

is a continuous function. Then $\gamma = g(\xi, \eta)$ will be a continuous random variable. The objective is to find the distribution function associated with γ given $f_{\xi\eta}(x, y)$. Often, a convenient way to find $f_\gamma(z)$ is to introduce an additional random variable $\nu = h(\xi, \eta)$ for some continuous function

$$h: R^2 \to R^1$$

The function h is usually heuristically determined. The joint density function $f_{\gamma\nu}(z, u)$ is found, and from this we obtain the desired (marginal) density

$$f_\gamma(z) = \int_{-\infty}^{\infty} f_{\gamma\nu}(z, u) \, du$$

Generalization to n dimensions is immediate: if the random vector $\mathbf{v}: S \to R^n$, the continuous function $g_1: R^n \to R^1$, and the density of $g_1(\mathbf{v})$ is desired, then introduce $n - 1$ other continuous functions $g_2, g_3, \ldots, g_n$, each mapping R^n into R^1, and find the joint density of $g_1(\mathbf{v}), g_2(\mathbf{v}), \ldots, g_n(\mathbf{v})$. Integration will yield the desired marginal density.

Returning to the two-dimensional case, we have $\gamma = g(\xi, \eta)$ and $\nu = h(\xi, \eta)$, the latter yet to be determined. If the equations $z = g(x, y)$ and $u = h(x, y)$ are uniquely solvable for x and y in terms of z and u—that is, $x = A(z, u)$ and $y = B(z, u)$—and if the partial derivatives

$$\frac{\partial x}{\partial z}, \frac{\partial x}{\partial u}, \frac{\partial y}{\partial z}, \frac{\partial y}{\partial u}$$

exist and are continuous, then the joint probability density function for γ and ν is given by

$$f_{\gamma\nu}(z, u) = f_{\xi\eta}(A(z, u), B(z, u)) \, | J(z, u) |$$

where $| J(z, u) |$ is the absolute value of the Jacobian determinant[15]

$$J(z, u) = \det \begin{pmatrix} \dfrac{\partial A}{\partial z} & \dfrac{\partial A}{\partial u} \\ \dfrac{\partial B}{\partial z} & \dfrac{\partial B}{\partial u} \end{pmatrix}$$

Some similarity exists between this result and Theorem 6.4, which holds in the one-dimensional case. Several illustrations of the Jacobian method will now be given.

A most important instance of this method is for

$$\gamma = \xi + \eta = g(\xi, \eta)$$

A second equation $\nu = h(\xi, \eta)$ must be employed to satisfy the preceding requirements. Perhaps the most straightforward possibility is to let $\nu = \xi$. Then, since $z = x + y$ and $u = x$, we obtain

$$x = A(z, u) = u$$
$$y = B(z, u) = z - u$$

The Jacobian of the transformation is

$$J = \det \begin{pmatrix} 0 & 1 \\ 1 & -1 \end{pmatrix} = -1$$

---

[15] Cramer, p. 293.

so $|J| = 1$. Hence,
$$f_{\gamma\nu}(z, u) = f_{\xi\eta}(u, z - u)$$
The desired (marginal) density is $f_{\gamma\nu}(z, \infty) = f_\gamma(z)$, so
$$f_\gamma(z) = \int_{u=-\infty}^{\infty} f_{\gamma\nu}(z, u) \, du$$
$$= \int_{u=-\infty}^{\infty} f_{\xi\eta}(u, z - u) \, du$$

Moreover,
$$F_\gamma(z) = \int_{t=-\infty}^{z} \int_{u=-\infty}^{\infty} f_{\xi\eta}(u, t - u) \, du \, dt$$

Suppose now that ξ and η are independent. Then
$$f_\gamma(z) = \int_{u=-\infty}^{\infty} f_\xi(u) f_\eta(z - u) \, du = (f_\xi * f_\eta)(z)$$

is the convolution of the individual density functions. Proceeding inductively, the probability density function for the sum of the three independent random variables ξ_1, ξ_2, ξ_3 can be found by a double convolution

$$f_{\xi_1} * (f_{\xi_2} * f_{\xi_3}) = (f_{\xi_1} * f_{\xi_2}) * f_{\xi_3}$$

where the equality represents the associativity of the convolution.

Let us formally apply the convolution formula to the independent discrete densities
$$f_\xi(x) = \sum_i \alpha_i \delta(x - x_i)$$
and
$$f_\eta(y) = \sum_j \beta_j \delta(y - y_j)$$

each of which is comprised of a finite number of delta functions. Formally,
$$(f_\xi * f_\eta)(z) = \int_{-\infty}^{\infty} f_\xi(u) f_\eta(z - u) \, du$$
$$= \sum_{i,j} \alpha_i \beta_j \int_{-\infty}^{\infty} \delta(u - x_i) \delta(z - u - y_j) \, du$$
$$= \sum_{i,j} \alpha_i \beta_j \int_{-\infty}^{\infty} \delta(u - x_i, z - u - y_j) \, du$$

But the integral is identically one if there exists a u such that $u - x_i = 0$ and

$z - u - y_j = 0$; otherwise it is zero. But these two equations hold only for $z = x_i + y_j$. Hence,

$$(f_\xi * f_\eta)(z) = \sum \{\alpha_i \beta_j : x_i + y_j = z\}$$

where by this expression we mean that $(f_\xi * f_\eta)(z) = 0$ if there exists no pair (x_i, y_j) such that $x_i + y_j = z$. In terms of discrete impulses (delta functions),

$$(f_\xi * f_\eta)(z) = \sum_k (\sum_{x_i + y_j = z_k} \alpha_i \beta_j) \delta(z - z_k)$$

where the summation is taken over all possible sums z_k of x_i and y_j. Referring to Example 6.36, we had $z_k = -1, 0, 1, 2, 3,$ and 4. We also had

$$P(\gamma = z_k) = \sum_{x_i + y_j = z_k} P(\xi = x_i) P(\eta = y_j)$$

Consequently, the result checks out. In sum, the convolution formula

$$f_{\xi + \eta} = f_\xi * f_\eta$$

applies formally to finite discrete densities, as well as rigorously to continuous ones.

The evaluation of $f_\xi * f_\eta$ for the finite discrete case is known as the method of *serial products*. The terminology results from a specific graphical technique for evaluating $f_\xi * f_\eta$. As illustrated in Figure 6.20(a), according to this technique we place the graph of $f_\eta(z - x)$ to the left of the graph of $f_\xi(x)$. We then move z to the right to find values of z for which delta functions from $f_\xi(x)$ and $f_\eta(z - x)$ are on top of each other. At any such z for which this occurs, multiply the heights of those impulses which are superimposed on each other and then add the corresponding products to obtain the appropriate coefficient γ_k for the impulse $\delta(z - z_k)$ in the density $(f_\xi * f_\eta)(z)$. In particular, in terms of f_ξ and f_η,

$$\gamma_k = \sum_{x_i + y_j = z_k} \alpha_i \beta_j$$

Figure 6.20 illustrates the method as applied to the densities of Example 6.36.

As another example of the Jacobian method, suppose $\gamma = \xi/\eta$. Again, we let $\nu = \xi$, so that $z = x/y$ and $u = x$, implying that $x = u$ and $y = u/z$. Consequently, the Jacobian is

$$J = \det \begin{pmatrix} 0 & 1 \\ \frac{-u}{z^2} & \frac{1}{z} \end{pmatrix} = \frac{u}{z^2}, \text{ for } z \neq 0$$

Hence,

$$f_{\gamma\nu}(z, u) = f_{\xi\eta}(u, u/z) \left| \frac{u}{z^2} \right|$$

Sec. 6.11 Functions of Two or More Random Variables

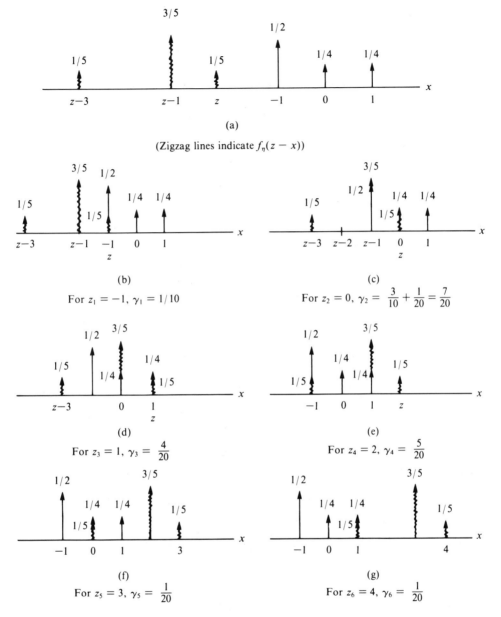

Figure 6.20 Method of Serial Products

and so

$$f_\gamma(z) = \int_{-\infty}^{\infty} f_{\xi\eta}(u, u/z) \left|\frac{u}{z^2}\right| du$$

Letting $w = u/z$, $dw = du/z$ and

$$f_\gamma(z) = \int_{-\infty}^{\infty} f_{\xi\eta}(zw, w) |w| dw$$

If ξ and η are independent, we obtain

$$f_\gamma(z) = \int_{-\infty}^{\infty} f_\xi(zw) f_\eta(w) |w| dw$$

Example 6.38

Let the gray level density be given by $f_\xi(x) = \frac{1}{20} I_{(-10,10)}(x)$, and suppose the gray level density is to be scaled by about 10 to reduce the dynamic range. More specifically, assume that $\gamma = \xi/\eta$, where η is a random variable with density $f_\eta(y) = \frac{1}{2} I_{(9,11)}(x)$. Suppose also that ξ and η are independent. We seek the probability density $f_\gamma(z)$ of γ. From the previous discussion,

$$f_\gamma(z) = \frac{1}{40} \int_{-\infty}^{\infty} I_{(-10,10)}(zw) I_{(9,11)}(w) |w| dw$$

Observing that $9 < w < 11$ and that $-10 < wz < 10$, we obtain

$$f_\gamma(z) = \begin{cases} \frac{1}{40} \int_{\max(-10/z,9)}^{\min(10/z,11)} w \, dw, \text{ for } z > 0 \\ \frac{1}{40} \int_{\max(10/z,9)}^{\min(-10/z,11)} w \, dw, \text{ for } z < 0 \end{cases}$$

and so, for $z \neq 0$,

$$f_\gamma(z) = \frac{1}{40} \left(\int_9^{10/z} w \, dw \right) I_{(10/11,10/9)}(z)$$
$$+ \frac{1}{40} \left(\int_9^{11} w \, dw \right) I_{(0,10/11)}(z)$$
$$+ \frac{1}{40} \left(\int_9^{-10/z} w \, dw \right) I_{(-10/9,-10/11)}(z)$$
$$+ \frac{1}{40} \left(\int_9^{11} w \, dw \right) I_{(-10/11,0)}(z)$$
$$= \frac{1}{80} \left(\frac{100}{z^2} - 81 \right) I_{(-10/9,-10/11) \cup (10/11,10/9)}(z)$$
$$+ \frac{1}{2} I_{(-10/11,10/11)}(z)$$

Sec. 6.11   Functions of Two or More Random Variables   359

Suppose now $\gamma = \xi\eta$. Then the probability density of γ is found by the Jacobian method to be

$$f_\gamma(z) = \int_{-\infty}^{\infty} f_{\xi\eta}(u, z/u) \frac{1}{|u|} du$$

If ξ and η are independent, then

$$f_\gamma(z) = \int_{-\infty}^{\infty} f_\xi(u) f_\eta(z/u) \frac{1}{|u|} du$$

A most interesting way of obtaining $f_\gamma(z)$ exists when, in addition to the independence of ξ and η, both $f_\xi(x)$ and $f_\eta(y)$ have support in $[0, \infty)$. In that case,

$$f_\gamma(z) = \int_0^{\infty} f_\xi(u) f_\eta(z/u) \frac{1}{u} du$$

which is the Mellin-type convolution of $f_\xi(x)$ and $f_\eta(y)$. (See Section 4.6.) As a

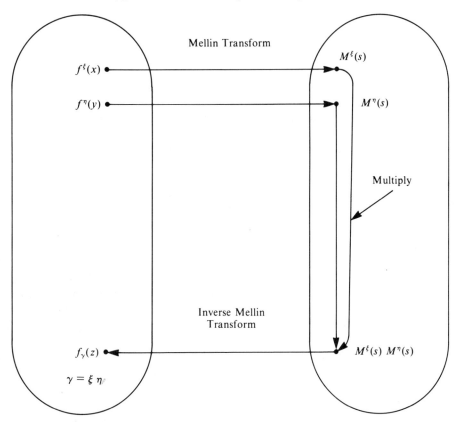

Figure 6.21 Mellin Transform for Determining Density of Product of Two Independent Random Variables

result, $f_\gamma(z)$ can be found using Mellin transform techniques as illustrated in Figure 6.21. As in the case of division, multiplication of random variables is used in image processing for scaling normalization, as well as for other purposes.

6.12 MULTIVARIATE GAUSSIAN DENSITY

In this section we introduce the *multivariate Gaussian,* or *multivariate normal,* density. Throughout,

$$\mathbf{v} = \begin{pmatrix} \xi_1 \\ \xi_2 \\ \vdots \\ \xi_n \end{pmatrix}$$

will denote a random vector.

Let M be an n by n positive definite matrix with determinant $|M|$, and let

$$\mathbf{u} = \begin{pmatrix} u_1 \\ u_2 \\ \vdots \\ u_n \end{pmatrix}$$

be a vector of real values. Then the multivariate normal density is

$$f_\mathbf{v}(\mathbf{x}) = \frac{1}{\sqrt{(2\pi)^n |M|}} e^{-\frac{1}{2}[(\mathbf{x} - \mathbf{u})'M^{-1}(\mathbf{x} - \mathbf{u})]}$$

where

$$\mathbf{x} = \begin{pmatrix} x_1 \\ x_2 \\ \vdots \\ x_n \end{pmatrix}$$

for x_i real. When $n = 1$, the usual univariate Gaussian density results, with $|M| = \sigma^2$ and $u = \bar{v} = \xi$. By employing standard techniques of linear algebra and calculus, the following theorem is easily shown to hold.

Theorem 6.7.[16] Let $f_\mathbf{v}$ be a multivariate normal density with $n \geq 2$. Then
(a) The vector $\mathbf{u}$ is the mean vector, i.e., $\mathbf{u} = \bar{\mathbf{v}}$.
(b) The matrix M is the covariance matrix of $\mathbf{v}$, i.e.,

$$\text{Cov}(\mathbf{v}) = M = E((\mathbf{v} - \mathbf{u})(\mathbf{v} - \mathbf{u})')$$

[16] Cramer, p. 310.

Sec. 6.12   Multivariate Gaussian Density   361

(c) The random variables $\xi_1, \xi_2, \ldots, \xi_n$ in **v** are independent if and only if M is a diagonal matrix, i.e., if and only if

$$M = \begin{pmatrix} \sigma_1^2 & 0 & \cdots & 0 \\ 0 & \sigma_2^2 & \cdots & \cdot \\ \vdots & \vdots & \ddots & \vdots \\ 0 & 0 & \cdots & \sigma_n^2 \end{pmatrix}$$

In this, the Gaussian case, uncorrelatedness and independence are equivalent. Should $\xi_1, \xi_2, \ldots, \xi_n$ be independent, then

$$f_{\mathbf{v}} = \frac{1}{\sqrt{(2\pi)^n}\, \sigma_1 \sigma_2 \cdots \sigma_n} e^{-1/2 \sum_{i=1}^{n} \left(\frac{x_i - u_i}{\sigma_i}\right)^2}$$

(d) If $\eta_i = a_{1i}\xi_1 + a_{2i}\xi_2 + \cdots + a_{ni}\xi_n$ for $i = 1, 2, \ldots, k$, $k \leq n$, and the η_i are linearly independent, then $f_{\mathbf{w}}$ is multivariate normal, where

$$\mathbf{w} = \begin{pmatrix} \eta_1 \\ \eta_2 \\ \vdots \\ \eta_k \end{pmatrix}$$

In particular, the marginal densities of a multivariate normal density are normal.

One of the more important of the multivariate normal densities is the bivariate normal density. Here,

$$M = \begin{pmatrix} \sigma_1^2 & \rho\sigma_1\sigma_2 \\ \rho\sigma_1\sigma_2 & \sigma_2^2 \end{pmatrix}$$

where $\rho = \sigma_{12}/\sigma_1\sigma_2$ is the correlation coefficient. We have

$$f_{\xi_1\xi_2}(x_1, x_2) = \frac{1}{2\pi\sigma_1\sigma_2\sqrt{1-\rho^2}} e^{-\left[\left(\frac{x_1-u_1}{\sigma_1}\right)^2 - 2\rho\left(\frac{x_1-u_1}{\sigma_1}\right)\left(\frac{x_2-u_2}{\sigma_2}\right) + \left(\frac{x_2-u_2}{\sigma_2}\right)^2\right]/2(1-\rho^2)}$$

From Theorem 6.7, it is seen that u_1 and u_2 are the mean values of ξ_1 and ξ_2, respectively. The variances are $\sigma_1^2 = E(\xi_1 - u_1)^2$ and $\sigma_2^2 = E(\xi_2 - u_2)^2$. Moreover, ξ_1 and ξ_2 are independent if and only if $\rho = 0$, and hence, in the independent case, ρ is not a parameter in the joint density $f_{\xi_1\xi_2}(x_1, x_2) = f_{\xi_1}(x_1) f_{\xi_2}(x_2)$.

The last part of Theorem 6.7 shows that if $\eta_1 = a\xi_1 + b\xi_2$ and $\eta_2 = c\xi_1 + d\xi_2$, with

$$\det \begin{pmatrix} a & b \\ c & d \end{pmatrix} \neq 0$$

then $f_{\eta_1\eta_2}(y_1, y_2)$ is also normal. Notice that the mean values are easily found:

$E(\eta_1) = au_1 + bu_2$ and $E(\eta_2) = cu_1 + du_2$. As regards the variance,

$$E(\eta_1 - \overline{\eta}_1)^2 = E(a(\xi_1 - u_1) + b(\xi_2 - u_2))^2$$
$$= a^2\sigma_1^2 + 2ab\sigma_{12} + b^2\sigma_2^2$$

The other variance and the covariance can be found similarly. Consequently, $f_{\eta_1\eta_2}(y_1, y_2)$ is easily specified since only five parameters are needed for the bivariate normal density. The marginal densities $f_{\eta_1}(y_1)$ and $f_{\eta_2}(y_2)$ are readily found, for they are also normal and involve only their respective means and variances.

The following example of a nonnormal bivariate density function illustrates the fact that even if the marginal densities are normal, the joint density need not be normal. This same example shows that even if random variables are uncorrelated and each has a normal distribution, they need not be independent.

Example 6.39

Consider two joint Gaussian density functions, each with mean zero and variance one, with the first having correlation coefficient $\sqrt{3/4}$ and the second having correlation coefficient $-\sqrt{3/4}$. We have

$$f_{\xi\eta}(x_1, x_2) = \frac{1}{\pi} e^{-2(x_1^2 - 2\sqrt{3/4}x_1x_2 + x_2^2)}$$

and

$$f_{\gamma\delta}(x_1, x_2) = \frac{1}{\pi} e^{-2(x_1^2 - 2\sqrt{3/4}x_1x_2 + x_2^2)}$$

The function

$$f(x_1, x_2) = \frac{1}{2} f_{\xi\eta}(x_1, x_2) + \frac{1}{2} f_{\gamma\delta}(x_1, x_2)$$

is nonnegative and of area one, and hence is a joint density function of a pair of random variables. From its form, however, we can see that this bivariate density is not normal. The marginal density functions associated with it are both Gaussian of mean zero and variance one; the new random variables associated with it are surely not independent, since the joint density is not the product of the marginal densities. However, these random variables are uncorrelated since

$$\int_{-\infty}^{\infty} \int_{-\infty}^{\infty} x_1 x_2 f(x_1, x_2) \, dx_1 \, dx_2 = 0$$

6.13 MOMENT TRANSFORMATION FOR FEATURE REPRESENTATION

Moment generation techniques (see Section 6.5) involving probability distributions can be employed for feature representation of images. In this application, the first $p + q$ moments associated with an image are used as a *signature* of that

Sec. 6.13 Moment Transformation for Feature Representation   363

image. For an observed image, after preprocessing for noise and other undesirable effects, the first $p + q$ moments are calculated, and from these, a normalized set of moments is developed. Usually, the normalization involves a translation, rotation, and scale factor magnification or attenuation. These steps are explained herein using the *ellipse of concentration*. Briefly, this ellipse centers itself about the digital image with its major axis in conformity with the greatest density of gray. Every digital image possesses an ellipse of concentration.

In what follows, we presuppose that the image f for which the moments are to be found has gray values between zero and one. In practice, a mapping can always be devised in a consistent, one-to-one, onto manner to produce a zero–one range, since, for practical models, gray values come from a bounded set of real numbers. In theory, some type of mapping can always be devised to obtain an image f in $([0, 1] \cup \{*\})^{Z \times Z}$.

Example 6.40

Consider the bound matrix

$$a = \begin{pmatrix} -5 & -5 & 5 \\ * & \underline{0} & * \\ * & 0 & * \end{pmatrix}$$

and let $g: R \cup \{*\} \to [0, 1] \cup \{*\}$ be given by

$$g(r) = \begin{cases} 1 & \text{if } r > 10 \\ \dfrac{x}{20} + \dfrac{1}{2} & \text{if } -10 \leq r \leq 10 \\ 0 & \text{if } r < -10 \\ * & \text{if } r = * \end{cases}$$

Letting $f(x, y) = g(a(x, y))$ produces the image

$$f = \begin{pmatrix} \frac{1}{4} & \frac{1}{4} & \frac{3}{4} \\ * & \underline{\frac{1}{2}} & * \\ * & \frac{1}{2} & * \end{pmatrix}$$

Other choices of g are also possible.

The p, q moment $m_{p,q}$ is given by

$$m_{p,q} = \frac{\sum_{i=-\infty}^{\infty} \sum_{j=-\infty}^{\infty} i^p j^q f(i, j)}{\sum_{i=-\infty}^{\infty} \sum_{j=-\infty}^{\infty} f(i, j)}$$

where, of course, the denominator cannot be zero. Note that all sums are finite and, consequently, there need not be any mention of convergence. The zero-order moment m_{00} is defined to be unity, as is 0^0.

Example 6.41

Using the image f from Example 6.40, we have

$$m_{p,q} = \frac{(-1)^p 1/4 + 0^p 1/4 + 0^p 0^q 1/2 + 0^p(-1)^q 1/2 + 3/4}{2.25}$$

and:

The first-order moments are $m_{10} = 2/9$, $m_{01} = 1/3$
The second-order moments are $m_{20} = 4/9$, $m_{02} = 7/9$, $m_{11} = 2/9$
The third-order moments are $m_{30} = 2/9$, $m_{03} = 1/3$, $m_{21} = 4/9$, $m_{12} = 2/9$
etc.

The *mean*, or center-of-gravity, vector $\bar{\mathbf{v}}$ is defined as in Section 6.10:

$$\bar{\mathbf{v}} = \begin{pmatrix} m_{10} \\ m_{01} \end{pmatrix} = \begin{pmatrix} \bar{x} \\ \bar{y} \end{pmatrix}$$

The second-order moment matrix S and the covariance matrix C are respectively defined as

$$S = \begin{pmatrix} m_{20} & m_{11} \\ m_{11} & m_{02} \end{pmatrix}$$

and

$$C = S - \bar{\mathbf{v}}\bar{\mathbf{v}}'$$

where $\bar{\mathbf{v}}'$ denotes the transpose of $\bar{\mathbf{v}}$. We could also write

$$C = \begin{pmatrix} \sigma_x^2 & \sigma_{xy} \\ \sigma_{xy} & \sigma_y^2 \end{pmatrix}$$

where

$$\sigma_x^2 = m_{20} - m_{10}^2$$
$$\sigma_y^2 = m_{02} - m_{01}^2$$

and

$$\sigma_{xy} = m_{11} - m_{01}m_{10} = \rho\sigma_x\sigma_y$$

The quantities in the C matrix correspond to moments of inertia in mechanics or variances in probability, and ρ is the familiar correlation coefficient. These quantities are instrumental in determining the rotational, translational, and scale factor normalization procedure, to be discussed shortly. Before doing so, however, we consider the uniform distribution in two dimensions.

Sec. 6.13   Moment Transformation for Feature Representation

A continuous random vector $\binom{\xi}{\eta}$ is said to be *uniformly distributed* on a finite region D of $R \times R$ if

$$f_{\xi\eta}(x, y) = \frac{1}{\text{Area }(D)} I_D(x, y)$$

where the area of D is nonzero.

We are most interested in the case where the support region D is elliptical. In particular, associated with an arbitrary joint distribution function, there is a uniformly distributed probability density with elliptical domain possessing the same first- and second-order moments as the original distribution. (See Theorem 6.8, to follow.) This ellipse is called the *ellipse of concentration* corresponding to the given distribution. In conjunction with the method of moments, the ellipse of concentration is used as a feature normalization technique for image recognition.

Theorem 6.8.[17]   Let $f_{\xi\eta}$ be the probability density function for random variables ξ and η, with respective variances $\sigma_\xi^2 \neq 0$ and $\sigma_\eta^2 \neq 0$, and respective means $\bar{\xi}$ and $\bar{\eta}$. Assume that $|\rho| \neq 1$. Then there is a density function of the form

$$g(x, y) = \frac{1}{\text{Area }(D)} I_D(x, y)$$

where the associated random variables have the same means, variances, and correlation coefficient. Moreover, D is the interior of the ellipse defined by

$$\left(\frac{x - \bar{\xi}}{\sigma_\xi}\right)^2 - 2\rho \frac{(x - \bar{\xi})(y - \bar{\eta})}{\sigma_\xi \sigma_\eta} + \left(\frac{y - \bar{\eta}}{\sigma_\eta}\right)^2 = 4(1 - \rho^2)$$

This ellipse is the ellipse of concentration.

When applied to images (viewed as point densities), the ellipse of concentration judiciously locates itself amidst the image in question and therefore can be perceived as one of the signatures of the image.

Example 6.42

Using the moments calculated from the image in Example 6.41, we obtain the ellipse of concentration

$$\frac{(9x - 2)^2}{32} - \frac{(9x - 2)(3y - 1)}{24} + \frac{(3y - 1)^2}{6} = \frac{11}{3}$$

This ellipse is depicted, together with the original image f, in Figure 6.22. It has a major axis located at an angle of about 67° from the abscissa. The length of the major axis is about 1.71, and the length of the minor axis is about 1.15.

---

[17] Cramer, p. 283.

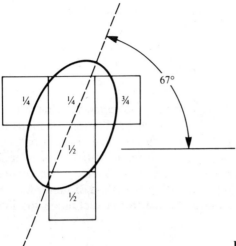

Figure 6.22 Image with Ellipse of Concentration

The fact that each image has an ellipse naturally associated with it helps to provide an easy-to-use translation-, rotation-, and scale-factor-invariant image recognition scheme. In all cases where the ellipse does not degenerate into a circle, straight line, or point, its center is located at $(\bar{x}, \bar{y})$, the angle of its major axis with respect to the x-axis is

$$\theta = \frac{\tan^{-1}\left(\frac{2\sigma_{xy}}{\sigma_x^2 - \sigma_y^2}\right)}{2} + \frac{\pi}{2}$$

and the two elliptical radii are

$$\ell_1 = \sqrt{\frac{4(1-\rho^2)}{\frac{\cos^2\theta}{\sigma_x^2} - \frac{2\rho\cos\theta\sin\theta}{\sigma_x\sigma_y} + \frac{\sin^2\theta}{\sigma_y^2}}}$$

and

$$\ell_2 = \sqrt{\frac{4(1-\rho^2)}{\frac{\sin^2\theta}{\sigma_x^2} + \frac{2\rho\cos\theta\sin\theta}{\sigma_x\sigma_y} + \frac{\cos^2\theta}{\sigma_y^2}}}$$

In theory, an image normalization and recognition scheme based on the ellipse of concentration is rather obvious for nondegenerate cases. For each pattern class in the database, calculate the appropriate number of moments $m_{p,q}$. In particular, obtain $(\bar{x}, \bar{y})$, the angle θ, and the length of the major axis

$\ell = \max(\ell_1, \ell_2)$. Then translate this ellipse so that the origin becomes the center of gravity, and translate all other moments accordingly. This translation yields

$$m_{p,q}^1 = \sum_{i=0}^{p} \sum_{j=0}^{q} \binom{p}{i} \binom{q}{j} (-\bar{x})^{p-i}(-\bar{y})^{q-j} m_{ij}$$

where the m_{ij} are the original moments and $m_{p,q}^1$ is called the p, q *moment about the mean*. The major axis of the ellipse of concentration is then scaled to unity by dividing by ℓ. Concurrently, this scale factor is used on all the moments; that is, the normalized moments $m_{p,q}^{11}$ with respect to length must be found. These are given by

$$m_{p,q}^{11} = \frac{m_{p,q}^1}{\ell^{p+q}}$$

Next, the new ellipse centered at the origin and of unit length should be rotated through the angle θ so that the major axis is the abscissa. Finally, the same rotation should be performed on the moments $m_{p,q}^{11}$. This latter rotation gives

$$m_{p,q}^{111} = \sum_{i=0}^{p} \sum_{j=0}^{q} \binom{p}{i} \binom{q}{j} (-1)^{q-j}(\cos \theta)^{p-i+j}(\sin \theta)^{q+i-j} m_{p+q-i-j,i+j}^{11}$$

The moments thus obtained are stored as features for each pattern class. Moreover, the rotation is often conducted using $\theta + \pi$ as well as θ. (See Section 5.12 for a similar methodology.) When the observed image is obtained and the associated moments found, the same normalization procedure is applied to these moments, first for translation, then for scale factor, and finally for rotation. These normalized moments are then compared with those stored in the database by using some norm, correlation technique, or functional. Recognition or lack of recognition occurs on the basis of the comparison.

EXERCISES

6.1. As in the procedures established in Example 6.2, let $X = \{(i, j): i, j = 0, 1\}$ and $Y = \{-1, 0, 1\}$. How many images are in the sample space? How many constant images are there in S? If $\mathcal{F} = 2^S$, what is the probability of occurrence of a singleton event? What is

$$P\left(\left\{\begin{pmatrix} 0 & 1 \\ 0 & 1 \end{pmatrix}, \begin{pmatrix} 1 & 1 \\ 0 & 1 \end{pmatrix}\right\}\right)?$$

Can the probability of the set

$$\left\{\begin{pmatrix} 0 & * \\ 1 & 1 \end{pmatrix}\right\}$$

be found?

6.2. In the previous exercise, let A be the event that the image contains zero and one among its gray values. Let B be the event that the gray value at $(0, 0)$ is zero. Find $P(A \mid B)$ and $P(B \mid A)$.

6.3. Show that a constant is a random variable for any probability space $(S, \mathscr{F}, P)$. Show also that when $\mathscr{F} = 2^S$, any function $\xi: S \to R$ is a random variable.

6.4. If

$$F_\xi(x) = \begin{cases} 0 & \text{if } x < 0 \\ \frac{1}{4} & \text{if } 0 \leq x < 1 \\ \frac{1}{2} & \text{if } 1 \leq x < 2 \\ \frac{1}{2} + \frac{x-2}{3} & \text{if } 2 \leq x < 3 \\ 1 & \text{if } 3 \leq x \end{cases}$$

is a gray level distribution function associated with a certain family S of images, describe a "typical" image within S.

6.5. Find the gray level density function $f_\xi(x)$ corresponding to $F_\xi(x)$ given in Exercise 6.4.

6.6. Using the distribution of the random variable given in Exercise 6.4, find:
  (a) $E(\xi) = \mu$
  (b) σ_ξ^2
  (c) $E|\xi - \mu|$
  (d) $P(|\xi - \mu| \geq 2\sigma)$, and compare with the results given by Chebyshev's inequality, Theorem 6.2.
  (e) the median.

6.7. Consider the image

$$a = \begin{pmatrix} 0 & 0 & 0 \\ -1 & 0 & * \\ 1 & \circledast & * \end{pmatrix}$$

If ξ is the gray level random variable associated with a, find $F_\xi(x)$ and answer (a), (b), (c), (d), and (e) as in Exercise 6.6.

6.8. Verify the entries given for the mean and variance in Table 6.1.

6.9. Verify the entries given for the mean and variance in Table 6.2.

6.10. For the image given in Exercise 6.7, find the gray level distribution and density associated with the "clipped" image obtained using the random variable

$$\eta = g(\xi) = \begin{cases} \xi & \text{if } \xi \geq \frac{1}{2} \\ 0 & \text{if } \xi < \frac{1}{2} \end{cases}$$

6.11. Answer the same question asked in Exercise 6.10 using $\eta = |\xi|$.

6.12. Refer to Example 6.26 involving an extended random field where S contains the eight illustrated images $\mathscr{E}_1, \mathscr{E}_2, \ldots, \mathscr{E}_8$. Assume that $P\{\mathscr{E}_1\} = P\{\mathscr{E}_2\} = P\{\mathscr{E}_3\} = P\{\mathscr{E}_4\} = \frac{1}{6}$ and $P\{\mathscr{E}_5\} = P\{\mathscr{E}_6\} = P\{\mathscr{E}_7\} = P\{\mathscr{E}_8\} = \frac{1}{12}$. Find $F_{\xi\eta}(x, y)$ and provide a "plot" similar to the one given in Figure 6.15.

6.13. Find the joint density function $f_{\xi\eta}(x, y)$ corresponding to the joint distribution given in Exercise 6.12.

6.14. Find the marginal and conditional density functions whose joint density is that of Exercise 6.13.

6.15. Using the random variables given in Exercise 6.12, find:
  (a) $E(\xi)$
  (b) $E(\eta)$
  (c) σ_ξ^2
  (d) σ_η^2
  (e) $\sigma_{\xi\eta}$
  (f) $E(\xi \mid \eta = y)$
  (g) $E(\eta \mid \xi = x)$

6.16. Repeat Example 6.36 using $\gamma = g(\xi, \eta) = \text{minimum}(\xi, \eta) = \xi \wedge \eta$ instead of maximum(ξ, η).

6.17. Let

$$f_{\xi\eta}(x, y) = \begin{cases} Ax & \text{if } 0 < x < y < 1 \\ 0 & \text{otherwise} \end{cases}$$

  (a) Find A such that $f_{\xi\eta}$ is a joint density function of the random vector $\mathbf{v} = \begin{pmatrix} \xi \\ \eta \end{pmatrix}$.
  (b) Find $\bar{\mathbf{v}} = \begin{pmatrix} \bar{\xi} \\ \bar{\eta} \end{pmatrix}$.
  (c) Find the covariance matrix Cov$(\mathbf{v})$.

6.18. Let ξ and η be independent random variables, each having a mean zero Gaussian distribution with variance σ^2.
  (a) Find the density function of $\xi^2 + \eta^2$.
  (b) Find the density function of $\gamma = \sqrt{\xi^2 + \eta^2}$.
  (c) What is $P(\gamma \leq \tfrac{1}{2})$?

6.19. Calculate the ellipse of concentration associated with the image

$$f = \begin{pmatrix} \tfrac{1}{2} & 1 & \tfrac{1}{2} \\ \tfrac{1}{2} & * & * \\ 1 & \tfrac{1}{4} & * \\ \text{①} & * & * \\ 1 & * & * \end{pmatrix}$$

7

Inferential Statistical Techniques in Image Processing

7.1 ESTIMATORS AND ESTIMATES

In Section 6.3 it was seen that a random variable is nothing more than a real-valued (measurable) function on a probability space. A value attained by this function, i.e., an observation of a random variable, is called a *sample value*. Statistical inference is the process of utilizing several sample values for the purpose of determining the underlying probability laws governing the random variables in question. For example, using observations of a random vector **v**, we might attempt to discover the joint distribution $F_\mathbf{v}$. Often, based upon expert opinion, heuristics, or hunches, it is assumed that the unknown distribution is a member of some family of distributions, such as the Gaussian, Poisson, or some other well-studied family. An assumption of this nature reduces the problem from determining a distribution to determining certain indexing parameters that characterize the distribution. For instance, if a member of the univariate normal family of distributions is to be determined, then only two parameters, the mean and variance, need be found. Once they are determined, a unique member of the Gaussian family is identified. The process of determining parameters used in characterizing families of distributions based on sample values is called *estimation*. When the family of distributions is assumed to be known, the process is called *parametric estimation*. When we seek parameters associated with a random variable whose specific distribution is unknown, *nonparametric*, or *distribution-free*, estimation is employed.

Suppose $\epsilon_1, \epsilon_2, \ldots, \epsilon_n$ are the parameters associated with a family of distributions. Let $D \subset R^n$ denote the set from which the parameter vector $(\epsilon_1, \epsilon_2, \ldots, \epsilon_n)$ is to be chosen. Then D is called the *parameter space*. For instance, if the mean of a Gaussian family is to be determined, then, since $\epsilon_1 = \mu$, $D = R$. On the other hand, if the mean and variance of a Gaussian family are desired, then $\epsilon_1 = \mu$, $\epsilon_2 = \sigma^2$, and $D = R \times (0, \infty)$. The same set D results if the median and standard deviation of the Gaussian density are desired.

The key to estimation is the formulation of a rule by which a set of sample values of a random variable might be used to estimate the values of the parameters to be determined. In practical terms, a function must be constructed which takes as input variables the outcomes of some experiment, and whose functional value, which might be vector valued, can be employed as a "reasonable" estimate for the vector of unknown parameters. The succeeding definitions formalize these intuitive notions.

An *estimation rule*, otherwise known as an *estimator*, is a function that maps a fixed-size, ordered collection of sample values into the parameter space D. In terms of the probability space $(S, \mathcal{F}, P)$, an ordered collection of sample values of size n is a vector-valued function $\mathbf{v}$ defined on the Cartesian product S^n. Rigorously, $\mathbf{v}: S^n \to R^n$ by

$$\mathbf{v}(s_1, s_2, \ldots, s_n) = \begin{pmatrix} \xi_1(s_1) \\ \xi_2(s_2) \\ \vdots \\ \xi_n(s_n) \end{pmatrix}$$

where each ξ_i is a random variable on S. An *estimator* is a mapping $\hat{\boldsymbol{\theta}}: S^n \to D$, where

$$\hat{\boldsymbol{\theta}}(\cdot) = \varphi(\mathbf{v}(\cdot)),$$

and where φ, the *estimation rule*, is defined on the range space of $\mathbf{v}$. In other words, the estimator $\hat{\boldsymbol{\theta}}$ is constructed from the component values of the random vector $\mathbf{v}$ by means of some rule φ. When we speak of the "goodness" of an estimator, we are referring to the degree (based on some measure of goodness) to which the estimation rule results in a satisfactory estimator.

For fixed $\mathbf{s} = (s_1, s_2, \ldots, s_n)$, we call $\mathbf{v}(\mathbf{s})$ an *observation*. Intuitively, an observation consists of n empirical real values, $\xi_1(s_1), \xi_2(s_2), \ldots, \xi_n(s_n)$. The real-valued vector $\varphi(\mathbf{v}(\mathbf{s}))$ is then called an *estimate*. Thus, an estimator is a function and an estimate is a value of such a function. Certain well-known notations and conventions have been established regarding estimates. If $\boldsymbol{\theta}$ is a parameter vector for the underlying distribution, then an estimate of $\boldsymbol{\theta}$ is denoted by $\overset{*}{\boldsymbol{\theta}}$. For instance, suppose that the mean μ and variance σ^2 of a Gaussian density are desired. Then, in this case,

$$\boldsymbol{\theta} = \begin{pmatrix} \mu \\ \sigma^2 \end{pmatrix}$$

and an estimate based on sampled values would be denoted by

$$\overset{*}{\boldsymbol{\theta}} = \begin{pmatrix} \overset{*}{\mu} \\ \overset{*}{\sigma}^2 \end{pmatrix}$$

An abuse of notation often occurs by calling $\hat{\boldsymbol{\theta}}$ an estimate. We must keep in mind, however, that an estimator is a fixed rule and an estimate is one of the many (vector) values returned by this rule. Much confusion arises over this point. The estimator $\hat{\boldsymbol{\theta}}$ is a random phenomenon, and hence is a random variable (or random vector); an estimate is a particular value (vector value) taken on by the estimator. Remember that the unknown parameters are to be estimated through the utilization of sample values. As a consequence, we must keep in mind that $\hat{\boldsymbol{\theta}}$ is a function, $\varphi(\mathbf{v}(s))$, defined on the sample space and, hence, is a random quantity. Once the experiment has been run and actual empirical values observed, a numerical value is computed from $\hat{\boldsymbol{\theta}}$ and an estimate results. Finally, note that $\mathbf{v}$ is a random vector on the product space S^n.

One of the most important cases in estimation occurs when the random variables $\xi_1, \xi_2, \ldots, \xi_n$ in $\mathbf{v}$ are independent and each possesses a marginal distribution identical to the random variable for the population being sampled. When this occurs, the method of sampling is called *random sampling* and, in that the random variables possess identical distributions, they are said to be *identically distributed*.

In practice, one performs an experiment and the outcome of the experiment corresponds to a particular sample value of the random vector $\mathbf{v}$. If the methodology of random sampling has been employed, then, for a fixed element $\mathbf{s} = (s_1, s_2, \ldots, s_n)$ in S^n,

$$\begin{pmatrix} \xi_1(s_1) \\ \xi_2(s_2) \\ \vdots \\ \xi_n(s_n) \end{pmatrix}$$

is called a *random (vector) sample of size n*.

For random sampling, the distribution associated with a given estimator, as a function of $\xi_1, \xi_2, \ldots, \xi_n$, is more easily found, since the joint distribution satisfies the relation

$$F_{\xi_1\xi_2\cdots\xi_n}(x_1, x_2, \ldots, x_n) = F_{\xi_1}(x_1) F_{\xi_1}(x_2) \cdots F_{\xi_1}(x_n)$$

(See Section 6.9, where functions of several random variables are discussed and associated marginal distributions are obtained.) The distribution function $F_{\xi_1}(x)$ may not be known at all, or it may be known except for several characterizing parameters.

Example 7.1

Suppose that the number of pixels having a nonnegative gray value in a certain set of scenes satisfies a Poisson density

$$f_\xi(x) = \sum_{n=0}^{\infty} \frac{\lambda^n}{n!} e^{-\lambda} \delta(x - n)$$

Let ξ_1, ξ_2, and ξ_3 denote independent identically distributed random variables with the same density as ξ. We use these random variables, which form a random sample of size 3, to construct an estimator $\hat{\lambda}$ for λ. In so doing, we employ the estimation rule

$$\varphi(\xi_1, \xi_2, \xi_3) = \frac{\xi_1 + \xi_2 + \xi_3}{3}$$

where each ξ_i denotes the number of nonnegative gray values in an image. Note that

$$\mathbf{v} = \begin{pmatrix} \xi_1 \\ \xi_2 \\ \xi_3 \end{pmatrix}$$

Hence,

$$\hat{\lambda}(s_1, s_2, s_3) = \varphi(\mathbf{v}(s_1, s_2, s_3)) = \frac{1}{3}[\xi_1(s_1) + \xi_2(s_2) + \xi_3(s_3)]$$

which, for a particular set of sample elements $\overset{*}{s}_1$, $\overset{*}{s}_2$, and $\overset{*}{s}_3$, gives the numerical mean of $\xi_1(\overset{*}{s}_1)$, $\xi_2(\overset{*}{s}_2)$, and $\xi_3(\overset{*}{s}_3)$.

Now suppose that a particular set of three sample images

$$\overset{*}{s}_1 = a = \begin{pmatrix} ③ & 2 \\ -7 & * \end{pmatrix}$$

$$\overset{*}{s}_2 = b = \begin{pmatrix} ④ & -2 \\ -3 & 0 \\ 1 & 8 \end{pmatrix}$$

$$\overset{*}{s}_3 = c = \begin{pmatrix} * & * \\ -2 & -6 \end{pmatrix}$$

is randomly chosen. The numbers of nonnegative gray values in a, b, and c are, respectively, $\xi_1(a) = 2$, $\xi_2(b) = 4$, and $\xi_3(c) = 0$. Therefore, the observed value of the random vector $\mathbf{v}$ is given by

$$\mathbf{v}(a, b, c) = \begin{pmatrix} 2 \\ 4 \\ 0 \end{pmatrix}$$

and the empirical estimate for λ, obtained from the particular sample in question, is

$$\overset{*}{\lambda} = \frac{2 + 4 + 0}{3} = 2$$

In the preceding example, the estimator

$$\hat{\theta} = \frac{1}{n}[\xi_1 + \xi_2 + \cdots + \xi_n]$$

was employed with $n = 3$. This estimator is known as the *sample mean*. Like ξ_1, $\xi_2, \ldots, \xi_n$, it is a random variable. Due to its frequent use, we denote the sample mean by $\overline{\theta}$. As a random variable, $\overline{\theta}$ possesses its own distribution function and associated parameters. For instance, assuming random sampling, the linearity of the expected value yields

$$E(\bar{\theta}) = \frac{1}{n} \sum_{i=1}^{n} E(\xi_i) = E(\xi)$$

Moreover, $\bar{\theta}$ has variance

$$\sigma_{\bar{\theta}}^2 = E(\bar{\theta}^2) - E(\bar{\theta})^2$$

But, due to the independence and identical distribution of the random variables $\xi_1, \xi_2, \ldots, \xi_n$,

$$E(\bar{\theta}^2) = 1/n^2 E((\xi_1 + \xi_2 + \cdots + \xi_n)^2)$$

$$= 1/n^2 \left[\sum_{i=1}^{n} E(\xi_i^2) + \sum_{i \ne j} E(\xi_i) E(\xi_j) \right]$$

$$= 1/n E(\xi^2) + \frac{n-1}{n} E(\xi)^2$$

Consequently, the variation of the sample mean is given by

$$\sigma_{\bar{\theta}}^2 = \frac{1}{n} E(\xi^2) - \frac{1}{n} E(\xi)^2 = \frac{1}{n} \sigma_{\xi}^2$$

where σ_{ξ}^2 denotes the variance of the population variable ξ. In words, the mean of the sample mean is the same as the population mean, while the variance of the sample mean is the population variance divided by the size of the random sample. Though we will not pursue applications of these two facts here, let us point out that, as a consequence of these relationships, the limit of the sample mean variance is zero as n goes to ∞. Hence, for large n, the distribution of $\bar{\theta}$ is spread tightly about the original population mean. For this reason, the sample mean estimator is often used to estimate the population mean, though numerous other estimation rules might be employed.

Before proceeding, a further comment on terminology is in order. When used alone, the term *mean* refers to the mean of the underlying probability distribution. It is the population, or theoretical mean, and it is a parameter. On the other hand, the sample mean is an estimator, a random variable. Finally, an observation of the sample mean is a number computed by applying the estimation rule for the sample mean to actual numerical values obtained from an experiment. This number, derived from observation, is an estimate. It is called an *empirical mean*. But be aware! It is common practice to use the term *sample mean* when referring to the empirical mean, and therefore one must pay careful attention to the context to see if it is an estimator or an estimate which is under discussion.

Returning to the situation in Example 7.1, if the sample mean is used, we obtain $\overset{*}{\lambda} = 2$ as an estimate of the average number of pixels with nonnegative gray value. The corresponding density is given by

$$f_\xi(x) = e^{-2} \sum_{n=0}^{\infty} \frac{2^n}{n!} \delta(x - n)$$

Before several estimation techniques, classification methods, and criteria are given, it should be mentioned that all estimators discussed herein are acceptable. An estimator is *acceptable* when it is not a function of any unknown parameter and the range of the estimator is a subset of D. By saying that the estimator is not dependent upon unknown parameters, we do not mean to imply that its distribution is free of unknown parameters. For instance, in Example 7.1 we utilized the sample mean to estimate the unknown parameter λ. What is of consequence is that there are no unknown parameters in the estimation rule for the sample mean. Finally, an acceptable estimator is often called a *statistic*.

7.2 ESTIMATION CLASSIFICATION CRITERIA

An estimator $\hat{\theta}$ is *linear* if it can be expressed as

$$\hat{\theta} = \varphi(\xi_1, \xi_2, \ldots, \xi_n) = B \begin{pmatrix} \xi_1 \\ \xi_2 \\ \vdots \\ \xi_n \end{pmatrix}$$

where B is an m by n matrix of real numbers.

Example 7.2

In Example 7.1, the sample mean for the average number of pixels with nonnegative gray value is a linear estimator. Indeed,

$$\bar{\theta} = \varphi(\xi_1, \xi_2, \xi_3) = (\tfrac{1}{3} \ \tfrac{1}{3} \ \tfrac{1}{3}) \begin{pmatrix} \xi_1 \\ \xi_2 \\ \xi_3 \end{pmatrix}$$

More generally, the sample mean

$$\bar{\theta} = (1/n \ 1/n \ \cdots \ 1/n) \begin{pmatrix} \xi_1 \\ \xi_2 \\ \xi_3 \\ \vdots \\ \xi_n \end{pmatrix}$$

is linear.

Example 7.3

An example of a nonlinear estimator is the *sample median*, which is the middle sample value of all values arranged in ascending order. In Example 7.1, since $\xi_1(a) = 2$, $\xi_2(b) = 4$, and $\xi_3(c) = 0$, the median estimate is $\overset{*}{\theta} = 2$.

Sec. 7.2 Estimation Classification Criteria

Assuming that the first moment exists, an important criterion of an estimator is unbiasedness. A real-valued estimator $\hat{\theta}$ is an *unbiased* estimator if $E(\hat{\theta}) = \theta$, the desired parameter. An estimator that is not unbiased is called a *biased* estimator. In general, the *bias* of an estimator is defined by

$$b = E(\hat{\theta}) - \theta$$

The sample mean is an unbiased estimator of the theoretical mean. That is, if $E(\xi) = \mu$ and $\xi_1, \xi_2, \ldots, \xi_n$ are independent random variables with the same distribution as ξ, then

$$E(\bar{\theta}) = (1/n \; 1/n \; \cdots \; 1/n) E \begin{pmatrix} \xi_1 \\ \xi_2 \\ \vdots \\ \xi_n \end{pmatrix} = \mu$$

A very important biased estimator is the *sample variance* s^2. Suppose the population mean (mean of the random variable) is μ, the variance σ^2 exists, and $\bar{\theta}$ is the sample mean. If $\xi_1, \xi_2, \ldots, \xi_n$ are obtained by random sampling, we define

$$s^2 = \frac{1}{n} \sum_{i=1}^{n} (\xi_i - \bar{\theta})^2$$

which can be written as

$$s^2 = \frac{1}{n} \left[\sum_{i=1}^{n} (\xi_i - \mu)^2 - n(\mu - \bar{\theta})^2 \right]$$

Taking expected values of both sides gives

$$E(s^2) = \sigma^2 - E\left[\frac{(\mu - \xi_1) + (\mu - \xi_2) + \cdots + (\mu - \xi_n)}{n} \right]^2$$

Using the fact that ξ_i and ξ_j are independent for $i \neq j$ gives $E((\mu - \xi_i)(\mu - \xi_j)) = 0$ for $i \neq j$. Consequently,

$$E(s^2) = \sigma^2 \frac{n-1}{n}$$

which makes s^2 a biased estimate with bias $-\sigma^2/n$. However, as $n \to \infty$, $E(s^2) \to \sigma^2$, and so the sample variance is said to be *asymptotically unbiased*, i.e., its bias tends to zero as $n \to \infty$.

An estimator $\hat{\theta}$ of a real-valued parameter θ is said to be *consistent* if there is a high probability that the estimate will be near the parameter when n is large. More precisely, $\hat{\theta}_n = \varphi(\xi_1, \xi_2, \ldots, \xi_n)$ is a consistent estimator of θ if for any $\epsilon > 0$,

$$P(|\hat{\theta}_n - \theta| \geq \epsilon) \to 0 \text{ as } n \to \infty$$

The sample mean $\bar{\theta} = \hat{\mu}$ is a consistent estimator of the population mean μ, and the sample variance $s^2 = \hat{\sigma}^2$ is a consistent estimator of the population variance σ^2. More generally, we have the following theorem.

Theorem 7.1.[1] Suppose $E(\xi^k)$ exists. Then

(a) $\dfrac{1}{n} \sum\limits_{i=1}^{n} \xi_i^k$ is a consistent estimator of $E(\xi^k)$.

(b) $\dfrac{1}{n} \sum\limits_{i=1}^{n} (\xi_i - \bar{\theta})^k$ is a consistent estimator of $E((\xi - \mu)^k)$.

The definition of consistency can be extended to the case where θ is a vector.

Of paramount importance is the concept of the *mean square error* associated with the estimator $\hat{\theta}$ of a real-valued parameter θ. This error is defined as $E((\hat{\theta} - \theta)^2)$, where it is assumed that the second moments exist. For notational clarity, we often write $E(\hat{\theta} - \theta)^2$. The mean square error can be written as $E(\hat{\theta} - \theta)^2 = \sigma_{\hat{\theta}}^2 + b^2$, where b is the bias and $\sigma_{\hat{\theta}}^2 = E(\hat{\theta} - E(\hat{\theta}))^2$ is the variance of $\hat{\theta}$. Written in that form, whenever $\hat{\theta}$ is an unbiased estimator of θ, the mean square error is simply $\sigma_{\hat{\theta}}^2$.

If $\hat{\theta}_1$ and $\hat{\theta}_2$ are both estimators of θ, and if $E(\hat{\theta}_1 - \theta)^2 < E(\hat{\theta}_2 - \theta)^2$, then $\hat{\theta}_1$ is said to be a *better estimator* of θ than $\hat{\theta}_2$. Sometimes there exists an estimator $\hat{\theta}$ of θ such that $E(\hat{\theta} - \theta)^2 \leq E(\tilde{\theta} - \theta)^2$ for any estimator $\tilde{\theta}$ of θ. In that case, $\hat{\theta}$ is said to be the *best estimator* of θ. Intuitively, a best estimator is one that minimizes the mean square error. In a later section, we concern ourselves with finding the *best linear unbiased estimator (BLUE)* of θ for a specific matrix model. This amounts to finding, for a certain class of problems, an estimator $\hat{\theta}$ of θ which is linear, and unbiased, and which minimizes $E(\hat{\theta} - \theta)^2$. A simple illustration is given in the next example.

Example 7.4

Suppose that a random sample consisting of the gray values of two pixels is chosen in an attempt to find the average gray value. Let the associated random variables be ξ_1 and ξ_2. Based on this sample, a BLUE estimator $\hat{\mu}$ of the average population gray value μ is desired. Suppose that second moments exist, and let $\sigma^2 = E(\xi - E(\xi))^2$. Then, since $\hat{\mu}$ is linear,

$$\hat{\mu} = B \begin{pmatrix} \xi_1 \\ \xi_2 \end{pmatrix}$$

where B is the one-by-two matrix $(a\ c)$. Hence, $\hat{\mu} = a\xi_1 + c\xi_2$.

Next, using the fact that $\hat{\mu}$ is an unbiased estimator of μ, we have $E(\hat{\mu}) = (a + c)\mu = \mu$, so that $c = 1 - a$, and therefore $\hat{\mu} = a\xi_1 + (1 - a)\xi_2$.

---

[1] Marek Fisz, *Probability Theory and Mathematical Statistics* (New York: John Wiley & Sons, Inc., 1963), pp. 238, 366.

Sec. 7.2 Estimation Classification Criteria

Finally, since $\hat{\mu}$ is best, it minimizes the mean square error. Consequently, we must find an a such that $E(\hat{\mu} - \mu)^2$ is minimized. Now,

$$E(\hat{\mu} - \mu)^2 = E(a(\xi_1 - \mu) + (1 - a)(\xi_2 - \mu))^2$$
$$= a^2\sigma^2 + (1 - a)^2\sigma^2$$
$$= (2a^2 - 2a + 1)\sigma^2$$

Taking the derivative of the last expression with respect to a and setting the result equal to zero gives $a = \frac{1}{2}$, so that the BLUE of μ is the sample mean $\bar{\theta} = \hat{\mu} = \frac{1}{2}\xi_1 + \frac{1}{2}\xi_2$. The associated mean square error is $E(\bar{\theta} - \theta)^2 = \sigma^2/2$.

More generally, if a random sample of size n were to be employed to find the average gray value, then again, the sample mean $\bar{\theta}$ would be the BLUE, where

$$\bar{\theta} = \frac{1}{n}\xi_1 + \frac{1}{n}\xi_2 + \cdots + \frac{1}{n}\xi_n$$

and the associated mean square error would be

$$E(\bar{\theta} - \theta)^2 = \frac{\sigma^2}{n}$$

As the number of samples increases, so does the accuracy of the estimate.

The mean square error in the vector case is described in Section 7.4. For unbiased estimators, it is often called the *output covariance matrix*.

When they exist, BLUEs are usually quite easy to obtain and analyze. If the linearity assumption is omitted, however, it is more difficult, and sometimes impossible, to find a best unbiased estimator $\hat{\theta}$. On the other hand, if the estimator $\tilde{\theta}$ is the BLUE, it cannot be better than $\hat{\theta}$. Using the unbiasedness assumption implies that $E(\hat{\theta} - \theta)^2 = \sigma_{\hat{\theta}}^2$, and therefore $\sigma_{\tilde{\theta}}^2 \geq \sigma_{\hat{\theta}}^2$.

When θ is known to lie in the interval (α, β), the *Cramer-Rao inequality*[2] provides a lower mean square error bound to all unbiased estimators $\hat{\theta}$ of θ. It is convenient in this inequality to denote the underlying density function for the population random variable ξ as $f_\xi(x; \theta)$ or $f(x; \theta)$. Letting $\xi_1, \xi_2, \ldots, \xi_n$ be a random sample of size n from this distribution, if $\hat{\theta} = \varphi(\xi_1, \xi_2, \ldots, \xi_n)$ is any acceptable estimator of θ, then the Cramer-Rao inequality is

$$\sigma_{\hat{\theta}}^2 \geq \frac{1}{nE\left[\dfrac{\partial \ln f(\xi; \theta)}{\partial \theta}\right]^2}$$

and is valid under regularity conditions to be subsequently presented. When the lower bound in the Cramer-Rao inequality is obtained by the estimator $\hat{\theta}$, the estimator is without question the best, and is called the *most efficient estimator* of θ.

---

[2] Ibid., p. 468.

Example 7.5

Suppose the population random variable ξ has an exponential density function $f_\xi(x, 1/\lambda) = \lambda e^{-\lambda x} I_{(0,\infty)}(x)$. Let $\xi_1, \xi_2, \ldots, \xi_n$ be a random sample, and let $\bar{\theta}$ be the sample mean estimator of the mean $E(\xi) = 1/\lambda$. Then,

$$\ln(f(x; 1/\lambda)) = \ln \lambda + (-\lambda x) + \ln(I_{(0,\infty)}(x))$$

$$= -\ln(1/\lambda) - \frac{x}{1/\lambda} + \ln(I_{(0,\infty)}(x))$$

Therefore,

$$\frac{\partial \ln(f(x; 1/\lambda))}{\partial(1/\lambda)} = -\lambda + x\lambda^2$$

and, since $\sigma_\xi^2 = \lambda^{-2}$,

$$E\left(\frac{\partial \ln f(\xi; 1/\lambda)}{\partial(1/\lambda)}\right)^2 = E(\lambda^2 \xi - \lambda)^2 = \lambda^2$$

Consequently, the Cramer-Rao lower bound is $(1/n)\lambda^{-2}$, which equals $\sigma_{\bar{\theta}}^2$ and this shows that $\bar{\theta}$ is the most efficient estimator of $1/\lambda$.

The regularity criteria which provide sufficient conditions for the Cramer-Rao inequality to hold are as follows:

1. $\partial f(x; \theta)/\partial \theta$ must exist for all real x and for θ in (α, β), the interval in which θ is known to lie.

2. $\dfrac{\partial}{\partial \theta} \int \cdots \int \prod_{i=1}^{n} f(x_i; \theta) \, dx_1 \ldots dx_n = \int \cdots \int \dfrac{\partial}{\partial \theta} \prod_{i=1}^{n} f(x_i; \theta) \, dx_1 \cdots dx_n$

3. $\dfrac{\partial}{\partial \theta} \int \cdots \int \varphi(x_1, x_2, \ldots, x_n) \prod_{i=1}^{n} f(x_i; \theta) \, dx_1 \cdots dx_n$

   $= \int \cdots \int \varphi(x_1, x_2, \ldots, x_n) \dfrac{\partial}{\partial \theta} \prod_{i=1}^{n} f(x_i; \theta) \, dx_1 \cdots dx_n$

4. $0 < E\left(\dfrac{\partial}{\partial \theta} \ln f(\xi; \theta)\right)^2 < \infty$ for all θ in (α, β).

Notice that conditions 2 and 3 have to do with bringing the partial derivative operator $\partial/\partial \theta$ inside the integral.

The final criterion of estimators to be given is *sufficiency*. As with the other criteria, sufficient statistics need not exist. However, when they do, it is often possible to find an unbiased efficient estimator based on the sufficient estimator. Intuitively, an estimator $\hat{\theta}$ is said to be sufficient whenever all information relevant

Sec. 7.2   Estimation Classification Criteria

to the problem of estimating the parameter θ is contained in $\overset{*}{\hat\theta}$. In other words, the sample values themselves provide no more information about θ once the value of $\hat\theta$ is known. Consequently, utilization of other estimators of θ which might be obtained from the sample would not improve the results since $\hat\theta$ contains all the information pertaining to θ that the sample has to offer. More precisely, $\hat\theta$ is sufficient if the conditional distribution of the sample random vector for a given value of $\hat\theta$ does not depend on the parameter θ. Symbolically, $\hat\theta$ is sufficient if

$$f_{\xi_1\xi_2\ldots\xi_n}(x_1, x_2, \ldots, x_n \mid \hat\theta = \overset{*}{\hat\theta})$$

does not depend on the parameter θ.

To appreciate the importance of the preceding definition, recall that all information regarding the sample is contained in the density $f_{\xi_1\ldots\xi_n}(x_1, \ldots, x_n)$. Thus, if $\hat\theta$ is sufficient, then once it has a known value, i.e., once it is no longer random, the value of the sample vector

$$\mathbf{v} = \begin{pmatrix} \xi_1 \\ \xi_2 \\ \vdots \\ \xi_n \end{pmatrix}$$

does not depend on θ. Consequently, no knowledge concerning θ can be obtained by observing $\mathbf{v}$. Or, to put it crudely, $\hat\theta$ extracts all information from the sample regarding θ.

The preceding remarks may be interpreted directly in terms of estimation. Suppose $\hat\theta_1, \hat\theta_2, \ldots, \hat\theta_n$ are n estimators with respective estimation rules $\varphi_1, \varphi_2, \ldots, \varphi_n$. Then, for any $k = 1, 2, \ldots, n$, $\hat\theta_k = \varphi_k(\xi_1, \xi_2, \ldots, \xi_n)$, where the ξ_j denote a random sample. Moreover, suppose the system of equations for the rules,

$$y_1 = \varphi_1(x_1, x_2, \ldots, x_n)$$
$$y_2 = \varphi_2(x_1, x_2, \ldots, x_n)$$
$$\vdots \qquad \vdots$$
$$y_n = \varphi_n(x_1, x_2, \ldots, x_n)$$

is uniquely solvable for $x_1, x_2, \ldots, x_n$. Then, as discussed in Section 6.11 for the case of two variables,

$$f_{\hat\theta_1\hat\theta_2\cdots\hat\theta_n}(y_1, y_2, \ldots, y_n; \theta) = |J| f_{\xi_1\xi_2\cdots\xi_n}(x_1, x_2, \ldots, x_n; \theta)$$

where $|J|$ is the absolute value of the Jacobian determinant. Now, suppose $\hat\theta_1$ is a sufficient estimator for θ. Then conditioning by $\hat\theta_1 = \overset{*}{\hat\theta}$ leaves both sides of the preceding equation independent of θ. In terms of the estimators $\hat\theta_1, \hat\theta_2, \ldots, \hat\theta_n$, this means that the conditional density

$$f_{\hat\theta_2\hat\theta_3\cdots\hat\theta_n}(y_2, y_3, \ldots, y_n \mid \hat\theta_1 = \overset{*}{\hat\theta})$$

does not depend upon the parameter θ.

Intuitively, estimation of a parameter depends upon some relationship between the parameter and the estimator in question. To say that the conditional density of $\hat{\theta}_2$ given $\hat{\theta}_1 = \overset{*}{\theta}$ is independent of θ means that once an estimate based upon $\hat{\theta}_1$ has been fixed, there is no further dependency between $\hat{\theta}_2$ and θ. Consequently, no information can be attained regarding θ by observation of $\hat{\theta}_2$.

Determining whether or not an estimator is sufficient by utilizing the definition is tedious, since joint densities must usually be found prior to the determination of the conditional density. Fortunately, a factorization procedure known as the *Neyman criterion for sufficiency* has been developed that is often easier to use than the definition.[3] This criterion states that $\hat{\theta}$ is sufficient if the joint density of the random sample $\xi_1, \xi_2, \ldots, \xi_n$, $f(x_1; \theta)f(x_2; \theta) \cdots f(x_n; \theta)$, is capable of being written as the product of the density of $\hat{\theta}$ with a function $h(x_1, x_2, \ldots, x_n)$, which does not depend upon θ. Perhaps an illustration might help clarify these somewhat abstruse notions.

Example 7.6

The estimator given in Example 7.5 is a sufficient estimator of the mean $1/\lambda$. The joint density of the random sample is

$$\prod_{i=1}^{n} f(x_i; \theta) = \lambda^n e^{-\lambda \left(\sum_{i=1}^{n} x_i\right)} I_{(0,\infty) \times (0,\infty) \times \cdots \times (0,\infty)}(x_1, x_2, \ldots, x_n)$$

The density function associated with the estimator

$$\hat{\theta} = \frac{\xi_1 + \xi_2 + \cdots + \xi_n}{n}$$

can be found in two steps. First, the density function associated with the sum of the independent random variables $\xi_1 + \xi_2 + \cdots + \xi_n$ is found by convolution or transform techniques. (See Section 6.11.) Either way, this density is seen to be in the gamma family, i.e.,

$$f_{\xi_1 + \xi_2 + \cdots + \xi_n}(x) = \frac{x^{\alpha-1} e^{-x/\beta}}{\Gamma(\alpha)\beta^\alpha} I_{(0,\infty)}(x)$$

with $\alpha = n$ and $\beta = 1/\lambda$. A change in variables (see Section 6.7) shows the density function of $\hat{\theta}$ to be

$$f_{\hat{\theta}}(\overset{*}{\theta}; \theta) = \frac{\lambda^n n^n \overset{*}{\theta}^{n-1} e^{-\lambda n \overset{*}{\theta}}}{(n-1)!} I_{(0,\infty)}(\overset{*}{\theta})$$

where we let $\overset{*}{\theta} = (x_1 + x_2 + \cdots + x_n)/n$ serve as the variable in the density for $\hat{\theta}$. Dividing the joint density of $\xi_1, \ldots, \xi_n$ by $f_{\hat{\theta}}(\overset{*}{\theta}; \theta)$ gives

$$h(x_1, x_2, \ldots, x_n) = \frac{(n-1)!}{n(x_1 + x_2 + \cdots + x_n)^{n-1}} I_{(0,\infty) \times (0,\infty) \times \cdots \times (0,\infty)}(x_1, x_2, \ldots, x_n)$$

which is not a function of λ. Consequently, $\hat{\theta}$ is sufficient.

Note that $h(x_1, \ldots, x_n)$ is not a density, nor should one expect it to be.

[3] Robert Hogg and Allen Craig, *Introduction to Mathematical Statistics* (London: Macmillan Company, 1970), p. 216.

7.3 ESTIMATION PROCEDURES

Among the most common estimation procedures is the *method of moments*, according to which parameters are determined by equating moments $E(\xi^n)$ of the population random variable ξ with the sample moments. Specifically, for the random sample $\xi_1, \xi_2, \ldots, \xi_p$, the nth *sample moment* M_n is given by

$$M_n = \frac{\sum_{k=1}^{p} \xi_k^n}{p}$$

The quantities M_n are themselves random variables, and, due to random sampling, $E(M_n) = E(\xi^n)$. The population moments are constants which usually involve some unknown parameters $\theta_1, \theta_2, \ldots, \theta_N$. Thus, $E(\xi^n) = h_n(\theta_1, \theta_2, \ldots, \theta_N)$.

For the observed values $x_1, x_2, \ldots, x_p$ of the random sample $\xi_1, \xi_2, \ldots, \xi_p$, the observed values m_n of the sample moments M_n are given by

$$m_n = \frac{\sum_{k=1}^{p} x_k^n}{p}$$

Estimation occurs by setting $E(\xi^n) = m_n$. This leads to the system of equations

$$h_1(\theta_1, \theta_2, \ldots, \theta_N) = E(\xi) = m_1$$
$$h_2(\theta_1, \theta_2, \ldots, \theta_N) = E(\xi^2) = m_2$$
$$\vdots$$
$$h_n(\theta_1, \theta_2, \ldots, \theta_N) = E(\xi^n) = m_n$$

where n, the number of moments used, is chosen such that a unique solution for $\theta_1, \theta_2, \ldots, \theta_N$ can be found. When a solution for $\theta_1, \theta_2, \ldots, \theta_N$ exists, this solution is denoted (as usual) by $\theta_1^*, \theta_2^*, \ldots, \theta_N^*$, and these are called *estimates* of $\theta_1, \theta_2, \ldots, \theta_N$. The corresponding *estimators* $\hat{\theta}_l$, $l = 1, 2, \ldots, N$, are found by solving the set of equations $E(\xi^k) = M_k$, $k = 1, 2, \ldots, n$, for $\theta_1, \theta_2, \ldots, \theta_N$ (if a solution exists).

Example 7.7

Suppose that, for a certain class of discrete images, it is known that the sum of the squares of consecutive gray values in each horizontal row satisfies a gamma density. Let ξ be the random variable associated with the sum of the squares of consecutive gray values in a horizontal row. The method of moments will be employed to find the two parameters α and β for this distribution. The first two sample moments will suffice. Using the fact (see Section 6.6) that

$$E(\xi) = \alpha\beta$$

and

$$E(\xi^2) = (\alpha + 1)\alpha\beta^2$$

we obtain the system of equations

$$M_1 = \frac{1}{p} \sum_{i=1}^{P} \xi_i = \alpha\beta$$

$$M_2 = \frac{1}{p} \sum_{i=1}^{P} \xi_i^2 = (\alpha + 1)\alpha\beta^2$$

This system of equations can be solved for α and β to obtain the estimators

$$\hat{\alpha} = \frac{M_1^2}{M_2 - M_1^2}$$

and

$$\hat{\beta} = \frac{M_2 - M_1^2}{M_1}$$

If $\bar{\mu}$ and s^2 denote the sample mean and variance, respectively, then

$$\hat{\alpha} = \frac{\bar{\mu}^2}{s^2}$$

and

$$\hat{\beta} = \frac{s^2}{\bar{\mu}}$$

Now suppose the four gray value runs

$$s_1 = (3, -2)$$
$$s_2 = (2)$$
$$s_3 = (4, -1)$$
$$s_4 = (1, 2, 3)$$

have been obtained by random selection of four images and the random choice of a run from each of those images. These images might have been

$$\begin{pmatrix} 1 & * & 3 & -2 \\ * & \circled{4} & 0 & * \end{pmatrix}, \begin{pmatrix} 2 & * \\ \circled{1} & 3 \end{pmatrix}$$

$$\begin{pmatrix} \circled{4} & -1 & * \\ 2 & 3 & 4 \end{pmatrix}, \begin{pmatrix} 1 & 2 & 3 & \circledast \\ 2 & -2 & 0 & 4 \end{pmatrix}$$

Whatever the case, we have $\xi_1(s_1) = 3^2 + (-2)^2 = 13$, $\xi_2(s_2) = 2^2 = 4$, $\xi_3(s_3) = 17$, and $\xi_4(s_4) = 14$. The observed value of the sample mean $\bar{\mu}$ is, then, 12 and the variance s^2 is $\frac{94}{4}$. Substituting gives $\overset{*}{\alpha}$ and $\overset{*}{\beta}$.

Another common estimation technique is the *maximum likelihood estimation* (or *filtering*) method, often denoted by *MLF*. Suppose that the parameter vector

$$\boldsymbol{\theta} = \begin{pmatrix} \theta_1 \\ \theta_2 \\ \vdots \\ \theta_k \end{pmatrix}$$

is to be estimated and that $\boldsymbol{\theta}$ is in parameter space $D \subset R^k$. Suppose further that a random sample $\xi_1, \xi_2, \ldots, \xi_m$ is obtained with values $x_1, x_2, \ldots, x_m$. Let the density function associated with any ξ_i be denoted by $f(x_i; \theta_1, \theta_2, \ldots, \theta_k)$. The joint density function of $\xi_1, \xi_2, \ldots, \xi_m$ is given by

$$\prod_{i=1}^{m} f(x_i; \boldsymbol{\theta})$$

When the values of the random sample are substituted into this joint density function, we obtain the *likelihood function* associated with these sample values:

$$L(\theta_1, \theta_2, \ldots, \theta_k; x_1, \ldots, x_m) = \prod_{i=1}^{m} f(x_i; \boldsymbol{\theta})$$

To avoid cumbersome notation, we shall simply write $L_\mathbf{x}(\theta_1, \theta_2, \ldots, \theta_k)$ to denote the likelihood function associated with the *sample value* vector $\mathbf{x} = (x_1, x_2, \ldots, x_m)$. When no confusion is likely to arise, we can suppress the subscript $\mathbf{x}$. Note that

$$L: D \to R$$

If, for a particular set of sample values $\mathbf{x}$, a set of parameters $\theta'_1, \theta'_2, \ldots, \theta'_k$ can be found such that $L_\mathbf{x}(\theta'_1, \theta'_2, \ldots, \theta'_k) \geq L_\mathbf{x}(\theta_1, \theta_2, \ldots, \theta_k)$ for all $\theta_1, \theta_2, \ldots, \theta_k$ in D, then this set of parameters is called a *maximum likelihood estimate* for the given set of sample values. If this maximization holds for every set of sample values, then $\hat{\boldsymbol{\theta}} = (\theta'_1, \theta'_2, \ldots, \theta'_k)$ is called the *maximum likelihood estimator*.

In practical situations, maximum likelihood estimates are found by employing conventional optimization techniques, such as setting partial derivatives of the likelihood function equal to zero.

Example 7.8

Let $\xi_1, \xi_2, \ldots, \xi_n$ constitute a random sample of background gray values whose density is Gaussian, with mean μ and variance σ^2. We seek the maximum likelihood estimator of $\boldsymbol{\theta} = \binom{\mu}{\sigma^2}$. The likelihood function is

$$L(\mu, \sigma^2) = \frac{1}{(2\pi\sigma^2)^{n/2}} e^{-1/2 \left(\frac{\sum_{i=1}^{n}(x_i - \mu)^2}{\sigma^2} \right)}$$

Taking logs of both sides and partial derivatives with respect to μ and σ^2 gives

$$\frac{\partial}{\partial \mu}[\ln L(\mu, \sigma^2)] = \frac{1}{\sigma^2} \sum_{i=1}^{n} (x_i - \mu)$$

and

$$\frac{\partial}{\partial \sigma^2}[\ln L(\mu, \sigma^2)] = \frac{-n}{2\sigma^2} + \frac{1}{2\sigma^4} \sum_{i=1}^{n} (x_i - \mu)^2$$

Setting these partial derivatives equal to zero and solving gives

$$\overset{*}{\mu} = \frac{1}{n} \sum_{i=1}^{n} x_i$$

and

$$\overset{*}{\sigma}^2 = \frac{1}{n} \sum_{i=1}^{n} (x_i - \overset{*}{\mu})^2$$

Since the same procedure applies to any set of sample values, we conclude that

$$\hat{\mu} = \frac{1}{n} \sum_{i=1}^{n} \xi_i$$

and

$$\hat{\sigma}^2 = \frac{1}{n} \sum_{i=1}^{n} (\xi_i - \hat{\mu})^2$$

Example 7.9

Consider the previous example, but this time assume that the variance σ^2 of the background gray values is known. Hence, all that is desired is to find the maximum likelihood estimate of μ. In this case, we again obtain the estimator

$$\hat{\mu} = \frac{\sum_{i=1}^{n} \xi_i}{n}$$

The likelihood function looks the same as it did before; however, it is denoted by $L(\mu)$, since $L: R \to R$.

Example 7.10

Consider Example 7.8 again, but assume this time that the mean μ of the background gray values is known exactly and an estimate of σ^2 is desired. The resulting maximum likelihood estimator is given by

$$\hat{\sigma}^2 = \frac{\sum_{i=1}^{n} (\xi_1 - \mu)^2}{n}$$

Notice that this result differs from the estimate of the background gray variance found in Example 7.8.

Example 7.11

Let ξ and η be random variables denoting gray values in some relation to each other within a class of images. Such a relation might occur through co-occurrency. Suppose further that these random variables are bivariate normally distributed, and consider a random sample $(\xi_1, \eta_1), (\xi_2, \eta_2), \ldots, (\xi_n, \eta_n)$ with corresponding observed values $(x_1, y_1), (x_2, y_2), \ldots, (x_n, y_n)$. If nothing at all is known about this normal distribution, then we have the likelihood function

$L(\mu_1, \mu_2, \sigma_1^2, \sigma_2^2, \rho)$

$$= \frac{1}{(4\pi^2 \sigma_1^2 \sigma_2^2 (1 - \rho^2))^{n/2}} e^{-\frac{1}{2(1-\rho^2)} \sum_{i=1}^{n} \left[\left(\frac{x_i - \mu_1}{\sigma_1}\right)^2 - \frac{2\rho}{\sigma_1 \sigma_2}(x_i - \mu_1)(y_i - \mu_2) + \left(\frac{y_i - \mu_2}{\sigma_2}\right)^2 \right]}$$

Taking logs of both sides and setting the partial derivatives with respect to $\mu_1, \mu_2, \sigma_1^2, \sigma_2^2$, and ρ equal to zero yields the following results:

$$\hat{\mu}_1 = \frac{1}{n} \sum_{i=1}^{n} \xi_i$$

$$\hat{\mu}_2 = \frac{1}{n} \sum_{i=1}^{n} \eta_i$$

$$\hat{\sigma}_1^2 = \frac{1}{n} \sum_{i=1}^{n} (\xi_i - \hat{\mu}_1)^2$$

$$\hat{\sigma}_2^2 = \frac{1}{n} \sum_{i=1}^{n} (\eta_i - \hat{u}_2)^2$$

$$\hat{\rho} = \frac{1}{n} \frac{\sum_{i=1}^{n} (\xi_i - \hat{\mu}_1)(\eta_i - \hat{\mu}_2)}{\sqrt{\hat{\sigma}_1^2 \hat{\sigma}_2^2}}$$

If we consider the maximum likelihood methodology from an intuitive perspective, its genesis is quite apparent. Maximizing the density $L_\mathbf{x}(\theta_1, \theta_2, \ldots, \theta_n)$ over all choices of $\theta_1, \theta_2, \ldots, \theta_n$ for a given input observation $\mathbf{x} = (x_1, x_2, \ldots, x_n)$ simply maximizes the "probability" that the sample $x_1, x_2, \ldots, x_n$ actually obtained would be obtained. In other words, the methodology is *a posteriori* in that estimates are made which are those which would most likely have given the actual empirical sample obtained.

Suppose the parameter θ is an observed value of the uniform random variable η with support in $D = [a, b]$. The density of η is given by

$$f_\eta(\theta) = \left(\frac{1}{b-a}\right) I_{[a,b]}(\theta)$$

Suppose further that ξ is a random variable with a probability density function having θ as a parameter, and let $\xi_1, \xi_2, \ldots, \xi_n$ be a random sample with the same distribution as ξ. Then, from the joint density function

$f_{\xi_1\xi_2\cdots\xi_n\eta}(x_1, x_2, \ldots, x_n, \theta)$, the associated marginal and conditional densities can be found. In particular,

$$f_{\xi_1\xi_2\cdots\xi_n}(x_1, \ldots, x_n) = \int_{-\infty}^{\infty} f_{\xi_1\xi_2\cdots\xi_n\eta}(x_1, x_2, \ldots, x_n, \theta) \, d\theta$$

If this quantity is nonzero, then the conditional density of η given $\xi_1 = x_1$, $\xi_2 = x_2, \ldots, \xi_n = x_n$ can be found. Indeed, for $a \leq \theta \leq b$,

$$f_\eta(\theta \mid x_1, \ldots, x_n) = \frac{f_{\xi_1\xi_2\cdots\xi_n\eta}(x_1, \ldots, x_n, \theta)}{f_{\xi_1\xi_2\cdots\xi_n}(x_1, \ldots, x_n)}$$

$$= \frac{f_{\xi_1\cdots\xi_n}(x_1, \ldots, x_n \mid \eta = \theta) f_\eta(\theta)}{f_{\xi_1\cdots\xi_n}(x_1, \ldots, x_n)}$$

$$= \frac{f_{\xi_1\cdots\xi_n}(x_1, \ldots, x_n \mid \eta = \theta)}{(b - a) f_{\xi_1\cdots\xi_n}(x_1, \ldots, x_n)} \quad (1)$$

Formally, in the maximum likelihood method the values $x_1, x_2, \ldots, x_n$ are kept fixed and the likelihood function $L(\theta)$ is maximized with respect to θ. Recall that

$$L(\theta) = \prod_{i=1}^{n} f_\xi(x_i; \theta)$$

Using independence, we could write $L(\theta)$ as

$$L(\theta) = f_{\xi_1\xi_2\cdots\xi_n}(x_1, x_2, \ldots, x_n; \theta)$$

This formula looks almost exactly like the density function in the numerator of (1). Therefore, formally, maximizing the numerator with respect to θ will provide the maximum likelihood estimate of θ. Furthermore, since the denominator of the expression does not involve θ, this maximum likelihood estimate will be equal to the maximum of $f_\eta(\theta \mid x_1, x_2, \ldots, x_n)$ with respect to θ, which, by definition, is the mode of $f_\eta(\theta \mid x_1, x_2, \ldots, x_n)$. (See Section 6.5.)

Changing directions momentarily, suppose we wish to measure the efficacy of an estimator $\hat\theta$ of the parameter θ. One way to proceed is to define a *loss function* $\ell(\hat\theta, \theta)$. Given a particular value of the theoretical parameter, $\ell(\hat\theta, \theta)$ is a random variable that depends upon the outcome of the random variable $\hat\theta$. For instance, we might let $\ell(\hat\theta, \theta) = [\hat\theta - \theta]^2$. ℓ would then be known as a *quadratic loss function*. Another obvious choice would be $\ell(\hat\theta, \theta) = |\hat\theta - \theta|$, an *absolute value loss function*. In any event, ℓ should certainly be chosen so that it is nonnegative.

Since a loss function is a random variable, it is customary to investigate its expected value. This leads to the *risk function*

$$\rho(\hat\theta, \theta) = E(\ell(\hat\theta, \theta))$$

Notice that for a given estimator $\hat{\theta}$, $\rho(\hat{\theta}, \theta)$ is a function of θ. Given two estimators, $\hat{\theta}_1$ and $\hat{\theta}_2$, we might choose to compare them based upon which results in a minimal risk function for a given loss function. The problem is to define minimality as it applies to risk functions. One popular way is simply to declare $\rho(\hat{\theta}_1, \theta)$ *smaller* than $\rho(\hat{\theta}_2, \theta)$ if the maximum value of the former is less than the maximum value of the latter.

In the techniques studied thus far, the estimation of the underlying parameter (parameter vector) θ has been accomplished without assuming any prior knowledge concerning the parameter. In *Bayesian estimation*, it is assumed that we have some knowledge regarding the parameter prior to sampling. To wit, we assume that θ is a value of a random variable η with a known distribution, the *a priori* distribution. Based upon a sample $\xi_1, \xi_2, \ldots, \xi_n$, we wish to estimate that θ which applies to the sample at hand. Intuitively, at least, one should expect that the availability of prior knowledge will sharpen the estimation procedure in that the parameter to be estimated is restrained by the distribution of the probability mass of η. The methodology is to utilize knowledge of the distribution of η in conjunction with the data observed in the sampling procedure.

In Bayesian theory, a parameter θ in $D \subset R$ is modeled as an observed value of a random variable η. The density function of ξ with parameter θ, viz., $f_\xi(x; \theta) = f(x; \theta)$, is viewed as a conditional density $f_\xi(x \mid \theta) = f(x \mid \theta)$. We then consider the sample $\xi_1, \xi_2, \ldots, \xi_n$ to be jointly distributed with η, and we let

$$f_{\xi_1 \xi_2 \cdots \xi_n \eta}(x_1, x_2, \ldots, x_n, \theta) = f(x_1, x_2, \ldots, x_n, \theta)$$

be the joint density. Each ξ_i is assumed to have the same conditional density as ξ, and the ξ_i are conditionally independent with respect to η; specifically, the conditional density of $\xi_1, \xi_2, \ldots, \xi_n$ given $\eta = \theta$ is given by

$$f(x_1, x_2, \ldots, x_n \mid \theta) = \prod_{i=1}^{n} f(x_i \mid \theta)$$

Consequently, by the definition of a conditional density, if the a priori density of η is $f_\eta(\theta)$, then

$$f(x_1, \ldots, x_n, \theta) = f_\eta(\theta) \prod_{i=1}^{n} f(x_i \mid \theta) = f(\theta \mid x_1, \ldots, x_n) f(x_1, \ldots, x_n)$$

In the Bayesian case, for a given loss function ℓ, the risk function takes on a different interpretation. Since θ is a value of the random variable η, the risk function is also a random variable. As a consequence, we consider the *mean risk*, or *Bayes risk*, which is defined by

$$r(\varphi) = E[\rho(\hat{\theta}, \eta)] = E[E(\ell(\hat{\theta}, \eta))]$$

where φ denotes the estimation rule for $\hat{\theta}$. If a decision rule φ can be found which minimizes $r(\varphi)$, then the estimator corresponding to φ is called the *Bayesian es-*

timator with respect to the a priori distribution. Straightforward substitution of the appropriate variables into the defining equation for *r* yields

$$r(\varphi) = E[E(\ell(\varphi(\xi_1, \xi_2, \ldots, \xi_n), \eta))]$$

$$= E\left[\int_{-\infty}^{\infty} \cdots \int_{-\infty}^{\infty} \ell(\varphi(x_1, x_2, \ldots, x_n), \eta) \right.$$

$$\left. \times f(x_1, x_2, \ldots, x_n | \eta) \, dx_1 \ldots dx_n \right]$$

$$= \int_{-\infty}^{\infty} \cdots \int_{-\infty}^{\infty} \ell(\varphi(x_1, x_2, \ldots, x_n), \theta)$$

$$\times f(x_1, \ldots, x_n, \theta) \, d\theta \, dx_1 \, dx_2 \cdots dx_n$$

(Note that the loss function in the integrand is assumed to be zero outside of its domain in R^{n+1}.) Hence, assuming that the expectation exists, the objective is to minimize

$$r(\varphi) = \int_{-\infty}^{\infty} \cdots \int_{-\infty}^{\infty} f(x_1, \ldots, x_n)$$

$$\times \left[\int_{-\infty}^{\infty} \ell(\varphi(x_1, \ldots, x_n), \theta) f_\eta(\theta | x_1, \ldots, x_n) \, d\theta\right] dx_1 \cdots dx_n$$

In a manner similar to maximum likelihood estimation, the inner integral (in brackets) is minimized for each set of values $x_1, x_2, \ldots, x_n$.

If a quadratic loss function is employed, then $r(\varphi)$ is minimized by minimizing

$$\int_{\theta=-\infty}^{\infty} (\varphi(x_1, x_2, \ldots, x_n) - \theta)^2 f_\eta(\theta | x_1, x_2, \ldots, x_n) \, d\theta$$

Like the results given in Section 6.5, the solution is the mean. Hence, in this case,

$$\overset{*}{\theta} = \varphi(x_1, x_2, \ldots, x_n) = E(\eta | x_1, x_2, \ldots, x_n)$$

which is known as the *conditional mean*.

Suppose we consider a loss function of the form

$$\ell(\varphi(x_1, x_2, \ldots, x_n), \theta) = |\varphi(x_1, x_2, \ldots, x_n) - \theta|$$

Then the median of the conditional density of η given $x_1, x_2, \ldots, x_n$ is the Bayesian solution. We now illustrate some of these results.

Example 7.12

Suppose the mean gray value for a certain class of images is known a priori to possess a Gaussian distribution with known mean μ_0 and known variance σ_0^2. By this it is meant that our observations of the population will be conditioned by whatever particular value of the actual mean is presented to us, the observers. Nevertheless, that presentation will be Gaussian. One might think of a situation where a light source

Sec. 7.3   Estimation Procedures

whose intensity is normally distributed is shining on the images but we do *not* know the intensity at the moment our sample is chosen. In order to estimate the population mean, let our sample be $\xi_1, \xi_2, \ldots, \xi_n$, and suppose the values $x_1, x_2, \ldots, x_n$ are observed. Finally, suppose each ξ_i satisfies a Gaussian distribution with variance σ^2 and mean depending upon the mean of the population—as presented to us by the state of nature! We are given the a priori density

$$f_\eta(\mu) = \frac{1}{\sqrt{2\pi}\sigma_0} e^{-\frac{(\mu - \mu_0)^2}{2\sigma_0^2}}$$

with μ_0 and σ_0 known. We will find the Bayesian estimate of μ by minimizing the least quadratic loss; and hence find $\overset{*}{\mu}$, the conditional mean. Each ξ_i has density

$$f_\xi(x_i \mid \mu) = \frac{1}{\sqrt{2\pi}\sigma} e^{-\frac{(x_i - \mu)^2}{2\sigma^2}}$$

where σ is known and μ is to be found.

In finding the joint density, it is convenient to let

$$\bar{x} = \frac{1}{n}\sum_{i=1}^{n} x_i$$

We then have

$$f(x_1, x_2, \ldots, x_n, \mu) = \prod_{i=1}^{n} f_\xi(x_i \mid \mu) f_\eta(\mu)$$
$$= K e^{-\left\{\left(\frac{\sigma^2 + n\sigma_0^2}{2\sigma^2 \sigma_0^2}\right)\left[\mu - \frac{\mu_0 \sigma^2 + n\bar{x}\sigma_0^2}{\sigma^2 + n\sigma_0^2}\right]\right\}}$$

where

$$K = \frac{1}{(\sqrt{2\pi})^{n+1}\sigma_0 \sigma^n} e^{-\left\{\left(\frac{\sigma^2 + n\sigma_0^2}{2\sigma^2 \sigma_0^2}\right)\left[\left(\sigma^2 \mu_0^2 + \sigma_0^2 \sum_{i=1}^{n} x_i^2\right) - \left(\frac{\mu_0 \sigma^2 + n\bar{x}\sigma_0^2}{\sigma^2 + n\sigma_0^2}\right)\right]\right\}}$$

Note that K does not involve μ. We must find the a *postiori* density

$$f_\eta(\mu \mid x_1, \ldots, x_n) = \frac{f(x_1, \ldots, x_n, \mu)}{f(x_1, \ldots, x_n)}$$

But

$$f(x_1, x_2, \ldots, x_n) = \int_{-\infty}^{\infty} f(x_1, x_2, \ldots, x_n, \mu) \, d\mu$$

and this marginal density has the same K as does the joint density $f(x_1, x_2, \ldots, x_n, \mu)$. Hence, the K in the numerator and the K in the denominator of $f(\mu \mid x_1, \ldots, x_n)$ reduce to unity. Thus,

$$f_\eta(\mu \mid x_1, x_2, \ldots, x_n) = \frac{e^{-\left\{\left(\frac{\sigma^2 + n\sigma_0^2}{2\sigma^2 \sigma_0^2}\right)\left[\mu - \frac{\mu_0 \sigma^2 + n\bar{x}\sigma_0^2}{\sigma^2 + n\sigma_0^2}\right]^2\right\}}}{\int_{-\infty}^{\infty} e^{-\left\{\left(\frac{\sigma^2 + n\sigma_0^2}{2\sigma^2 \sigma_0^2}\right)\left[\mu - \frac{\mu_0 \sigma^2 + n\bar{x}\sigma_0^2}{\sigma^2 + n\sigma_0^2}\right]^2\right\}} \, d\mu}$$

Since the integral in the denominator of $f(\mu \mid x_1, \ldots, x_n)$ possesses a Gaussian-type integrand, it must equal $1/\sqrt{2\pi}s$, where s is the standard deviation of the Gaussian (conditional) density. By inspection of $f(\mu \mid x_1, x_2, \ldots, x_n)$, we obtain $s^2 = \sigma^2\sigma_0^2/(\sigma^2 + n\sigma_0^2)$. Consequently,

$$f_\eta(\mu \mid x_1, \ldots, x_n) = \frac{1}{\sqrt{2\pi}\sqrt{\dfrac{\sigma^2\sigma_0^2}{\sigma^2 + n\sigma_0^2}}} e^{-\left\{\frac{1}{2s^2}\left[\mu - \frac{\mu_0\sigma^2 + n\bar{x}\sigma_0^2}{\sigma^2 + n\sigma_0^2}\right]^2\right\}}$$

Since the conditional mean

$$\overset{*}{\mu} = \int_{-\infty}^{\infty} \mu f_\eta(\mu \mid x_1, \ldots, x_n)\, d\mu$$

provides the least quadratic loss, and since $f_\eta(\mu \mid x_1, \ldots, x_n)$ is Gaussian, a simple inspection of the density provides the result

$$\overset{*}{\mu} = \frac{\mu_0\sigma^2 + n\bar{x}\sigma_0^2}{\sigma^2 + n\sigma_0^2}$$

It is beneficial to consider limiting cases for this result. If the initial knowledge of the average gray value μ is close to exact, that is, $\mu \cong \mu_0$ and $\sigma_0 \cong 0$, then, from the preceding formula, it follows that $\overset{*}{\mu} \cong \mu_0$.

If the observed gray values $x_1, \ldots, x_n$ are "heavily concentrated about μ," then $\sigma \cong 0$ and the best estimate $\overset{*}{\mu} \cong \bar{x}$ is the empirical mean of the observed gray values. In any event, if the number of sampled gray values gets large, i.e., if $n \to \infty$, then $\overset{*}{\mu} \to \bar{x}$ again.

An interesting special case of the above formula for $\overset{*}{\mu}$ is when $\mu_0 = 0$ and $\sigma = \sigma_0 = 1$. Then

$$\overset{*}{\mu} = \frac{x_1 + x_2 + \cdots + x_n}{n + 1}$$

Moreover, since the variance associated with the conditional mean is

$$s^2 = \frac{\sigma^2\sigma_0^2}{\sigma^2 + n\sigma_0^2}$$

we obtain $1/(n + 1)$ as the variance for the conditional mean (as compared to $1/n$ for the variance of the sample mean).

7.4 MATRIX ESTIMATION TECHNIQUES

The object of this section is to give, for comparative purposes, the formulas of several different well-known matrix estimation procedures together with a quick synopsis of the conditions under which they can be used. Each of the estimators is to be employed in estimating an $n \times 1$ parameter vector **x** using the information obtained by observing a value of the $K \times 1$ vector **y**.

We shall assume that the estimator $\hat{\mathbf{x}}$ is linear, i.e.,

$$\hat{\mathbf{x}} = B\mathbf{y}$$

where B is an $n \times K$ matrix that is to be found. Furthermore, **x** and **y** are assumed to be related by

$$\mathbf{y} = H\mathbf{x} + \mathbf{u}$$

or, written out,

$$\begin{pmatrix} y_1 \\ y_2 \\ \vdots \\ y_K \end{pmatrix} = \begin{pmatrix} h_{11} & \cdots & h_{1n} \\ \vdots & & \vdots \\ h_{K1} & \cdots & h_{Kn} \end{pmatrix} \begin{pmatrix} x_1 \\ x_2 \\ \vdots \\ x_n \end{pmatrix} + \begin{pmatrix} u_1 \\ u_2 \\ \vdots \\ u_K \end{pmatrix}$$

where H is a known $K \times n$ *design* matrix and **u** is a $K \times 1$ "input noise" vector of random variables or errors.

The object in matrix estimation is to find B. It is assumed throughout that we are dealing with an overdetermined model, i.e., that $K \geq n$, and that the rank of H is n. The matrix B is to be found so that it is optimal with respect to some specified criterion. In doing so, it must be remembered that an estimator may not be optimal with respect to a given criterion unless *all* the information concerning the parameter to be estimated is used in formulating the estimate.

Suppose that nothing more is known about the input noise **u** than the equation $\mathbf{y} = H\mathbf{x} + \mathbf{u}$. Suppose also that it is desired to find the estimator which minimizes the sum of the squares of the deviations; that is, the optimization criterion is to find $\hat{\mathbf{x}} = B\mathbf{y}$ which minimizes

$$S = \mathbf{u}'\mathbf{u} = u_1^2 + u_2^2 + \ldots + u_K^2$$

when actual values $u_1, u_2, \ldots, u_K$ are employed.

The optimal solution for this case, commonly called the *least squares* (*LS*) *solution*, is given by

$$B = (H'H)^{-1}H'$$

where H' denotes the transpose of H and $(H'H)^{-1}H'$ is called the *pseudoinverse* of H.[4] Hence, the optimal solution is

$$\hat{\mathbf{x}} = (H'H)^{-1}H'\mathbf{y}$$

If it is possible to obtain some information on the noise vector **u**, then some other estimation procedure which utilizes this additional information should be employed. Notice that in forming a least squares estimator nothing other than the design matrix H is needed. (When H is not overdetermined, a pseudoinverse again provides the least squares solution; however, a different formula for the pseudoinverse is utilized.) Additional assumptions will now be imposed, thereby leading to better estimates.

---

[4] Thomas L. Boullion and Patrick L. Odell, *Generalized Inverse Matrices* (New York: John Wiley & Sons, 1971), p. 50.

An important case is when the input noise vector has mean zero, i.e.,

$$\bar{u} = E(u) = 0$$

and possesses a known covariance matrix

$$L = E(uu')$$

that is assumed to be positive definite. The optimization criterion is to find $\hat{x} = By$ which is a BLUE, that is, a best linear unbiased estimator. Recall that the estimator's being unbiased means that, on the average, the solution will be exact, i.e.,

$$E(\hat{x}) = x$$

The solution will also be best in that the variance will be minimized by finding $\hat{x}$ such that

$$S = E((\hat{x} - x)'(\hat{x} - x))$$

is minimized. Under these conditions, the BLUE is given by

$$\hat{x} = (H'L^{-1}H)^{-1}H'L^{-1}y$$

Note that if $L = \sigma^2 I$, then

$$\hat{x} = (H'H)^{-1}H'y$$

and x is not only the BLUE, but it is also the least squares estimator.[5] Moreover, if we know that the input noise has mean zero, and if L is known to be of the form $L = \sigma^2 I$, where σ may or may not be known, then the least squares solution is also the BLUE. However, in general, in order to use the BLUE, it is necessary to know the input covariance matrix L, and not simply its form.

If additional information on the input noise vector **u** is available, then an even better estimation procedure will result. For instance, if the joint frequency function f of the elements in the noise vector **u** is known, then an estimation procedure that utilizes this additional information should be employed.

As an example, suppose **u** is mean zero Gaussian with (positive definite) covariance matrix L, so that

$$f_u(w) = \frac{1}{\sqrt{(2\pi)^K}\sqrt{\det L}} e^{-\frac{w'L^{-1}w}{2}}$$

Then one may apply maximum likelihood techniques to find $\hat{x}$. Consequently, $\hat{x}$ is called the *maximum likelihood estimator* or *maximum likelihood filter estimator* (*MLF estimator*).

---

[5] T. O. Lewis and P. L. Odell, *Estimation in Linear Models* (Englewood Cliffs, NJ: Prentice-Hall, Inc., 1971), p. 55.

Sec. 7.4   Matrix Estimation Techniques   395

Under the preceding Gaussian assumptions,

$$f_{\mathbf{u}}(\mathbf{w}) = \frac{1}{(2\pi)^{K/2}\sqrt{\det L}} e^{-\frac{(\mathbf{v} - H\mathbf{x})'L^{-1}(\mathbf{v} - H\mathbf{x})}{2}}$$

by letting $\mathbf{v} = H\mathbf{x} + \mathbf{w}$. Upon taking the gradient with respect to $\mathbf{x}$, we obtain the MLF estimate of $\mathbf{x}$, namely,

$$\hat{\mathbf{x}} = (H'L^{-1}H)^{-1}H'L^{-1}\mathbf{y}$$

In addition to giving an MLF estimate, this solution also provides the best linear unbiased estimate of $\mathbf{x}$. Note that this is not true in general, but is a fundamental property of the Gaussian family.

The preceding facts concerning matrix estimation techniques are summarized in Table 7.1. Included is the output covariance matrix $E((\hat{\mathbf{x}} - \mathbf{x})(\hat{\mathbf{x}} - \mathbf{x})')$ of each estimator. In the table, it is seen that all three estimators of $\mathbf{x}$ are unbiased; thus, on the average, they yield the exact results.

The following example illustrates the formal manipulations involved in performing some of these estimates. Further applications are given in Section 7.6, where inter- and intrapixel gray level interference are briefly investigated.

Example 7.13

Consider the system of equations

$$2 = x_1 + x_2$$
$$0 = x_1 - x_2$$

which is an instance of $\mathbf{y} = H\mathbf{x} + \mathbf{u}$, where

$$H = \begin{pmatrix} 1 & 1 \\ 1 & -1 \end{pmatrix}$$

and

$$\mathbf{y} = \begin{pmatrix} 2 \\ 0 \end{pmatrix}$$

An exact solution,

$$x = \begin{pmatrix} x_1 \\ x_2 \end{pmatrix}$$

exists. Here,

$$\mathbf{x} = H^{-1}\mathbf{y} = \begin{pmatrix} 1 \\ 1 \end{pmatrix}$$

If an additional observation $y_3 = 1.1$ is observed for x_1, then the new system

$$2 = x_1 + x_2$$
$$0 = x_1 - x_2$$
$$1.1 = x_1$$

TABLE 7.1. COMPARISON OF LS, BLUE, AND MLF ESTIMATION TECHNIQUES

$$y = Hx + u$$

Name of Estimator $\hat{x}$	What Must Be Known to Use the Technique	$\hat{x} =$	Bias of $\hat{x}$	Output Covariance Matrix of $\hat{x}$, $M =$
Least-Squares Estimator (LS)	Nothing	$(H'H)^{-1}H'y$	0*	$(H'H)^{-1}H'LH(H'H)^{-1}$**
Best Linear Unbiased Estimator (BLUE)	$E(u) = 0$ covariance matrix of u	$(H'L^{-1}H)^{-1}H'L^{-1}y$	0	$(H'L^{-1}H)^{-1}$
Maximum Likelihood Filter Estimator (MLF)	Density of Noise Vector	*Gaussian case* $(H'L^{-1}H)^{-1}H'L^{-1}y$	*Gaussian case* 0	*Gaussian case* $(H'L^{-1}H)^{-1}$

* Must assume that $E(u) = 0$.
** Must assume that $E(u) = 0$ and L is the known covariance matrix of the noise.

Sec. 7.4   Matrix Estimation Techniques

arises. The same model $\mathbf{y} = H\mathbf{x} + \mathbf{u}$ is applicable, now, however, with

$$\mathbf{y} = \begin{pmatrix} 2 \\ 0 \\ 1.1 \end{pmatrix}, H = \begin{pmatrix} 1 & 1 \\ 1 & -1 \\ 1 & 0 \end{pmatrix}, \mathbf{x} = \begin{pmatrix} x_1 \\ x_2 \end{pmatrix}$$

As given, we have $\mathbf{y} = H\mathbf{x}$, and from elementary matrix analysis there cannot be an exact solution for $\mathbf{x}$ in this system because the rank, which happens to be two, of the design matrix H is strictly less than the rank, which happens to be three, of the augmented matrix

$$(H \colon \mathbf{y}) = \begin{pmatrix} 1 & 1 & 2 \\ 1 & -1 & 0 \\ 1 & 0 & 1.1 \end{pmatrix}$$

Hence, the system is contradictory, and we seek only an approximate solution to the new system.

Utilizing different terminology, "slack" variables u_1, u_2, and u_3 are introduced into the system so that a revised system that is not contradictory results. Thus, we consider $\mathbf{y} = H\mathbf{x} + \mathbf{u}$, with

$$\mathbf{u} = \begin{pmatrix} u_1 \\ u_2 \\ u_3 \end{pmatrix} \neq \begin{pmatrix} 0 \\ 0 \\ 0 \end{pmatrix}$$

As we have just seen, a "solution" $\hat{\mathbf{x}}$ can be found which minimizes $S = \mathbf{u}'\mathbf{u}$. In the present case,

$$\hat{\mathbf{x}} = (H'H)^{-1}H'\mathbf{y} = \begin{pmatrix} \frac{1}{3} & \frac{1}{3} & \frac{1}{3} \\ \frac{1}{2} & -\frac{1}{2} & 0 \end{pmatrix} \begin{pmatrix} 2 \\ 0 \\ 1.1 \end{pmatrix}$$

Consequently, $\hat{x}_1 = \dfrac{3.1}{3}$ and $\hat{x}_2 = 1$. In fact, from an intuitive point of view, values around this size should have been expected.

Example 7.14

Using the matrix H, which was employed in the latter part of the previous example, and assuming that $\mathbf{u}$ is a random variable with $E(\mathbf{u}) = 0$ and

$$E(\mathbf{u}\mathbf{u}') = L = \begin{pmatrix} \sigma_1^2 & 0 & 0 \\ 0 & \sigma_2^2 & 0 \\ 0 & 0 & \sigma_3^2 \end{pmatrix}$$

we have, from the last column, first row of Table 7.1, the output covariance matrix

$$M = E((\hat{\mathbf{x}} - \mathbf{x})(\hat{\mathbf{x}} - \mathbf{x})') = \begin{pmatrix} \frac{1}{3} & \frac{1}{3} & \frac{1}{3} \\ \frac{1}{2} & -\frac{1}{2} & 0 \end{pmatrix} \begin{pmatrix} \sigma_1^2 & 0 & 0 \\ 0 & \sigma_2^2 & 0 \\ 0 & 0 & \sigma_3^2 \end{pmatrix} \begin{pmatrix} \frac{1}{3} & \frac{1}{2} \\ \frac{1}{3} & -\frac{1}{2} \\ \frac{1}{3} & 0 \end{pmatrix}$$

$$= \begin{pmatrix} \dfrac{\sigma_1^2 + \sigma_2^2 + \sigma_3^2}{9} & \dfrac{\sigma_1^2 - \sigma_2^2}{6} \\ \dfrac{\sigma_1^2 - \sigma_2^2}{6} & \dfrac{\sigma_1^2 + \sigma_2^2}{4} \end{pmatrix}$$

In the preceding example, the values $\sigma_1, \sigma_2, \sigma_3 > 0$ are indicative of the accuracy of the equations $2 = x_1 + x_2$, $0 = x_1 - x_2$, and $1.1 = x_1$, respectively. If σ_1 is smaller than σ_2 or σ_3, then the first equation is the most accurate. The situation is similar with regard to other possible orderings. Of course, no matter what the ordering of these variances is, the LS estimator $\hat{x}$ is always the same. On the other hand, the "goodness" of this estimator, as measured by M, is a function of the values of σ_1^2, σ_2^2, and σ_3^2.

The BLUE procedure takes into account the covariance matrix L; therefore, it must yield a better estimate than the estimate obtained by the LS method.

Example 7.15

In the BLUE procedure, the output covariance matrix for the latter system of Example 7.13 is given by

$$V = \frac{\begin{pmatrix} \sigma_2^2 \sigma_3^2 + \sigma_1^2 \sigma_3^2 & \sigma_1^2 \sigma_3^2 - \sigma_2^2 \sigma_3^2 \\ \sigma_1^2 \sigma_3^2 - \sigma_2^2 \sigma_3^2 & \sigma_2^2 \sigma_3^2 + \sigma_1^2 \sigma_3^2 + \sigma_1^2 \sigma_2^2 \end{pmatrix}}{4\sigma_3^2 + \sigma_2^2 + \sigma_1^2}$$

Simple algebra shows that the (1, 1) and (2, 2) entries in the covariance matrix for the BLUE procedure are always less than the corresponding entries in the covariance matrix of the LS procedure. Using the results of Example 7.14, and supposing the values $\sigma_1^2 = \sigma_2^2 = .1$ and $\sigma_3^2 = .2$, gives, in the LS case, the output covariance matrix

$$M = \begin{pmatrix} \frac{.4}{9} & 0 \\ 0 & \frac{.2}{4} \end{pmatrix}$$

In the BLUE situation the same values yield the output covariance matrix

$$V = \begin{pmatrix} \frac{.4}{10} & 0 \\ 0 & \frac{.5}{10} \end{pmatrix}$$

The corresponding optimal estimate in the BLUE case is given by

$$\hat{x} = \begin{pmatrix} 1.02 \\ 1 \end{pmatrix}$$

Note that the first two equations in the system

$$\begin{cases} 2 = x_1 + x_2 \\ 0 = x_1 - x_2 \\ 1.1 = x_1 \end{cases}$$

have a smaller σ associated with them than the third ($\sigma_1^2, \sigma_2^2 < \sigma_3^2$), and, as a consequence, carry more influence in the BLUE solution.

Sec. 7.4 Matrix Estimation Techniques

At this juncture, a point regarding modeling seems appropriate. Why is it reasonable to model an overdetermined system $\mathbf{y} = H\mathbf{x} + \mathbf{u}$ with $\mathbf{u}$ being considered as a noise random variable with covariance matrix $\text{Cov}(\mathbf{u}) = E(\mathbf{uu'})$? At its deepest philosophic level, the matter has to do with the theory of measurement.

Assuming that the system equations in terms of the x_i result from measurement, it is appropriate to hypothesize that their incompatibility results from inherent measurement uncertainty. Each equation results from a measurement of some type and, consequently, had consistent, deterministic observations been recorded, the system would not be contradictory no matter how many equations result. The need for slack variables is a consequence of the inconsistency of the equational constraints implied by observation. Thus, it is natural to look upon these variables as random phenomena.

In the case when the covariance matrix of the slack variables is diagonal, these variables are uncorrelated. Moreover, the variance σ_i^2 is a measure of the uncertainty (error) in the ith observational equation: the more uncertain the measurement, the greater the variability of the slack. Finally, the output covariance matrix measures, in an expected value sense, the error of the measurement.

In terms of the foregoing noise model, the BLUE, in that it utilizes the covariance matrix of the noise $\mathbf{u}$, takes into account the variance of the slack variables. In doing so, it makes use of prior knowledge regarding experimental indeterminacy. The LS solution is independent of any such knowledge.

We next set out a method for representing the estimation formula

$$\hat{\mathbf{x}} = (H'L^{-1}H)^{-1}H'L^{-1}\mathbf{y}$$

in a well-known recursive fashion. This recursive algorithm will enable a "continuous" estimate of the parameter vector $\mathbf{x}$ to be found as more and more observations are taken. In other words, as the dimension of the vector $\mathbf{y}$ increases, improved estimates of $\mathbf{x}$ are found on the basis of previous estimates of $\mathbf{x}$.

We begin with a discussion on notation. Instead of writing the vector linear model $\mathbf{y} = H\mathbf{x} + \mathbf{u}$, we shall write it as

$$\mathbf{y}_K = H_K\mathbf{x} + \mathbf{u}_K$$

where $\mathbf{x}$ is the $n \times 1$-parameter vector to be estimated and

$$\mathbf{y}_K = \begin{pmatrix} y(1) \\ \vdots \\ y(K) \end{pmatrix}$$

is a $K \times 1$ vector of scalar observations. Next, let $\mathbf{u}_K$ be a $K \times 1$ vector of random variables $u(i)$ such that (1) each $u(i)$ has mean zero, (2) as a collection, all the $u(i)$ are uncorrelated, and (3) each $u(i)$ possesses nonzero variance σ_i^2; that is,

$$\mathbf{u}_K = \begin{pmatrix} u(1) \\ \vdots \\ u(K) \end{pmatrix}$$

$$E(u(i)) = 0$$

and

$$E(u(i)u(j)) = \sigma_i^2 \delta_{ij}, \text{ where } \delta_{ij} = \begin{cases} 1 & \text{if } i = j \\ 0 & \text{if } i \neq j \end{cases}$$

Let the covariance matrix for $\mathbf{u}_K$ be

$$L_K = \begin{pmatrix} \sigma_1^2 & 0 & \cdots & 0 \\ 0 & \sigma_2^2 & \cdots & \vdots \\ \vdots & \vdots & \ddots & \vdots \\ 0 & 0 & \cdots & \sigma_K^2 \end{pmatrix}$$

The design matrix H_K is then a $K \times n$ matrix given by

$$H_K = \begin{pmatrix} h_{11} & h_{12} & \cdots & h_{1n} \\ \vdots & \vdots & & \vdots \\ h_{K1} & h_{K2} & \cdots & h_{Kn} \end{pmatrix}$$

Finally, let the vector $\mathbf{h}_i$ be made up of the elements in the rows of H_K, i.e.,

$$\mathbf{h}_i = \begin{pmatrix} h_{i1} \\ h_{i2} \\ \vdots \\ h_{in} \end{pmatrix}$$

Then H_K can be expressed as

$$H_K = \begin{pmatrix} h_1' \\ h_2' \\ \vdots \\ h_K' \end{pmatrix}$$

where h_i' is the transpose of h_i. With this new notation, an optimal estimate based on the K observations $y(1), y(2), \ldots, y(K)$ is given by

$$\hat{\mathbf{x}}_K = V_K H_K' L_K^{-1} \mathbf{y}_K$$

where

$$V_K = (H_K' L_K^{-1} H_K)^{-1}$$

Now let an additional observation $y(K + 1)$ be received. If a new BLUE optimal estimator $\hat{\mathbf{x}}_{K+1}$ based on all observations, including the most recent one is desired, the formula for $\hat{\mathbf{x}}_K$ above can be employed with $K + 1$ in place of K. However, this formula is not recursive. A recursive optimal algorithm that yields the same results utilizes the pair of equations[6]

$$\hat{\mathbf{x}}_{K+1} = \hat{\mathbf{x}}_K + V_K h_{K+1} (h_{K+1}' V_K h_{K+1} + \sigma_{K+1}^2)^{-1} [y(K + 1) - h_{K+1}' \hat{\mathbf{x}}_K]$$

[6] Lewis and Odell, p. 75.

and

$$V_{K+1} = V_K - V_K h_{K+1}(h'_{K+1} V_K h_{K+1} + \sigma^2_{K+1})^{-1} h'_{K+1} V_K$$

The first equation is known as the *state update* equation, while the second is known as the *covariance update* equation.

Example 7.16

We consider again the problem of Example 7.13, except, for notational clarity, we use a and b in place of x_1 and x_2. The new system of equations for which a and b are desired is given by

$$2 = a + b$$
$$0 = a - b$$
$$1.1 = a$$
$$.9 = b$$

Writing this system in the form $\mathbf{y} = H\mathbf{x} + \mathbf{u}$, and assuming that $E(\mathbf{u}) = 0$ and

$$E(\mathbf{uu'}) = \begin{pmatrix} .1 & 0 & 0 & 0 \\ 0 & .1 & 0 & 0 \\ 0 & 0 & .2 & 0 \\ 0 & 0 & 0 & \sigma^2_4 \end{pmatrix}$$

the BLUE $\hat{\mathbf{x}}$ can then be found directly, as in the last example. However, if we notice that the new system consists of the system of Example 7.13 together with a single new observation, $b = .9$, we can use the recursive procedure just presented and make direct use of the optimal results obtained previously. The only thing to do is identify important parameters in the new notation. For instance,

$$\hat{\mathbf{x}}_3 = \begin{pmatrix} 1.02 \\ 1 \end{pmatrix}$$

and

$$\mathbf{V}_3 = \begin{pmatrix} .04 & 0 \\ 0 & .05 \end{pmatrix}$$

represent $\hat{\mathbf{x}}$ and V as given at the end of Example 7.15. Moreover, $h_4 = \begin{pmatrix} 0 \\ 1 \end{pmatrix}$, since h'_4 is the bottom row in the H_4 matrix; $y(4) = .9$ is the new observed value; and σ_4 will be left temporarily unspecified. Substituting into the recursive formula above gives

$$\hat{\mathbf{x}}_4 = \begin{pmatrix} 1.02 \\ 1 - \dfrac{.005}{.05 + \sigma^2_4} \end{pmatrix} = \begin{pmatrix} \hat{a} \\ \hat{b} \end{pmatrix}$$

Notice that if $\sigma_4 \to 0$, meaning that the last observation $b = .9$ is very exact, $\hat{b} \to .9$. On the other hand, if the last observation is poor, meaning that $\sigma_4 \to \infty$, we obtain $\hat{\mathbf{x}}_4 = \hat{\mathbf{x}}_3$, as expected, since $\sigma_4 \to \infty$ means that there is no information in the last observation.

If we wish, we can find the output covariance matrix, but we shall not do so here.

7.5 KALMAN-BUCY FILTERING

This section is written in the same applied spirit as the section on matrix estimation procedures. In a sense the material presented herein is a direct continuation of concepts presented there.

Kalman-Bucy filtering, or *KBF*, is an optimal estimation technique based on a linear model similar to the one in Section 7.4. As in that model, when more observations are given and a new (updated) estimate is required, there is no need to redo the entire computation. A new best estimate can be found on the basis of only a previously found best estimate and the current observation. Thus, the KBF is a recursive procedure.

Consider one of the simplest recursive procedures, calculation of the sample mean. If

$$\bar{x}_n = \frac{x_1 + x_2 + \cdots + x_n}{n}$$

and a new observation x_{n+1} is given, then $\bar{x}_{n+1}$ can be written as

$$\bar{x}_{n+1} = \frac{\frac{n(x_1 + x_2 + \cdots + x_n)}{n} + x_{n+1}}{n + 1}$$

Consequently,

$$\bar{x}_{n+1} = \frac{n}{n+1}\bar{x}_n + \frac{1}{n+1}x_{n+1}$$

The form of this recursive solution is important: the new sample mean is a weighted average of the old sample mean and the latest observation. The KBF has a similar form. It differs vastly from previous matrix estimation techniques in that more is given than a matrix linear model involving observations. The KBF requires a *system model* in the form of a matrix difference equation. In this model, the $n \times 1$-parameter vector, also called the *state vector*, is indexed on the set of nonnegative integers. This (unknown) state vector is denoted by $\mathbf{x}_k$. In signal processing applications k might denote $k\Delta t$ units in time, $\Delta t > 0$. In image processing, the indexing set can take on various other meanings. For instance, a total ordering can be imposed on the pixels or groups of pixels within or between images, or both. These values then form the indexing set. The principal purpose of the KBF is to find estimates of $\mathbf{x}_k$. In order to describe these estimates, a three-part model must be given.

Sec. 7.5 Kalman-Bucy Filtering

The difference equation that is the system model for the KBF is called the *model equation* and is given by

$$\mathbf{x}_k = \Phi_{k-1}\mathbf{x}_{k-1} + \mathbf{w}_{k-1}, \text{ for } k = 1, 2, 3, \ldots$$

where Φ_{k-1}, called the *transition matrix*, is a known n by n matrix that depends on the indexing set. The *noise vector* $\mathbf{w}_{k-1}$ is an n by 1 random vector depending on the indexing set. It is assumed that this noise vector is of mean zero, i.e., $E(\mathbf{w}_{k-1}) = 0$, and that $\text{Cov}(\mathbf{w}_{k-1}) = E(\mathbf{w}_{k-1}\mathbf{w}'_{k-1}) = Q_{k-1}$ is known. It is also assumed that $E(\mathbf{w}_i\mathbf{w}'_j) = 0$ for $i \ne j$. This last assumption on noise vectors is described by calling the vectors *white noise*. The matrix Φ_{k-1} indicates how the states change from one value of the index parameter to the next value. The white noise vector is employed to indicate uncertainty in the model. Although the covariance matrix Q_{k-1} need not be a diagonal matrix, the diagonal elements, being variances, are indicative of the "amount" of noise being added onto corresponding n-tuples of the state vector: the larger the variance, the more noise, and consequently, the greater the uncertainty.

The second part of the KBF model is similar to the matrix model given in Section 7.4. It involves measurements that are linear combinations of *state variables* corrupted by noise. The relation between these measurements is called the *observation equation* and is given by

$$\mathbf{z}_k = H_k\mathbf{x}_k + \mathbf{v}_k, \text{ for } k = 1, 2, \ldots$$

where $\mathbf{z}_k$ is a known m by 1 vector of measurements (obtained from some sensor) that depends on the indexing parameter k. If the indexing set involves some type of ordering involving the location of pixels, then the measurements are observed as a function of the location k.

The m by n design matrix H_k illustrates the linear relations the state variables bear to each other (as measurements) when no noise is present. This matrix, which might also change as k changes, is assumed to be known. The measurement noise vector $\mathbf{v}_k$ is an m by 1 random vector that can be used to model sensor noise as well as ambient-type noise. In any case, it is assumed to be of mean zero, i.e., $E(\mathbf{v}_k) = 0$. We also assume that the covariance $\text{Cov}(\mathbf{v}_k) = E(\mathbf{v}_k\mathbf{v}'_k) = R_k$ is known. Like the system noise vector, the observation, or measurement, noise vector is white; thus, $E(\mathbf{v}_i\mathbf{v}'_j) = 0$ for $i \ne j$. Finally, the second part of the KBF model assumes that

$$E(\mathbf{v}_i\mathbf{w}'_j) = 0 \text{ for all } i, j = 1, 2, 3, \ldots$$

The third part of the KBF model consists of initial conditions. Specifically, it is assumed that the initial n by 1 estimator of the state vector is known and equals the average value of the initial state vector, i.e., $\hat{\mathbf{x}}_0 = E(\mathbf{x}_0)$. Also, the initial n by n covariance matrix $P_0 = E((\mathbf{x}_0 - \hat{\mathbf{x}}_0)(\mathbf{x}_0 - \hat{\mathbf{x}}_0)')$ is assumed to be known.

The objective is to find an estimator $\hat{\mathbf{x}}_k$ of $\mathbf{x}_k$ which is best in certain respects to be spelled out subsequently. Moreover, this estimator must be given in a recursive manner, i.e., $\hat{\mathbf{x}}_{k+1}$ should be obtainable from only $\hat{\mathbf{x}}_k$ and $\mathbf{z}_k$, together with known quantities previously described. The criteria for which the KBF is optimal will be discussed in more depth after the actual algorithm is given. Suffice it to say here that in the most general case this algorithm is a BLUE. Furthermore, when the noise vectors are assumed to be multivariate Gaussian, the results can be seen to be best unbiased estimators as well as MLFs. The KBF also provides minimum variance Bayesian estimates.

The KBF is given by the following five equations,[7] each of which is subsequently explained:

k1. $\hat{\mathbf{x}}_k^- = \Phi_{k-1}\hat{\mathbf{x}}_{k-1}^+$.
k2. $P_k^- = \Phi_{k-1}P_{k-1}^+\Phi_{k-1}' + Q_{k-1}$.
k3. $C_k = P_k^- H_k'[H_k P_k^- H_k' + R_k]^{-1}$.
k4. $\hat{\mathbf{x}}_k^+ = \hat{\mathbf{x}}_k^- + C_k[\mathbf{z}_k - H_k\hat{\mathbf{x}}_k^-]$.
k5. $P_k^+ = [I - C_k H_k]P_k^-$.

A minus sign $(-)$ is placed after certain symbols to indicate that these quantities are calculated before the corresponding measurement is received. Hence, $\hat{\mathbf{x}}_k^-$ denotes an estimator of the state vector $\mathbf{x}_k$ based on all measurements $\mathbf{z}_1, \mathbf{z}_2, \ldots, \mathbf{z}_{k-1}$, but not on $\mathbf{z}_k$. Similarly, a plus sign $(+)$ is placed after various symbols to indicate that the quantity in question includes the most current measurement. Consequently, $\hat{\mathbf{x}}_k^+$ denotes an estimator of the state vector $\mathbf{x}_k$ based on all information up to and including $\mathbf{z}_k$.

Equation k1 is called the *extrapolation equation* of the state estimator. It relates the best estimator before a new observation is given with the most recent best estimate based on the last observation. For $k = 1$, we use $\hat{\mathbf{x}}_0^+ = \hat{\mathbf{x}}_0 = E(\mathbf{x}_0)$, which is known. This begins the recursive process, since $\hat{\mathbf{x}}_1^-$ is thereby obtained. The next estimator $\hat{\mathbf{x}}_1^+$ is found from equation k4, but first other values must be calculated.

Equation k2 is called the *error covariance extrapolation*. The covariance immediately before the kth measurement $\mathbf{z}_k$ occurs is

$$P_k^- = E((\hat{\mathbf{x}}_k^- - \mathbf{x}_k)(\hat{\mathbf{x}}_k^- - \mathbf{x}_k)')$$

Similarly, the covariance immediately after $\mathbf{z}_k$ is observed is

$$P_k^+ = E((\hat{\mathbf{x}}_k^+ - \mathbf{x}_k)(\hat{\mathbf{x}}_k^+ - \mathbf{x}_k)')$$

---

[7] Richard S. Bucy and Peter D. Joseph, *Filtering for Stochastic Processes with Applications to Guidance* (New York: John Wiley & Sons, 1968), p. 140.

From equation k2, it is seen that P_k^- depends, among other things, directly on the covariance matrix Q_{k-1} for the model noise. Consequently, if diagonal entries in Q_{k-1} are large (i.e., approach infinity), so will be the corresponding diagonal entries in P_k^-. In any case, P_k^- is a measure of how well $\hat{\mathbf{x}}_k^-$ estimates $\mathbf{x}_k$.

Equation k3 is called the *Kalman-Bucy gain equation*. Among the most interesting features of this equation is the covariance matrix of the measurement noise R_k, where R_k is assumed to be positive definite. If the diagonal entries of R_k are all large, this means that the measurements are not accurate; intuitively, the corresponding matrix C_k will have entries that are small in magnitude. The ramifications of this are better understood by investigating the role of C_k in the ensuing equations.

The *state estimate update* is given in equation k4. This equation is very similar to the "moving average" calculation of the sample mean given in the opening remarks of this section. Notice that $\hat{\mathbf{x}}_k^+$ is expressed as a linear combination of $\hat{\mathbf{x}}_k^-$ and the observation $\mathbf{z}_k$. This is seen by writing

$$\hat{\mathbf{x}}_k^+ = (I - C_k H_k)\hat{\mathbf{x}}_k^- + C_k \mathbf{z}_k$$

Moreover, as mentioned in the previous paragraph, if the measurement $\mathbf{z}_k$ is not very accurate, then C_k will have entries that are very small in magnitude. In this case, $\hat{\mathbf{x}}_k^+ \cong \hat{\mathbf{x}}_k^-$ and the estimator $\hat{\mathbf{x}}_k^+$ essentially ignores the measurement $\mathbf{z}_k$.

Equation k5 determines the error covariance matrices immediately after the measurement $\mathbf{z}_k$ is obtained by using the error covariance matrix calculated immediately before $\mathbf{z}_k$ is observed. This equation is called the *error covariance update equation*. Notice that if C_k has entries close to zero, then $P_k^+ \cong P_k^-$. Relating this to the discussion in the previous two paragraphs, if the measurement $\mathbf{z}_k$ is of no value, it is ignored and the error remains the same!

The KBF equations should be used in order, k1 → k5, repeatedly. A simple scalar example is given next.

Example 7.17

Suppose that it is known that the state variable x_k, which is in this example assumed to be scalar, satisfies the model equation

$$x_k = x_{k-1} + w_{k-1}$$

i.e., $\Phi_{k-1} = 1$. Let $Q_{k-1} = E(w_{k-1}^2) = \sigma_w^2$, and let the observation equation be

$$z_k = x_k + v_k$$

It follows that $H_k \equiv 1$. Let $R_k = E(v_k^2) = \sigma_v^2 > 0$. We assume that the variances are constant, i.e., not changing with the index parameter. Finally, let

$$P_0^+ = E(x_0 - \hat{x}_0^+)^2 = \sigma_0^2,$$

and recall that $\hat{x}_0^+ = E(x_0)$ is known.

The first iteration (and the only iteration we shall give) of the equations k1 → k5 follows:

k1. $\hat{x}_1^- = E(x_0)$.

k2. $P_1^- = \sigma_0^2 + \sigma_w^2$.

k3. $C_1 = \dfrac{\sigma_0^2 + \sigma_w^2}{\sigma_0^2 + \sigma_w^2 + \sigma_v^2}$.

k4. $\hat{x}_1^+ = E(x_0) + \left(\dfrac{\sigma_0^2 + \sigma_w^2}{\sigma_0^2 + \sigma_w^2 + \sigma_v^2}\right)[z_1 - E(x_0)]$.

k5. $P_1^+ = \dfrac{\sigma_v^2(\sigma_0^2 + \sigma_w^2)}{\sigma_0^2 + \sigma_w^2 + \sigma_v^2}$.

From equation k4, if $\sigma_v \to 0$, meaning that the measurement z_1 is very accurate, then $\hat{x}_1^+ \to z_1$. Also, in this case $P_1^+ \to 0$, as should be expected. On the other hand, if $\sigma_v \to \infty$, then $\hat{x}_1^+ \to E(x_0)$, $P_1^+ \to \sigma_0^2 + \sigma_w^2$, and the uncertainty in the error P_1^+ is the sum of the uncertainty in the initial condition σ_0^2 and the uncertainty in the model σ_w^2. Other limiting cases should also be investigated, but we will not do so now.

The KBF satisfies criteria similar to those of the BLUE method given in Section 7.4. Specifically, the estimator $\hat{x}_k^+$ is linear in $\hat{x}_k^-$ and z_k; that is, it can be expressed in the form

$$\hat{x}_k^+ = A_k \hat{x}_k^- + B_k z_k$$

It was seen earlier that $A_k = I - C_k H_k$ and $B_k = C_k$. The estimator $\hat{x}_k^+$ is unbiased, i.e., $E(\hat{x}_k^+) = E(x_k)$. Notice that unbiasedness is defined slightly differently from the way it was in previous sections since the state (parameter) vector is random. Finally, the KBF is best in the sense that it minimizes the sum of the variances in the error covariance matrix. Specifically, the KBF yields an estimator $\hat{x}_k^+$ that minimizes the trace of P_k^+.

A more useful version of the KBF can be obtained by employing the model equation

$$x_k = \Phi_{k-1} x_{k-1} + w_{k-1} + u_{k-1}$$

where u_{k-1} is an n by 1 input vector or driving function. The resulting KBF involves the same equations k2, k3, and k5; however, k1 and k4 together become[8]

k'1. $\hat{x}_k^+ = \Phi_{k-1}\hat{x}_{k-1}^+ + u_{k-1} + C_k[z_k - H_k \Phi_{k-1}\hat{x}_{k-1}^+]$

Kalman-Bucy filtering is currently very popular in image processing for the estimation of state vectors. Unfortunately, insofar as a text is concerned, worked-out examples are unpractical because of the inordinate amount of computation involved. Consequently, only a scalar example has been presented. Pragmatic

[8] Arthur Gelb, ed., *Applied Optimal Estimation* (Cambridge, MA: M.I.T. Press, 1974), p. 130.

computational considerations make the KBF a strictly machine-dependent procedure.

7.6. LEAST-SQUARES IMAGE RESTORATION

In this section we employ the least-squares techniques presented in Section 7.4 for image restoration, which is the process of removing errors introduced by imaging sensors in the observation process. Figure 7.1 depicts an intuitive description of image restoration, which is essentially an estimation procedure.

The first step in the restoration process is to convert the digital image into a form to which the matrix estimation techniques given in Section 7.4 are applicable. This is best done using the VECTRAN operation, where $\mathbf{f} = \text{VECTRAN}(f)$ is the transpose of $\text{RANGE}(f)$ for any bound matrix $f = (a_{pq})_{rt}$. Put simply, $\mathbf{f}$ consists of the columns of f from left to right stacked on top of each other with stars removed.

Example 7.18

If

$$f = \begin{pmatrix} 7 & 2 \\ * & 1 \\ 3 & * \end{pmatrix}_{4,8}$$

then

$$\mathbf{f} = \text{VECTRAN}(f) = \begin{pmatrix} 7 \\ 3 \\ 2 \\ 1 \end{pmatrix}$$

In utilizing the VECTRAN operation, it is convenient to simultaneously employ the domain extraction operation DOMAIN. This will allow an inverse operation similar to CREATE to be applied. Here, a CREATE-type operation will be employed using DOMAIN and a vector $\hat{\mathbf{g}}$ obtained from $\mathbf{f}$, where $\hat{\mathbf{g}}$ has the same dimension as $\mathbf{f}$. The vector $\hat{\mathbf{g}}$ will be an estimator of $\mathbf{g}$, which will be a vector representation of an image.

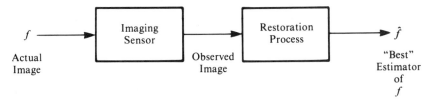

Figure 7.1 Restoration Procedure

An LS (least-squares) procedure for image restoration utilizing VECTRAN follows. The methodology is the same as that given in the early part of Section 7.4. The procedure employs both intra- and interpixel gray level information. In the former case, observed gray levels of a given pixel are a function of only the actual gray level of that same pixel. In the latter case, observed gray levels of a given pixel depend on the actual gray levels of other pixels. The goal in restoration is to make a best estimate of the true gray value of a pixel based upon observed gray values. Sometimes more than one image of a specific object is given, as, for instance, in the case when two different sensors are used in observing the same object, or when a fixed object is observed at two different times by the same sensor and from the same perspective. In any case, $n \geq 1$ observed images $f_1, f_2, \ldots, f_n$ might be given for the purpose of producing an estimate $\hat{g}$ of the image g. The procedure for finding $\hat{g}$ is:

1. Find DOMAIN(g)
2. Take VECTRAN(f_i) = $\mathbf{f}_i$, for $i = 1, 2, \ldots, n$, and form the vector

$$\mathbf{f} = \begin{pmatrix} \mathbf{f}_1 \\ \mathbf{f}_2 \\ \vdots \\ \mathbf{f}_n \end{pmatrix}$$

3. Take VECTRAN(g) = $\mathbf{g}$ and form $\mathbf{f} = H\mathbf{g}$, where the design matrix H is assumed to be an r by m matrix of full rank with $r \geq m$. (Note that $\mathbf{f}$ is known from the observations and H is known from our knowledge concerning the action of the sensors.)
4. Use the pseudoinverse relation $\hat{\mathbf{g}} = H^+ \mathbf{f}$, where $H^+ = (H'H)^{-1}H'$. This yields the least-squares estimate $\hat{\mathbf{g}}$ of $\mathbf{g}$. (See Section 7.4.)
5. Use the DOMAIN(g) information and $\hat{\mathbf{g}}$ to find $\hat{g}$.

The entire five-step procedure is illustrated in Figure 7.2.

Example 7.19

Suppose that an estimate of the two gray values in the image $(a\ b)_{0,0}$ is desired. Suppose further that two observed images

$$f_1 = \begin{pmatrix} 3 & 4 \\ 1 & * \end{pmatrix}_{0,0}$$

and

$$f_2 = \begin{pmatrix} 3 & * \\ * & 4 \end{pmatrix}_{0,0}$$

are given. The model to be presented shortly will be used in determining the design matrix H. It will be assumed that an observed value of gray at a given pixel equals

Sec. 7.6 Least-Squares Image Restoration

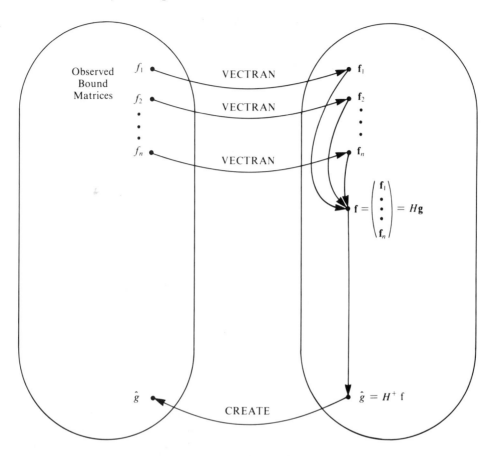

Figure 7.2 Least-Squares Solution Using VECTRAN

.6 times the true gray value at that pixel plus .1 times the value of gray at each strong neighbor. Then, since

$$\text{VECTRAN}(f_1) = \begin{pmatrix} 3 \\ 1 \\ 4 \end{pmatrix} = \mathbf{f}_1$$

and

$$\text{VECTRAN}(f_2) = \begin{pmatrix} 3 \\ 4 \end{pmatrix} = \mathbf{f}_2$$

the observed vector obtained by concatenation is

$$\mathbf{f} = \begin{pmatrix} 3 \\ 1 \\ 4 \\ 3 \\ 4 \end{pmatrix}$$

Utilizing the model for determining H gives $\mathbf{f} = H\mathbf{g}$:

$$\begin{pmatrix} 3 \\ 1 \\ 4 \\ 3 \\ 4 \end{pmatrix} = \begin{pmatrix} .6 & .1 \\ .1 & 0 \\ .1 & .6 \\ .6 & .1 \\ 0 & .1 \end{pmatrix} \begin{pmatrix} a \\ b \end{pmatrix}$$

where

$$\text{VECTRAN}((a \;\; b)_{0,0}) = \begin{pmatrix} a \\ b \end{pmatrix} = \mathbf{g}$$

The pseudoinverse of the overdetermined matrix H is

$$H^+ = \begin{pmatrix} .843 & .152 & -.268 & .843 & -.70 \\ -.133 & -.70 & 1.663 & -.133 & .288 \end{pmatrix}$$

Therefore,

$$\hat{\mathbf{g}} = H^+ \mathbf{f} = \begin{pmatrix} \hat{a} \\ \hat{b} \end{pmatrix} = \begin{pmatrix} 3.85 \\ 6.84 \end{pmatrix}$$

Utilizing the domain information gives the least-squares best estimate image

$$\hat{g} = (\hat{a} \;\; \hat{b})_{00} = (3.85 \;\; 6.84)_{0,0}$$

7.7 DETECTION

In this section we introduce hypothesis tests for solving detection problems. Like estimation theory, detection theory falls into the category of inferential statistics. However, detection problems differ from estimation problems in several ways. Estimation theory is used for finding parameters that are unknown, except perhaps for bounds or some knowledge of underlying probabilities, whereas detection theory is utilized for determining which of several known candidate parameters should be chosen based on a given sample. These parameters are known beforehand and, in imaging, often take the form of feature vectors. Sometimes both estimation and detection can be employed to solve a problem. When this occurs, the detection techniques often provide better results.

When dealing with detection problems, the range space D for the parameter is often called the *set of states of nature* and a random variable, or random vector, called the *observation*, has a probability density that depends on a parameter in D. The random variable (vector), or a random sample of the random variable, is observed and a choice of parameters is made.

Though the decision logic of hypothesis testing might at first appear somewhat abstruse, the intuition behind the methodology is quite straightforward. Suppose ξ is a random variable possessing a univariate Gaussian distribution with known variance σ^2 and unknown mean μ. However, suppose that, due to some

Sec. 7.7   Detection

form of prior knowledge, we have reduced the problem to choosing between two values, μ_0 and μ_1. In a sense, we must estimate μ, but in the present situation the parameter space D consists of only two points, μ_0 and μ_1. Since μ is the mean of the distribution, a reasonable way to proceed is to employ the sample mean and compute an empirical mean $\overset{*}{\mu}$ and, based on the empirical mean, decide between μ_0 and μ_1. There are essentially two hypotheses: $\mu = \mu_0$, and $\mu = \mu_1$. Based upon a function of the random variable, in this case the sample mean, a decision is made as to which hypothesis is to be selected.

An immediate variant of the preceding hypothesis problem concerns the situation where one simply desires to decide whether or not a particular value, say c, is a reasonable choice for some distributional parameter. Again, the decision is to be based upon some statistic resulting from a sample. If the parameter under consideration happens to be the mean μ, then the two competing hypotheses are: $\mu = c$, and $\mu \neq c$. Additional variations are possible, including several competing hypotheses.

The subsequent discussion provides a general decision theoretic development of hypothesis testing. Nevertheless, one should keep in mind the preceding motivating comments.

The simplest detection problem involves a partition $\{D_0, D_1\}$ of D where $D_0 \cup D_1 = D$ and $D_0 \cap D_1 = \varnothing$. The objective is to detect in which set, D_0 or D_1, the unknown parameter (the true state of nature) lies. The condition $\theta \in D_0$ is called the *null hypothesis* and is denoted by

$$H_0: \theta \in D_0$$

The condition $\theta \in D_1$ is referred to as the *alternative hypothesis* and is denoted by

$$H_1: \theta \in D_1$$

The decision as to whether H_0 or H_1 is true is made on the basis of observed values of the random variable (vector) in question. It is convenient to define the set A of possible actions to be taken. Here, the action will be either to *accept* H_0, or to *reject* it, and two real numbers a_0 and a_1 are used to denote the actions in A. Moreover, a decision rule is utilized in determining which action is to be taken. A *decision rule d* is a (Baire) function from the range space of the underlying random variable (vector) into A.

Assuming that ξ is the underlying random variable, let

$$\mathbf{v} = \begin{pmatrix} \xi_1 \\ \xi_2 \\ \vdots \\ \xi_n \end{pmatrix}$$

denote a random sample of size n. Based upon an observation of $\mathbf{v}$, an action is determined either to accept or to reject H_0; this is called a *test* of the null hypothesis H_0 versus the alternative H_1. If S is the underlying sample space, then the range space of the random vector $\mathbf{v}$ is denoted by $\mathbf{v}(S^n) = B \subset R^n$, and $d: B \to A$. Letting d^{-1} denote the inverse relation and letting $d^{-1}(\{a_0\}) = B_0$ and $d^{-1}(\{a_1\}) = B_1$ yields a partition of B, i.e., $B = B_0 \cup B_1$ and $B_0 \cap B_1 = \emptyset$. Since for every $b \in B_0$, $d(b) = a_0$, and therefore H_0 is accepted, we call B_0 the *acceptance region*. B_1 is called the *rejection* or *critical region*.

If action a_0 is taken when $\theta \in D_0$, then we have accepted a true hypothesis, and no error or loss occurs. Similarly, no loss occurs if we reject H_0, take action a_1, and it is true that $\theta \in D_1$. However, if action a_1 is taken (based on sampled evidence) when in fact $\theta \in D_0$, then an error known as a *Type I error* occurs. This error is sometimes called an *error of nondetection*. A *Type II error* occurs when action a_0 is taken when in fact $\theta \in D_1$. This error is sometimes called a *false-alarm-type error*. Type I and Type II errors are summarized in Figure 7.3, along with the decision rule and a diagram of the acceptance and critical regions.

If we let α and β denote the probabilities of Type I and Type II error, respectively, it is important to note that both α and β are conditional probabilities:

$$\alpha = P(\mathbf{v} \in B_1 \mid H_0: \theta \in D_0) = P(d = a_1 \mid H_0)$$

$$\beta = P(\mathbf{v} \in B_0 \mid H_1: \theta \in D_1) = P(d = a_0 \mid H_1)$$

Example 7.20

Let ξ be a random variable indicating the level of gray. Suppose that gray values are encoded using the code 0 = white, 1 = light gray, 2 = medium gray, 3 = black. Suppose also that there exists a density function for ξ which depends on a parameter θ which, depending on the state of nature, can attain one of two values, θ_0 or θ_1. Assume that $f_\xi(x; \theta_0)$ is the density function for the level of gray in the foreground image, and assume that $f_\xi(x; \theta_1)$ is associated with the gray levels in the background image. For convenience of understanding, we will posit that f_ξ possesses a binomial distribution:

$$f_\xi(x; \theta) = \sum_{k=0}^{3} \binom{3}{k} \theta^k (1 - \theta)^{3-k} \delta(x - k)$$

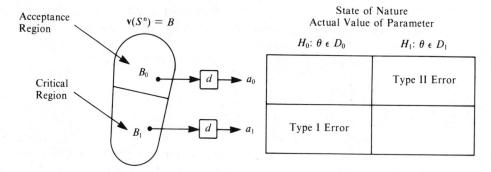

Figure 7.3 Type I and Type II Errors

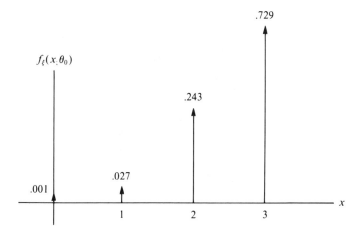
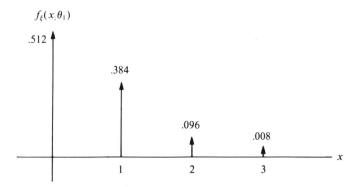

Figure 7.4 Densities Depending on a Parameter

where $\theta_0 = .9$ and $\theta_1 = .2$, so that the foreground has a larger amount of darker gray than the background. The two densities are illustrated in Figure 7.4.

Suppose we desire to test the hypothesis H_0: $\theta = \theta_0$ against the alternative H_1: $\theta = \theta_1$ based on a single observed value of ξ. In other words, based upon the gray value of a single pixel, we wish to determine whether the pixel lies in the foreground ($\theta = \theta_0$) or the background ($\theta = \theta_1$). Moreover, suppose that $B_0 = \{2, 3\}$; that is, if the value of the sample is 2 or 3, the decision would be a_0, i.e., accept H_0. Consequently, the critical region is $B_1 = \{0, 1\}$, and if the observed value lies in this set, then the decision would be a_1, i.e., accept H_1.

In this case the probability of a Type I error is $\alpha = .028$, which equals the "area" under $f_\xi(x; \theta_0)$ in the critical region. The probability of a Type II error is $\beta = .104$, which is the "area" under $f_\xi(x; \theta_1)$ in B_0.

Just as in estimation, it is possible to introduce a loss function into detection analysis, only here we employ a (random) decision d instead of an estimator.

Since $d: B \to A$, the action space, and since $d^{-1}(\{a_j\}) = B_j$, for $j = 0, 1$, we can look upon the loss function as either being defined on $B \times D$ or $A \times D$.

A simple choice for a loss function in the present situation is

$$\ell(a, \theta) = \begin{cases} 0 & \text{if } (a, \theta) \text{ is in } \{a_0\} \times D_0 \\ 0 & \text{if } (a, \theta) \text{ is in } \{a_1\} \times D_1 \\ 1 & \text{if } (a, \theta) \text{ is in } \{a_0\} \times D_1 \\ 1 & \text{if } (a, \theta) \text{ is in } \{a_1\} \times D_0 \end{cases}$$

Put simply, $\ell(a, \theta) = 0$ for a correct decision, while $\ell(a, \theta) = 1$ for an incorrect decision. Since $d: B \to A$, we can consider the composite function

$$\ell(d(\cdot), \cdot): B \times D \to R,$$

where, if values $x_1, x_2, \ldots, x_n$ of the observed sample are used, then the following risk function r results:

$$r(d, \theta) = \int_{-\infty}^{\infty} \cdots \int_{-\infty}^{\infty} \left[\ell(d(x_1, \ldots, x_n), \theta) \prod_{i=1}^{n} f(x_i; \theta) \right] dx_1 \cdots dx_n$$

A clear objective is to find d such that α and β are both small; however, a decrease in one of these probabilities usually results in an increase in the other. Consequently, our objective will be to find the critical region B_1 of a given size α which minimizes β among all critical regions of size $\leq \alpha$. If such a critical region exists, it is called the *best critical region* of size α. Usually the value α is given in advance, and all critical regions of size $\leq \alpha$ are investigated to see whether they are best. The corresponding test is called a *best test*.

Example 7.21

Suppose that ξ is a random variable indicating the "number of pixels on target." Suppose also that there are two targets and there is a density function associated with ξ depending on a parameter θ. For the first target, the density function is $f_\xi(x; \theta_0)$, and for the second it is $f_\xi(x; \theta_1)$, where

$$f_\xi(x; \theta_0) = .03\delta(x - 40) + .02\delta(x - 41) + .4\delta(x - 42)$$
$$+ .45\delta(x - 43) + .05\delta(x - 44) + .05\delta(x - 45)$$
$$f_\xi(x; \theta_1) = .09\delta(x - 40) + .08\delta(x - 41) + .23\delta(x - 42)$$
$$+ .3\delta(x - 43) + .29\delta(x - 44) + .01\delta(x - 45)$$

Let the hypothesis $H_0: \theta = \theta_0$ be tested against the alternative $H_1: \theta = \theta_1$ on the basis of a single observed value of ξ, with critical region of size $\alpha = .05$. The critical regions of size .05 and less correspond to values of x in the sets $\{40\}$, $\{41\}$, $\{44\}$, $\{45\}$, and $\{40, 41\}$. The values of β associated with each of these sets are .91, .92, .71, .99, and .83, respectively. Hence, the best critical region of size α is $\{44\}$.

The Neyman-Pearson Lemma, presented next, provides a method for constructing a best critical region of a given size.

Sec. 7.7 Detection

Theorem 7.2 (Neyman-Pearson Lemma).[9] Consider a random sample of size n with values $x_1, x_2, \ldots, x_n$, each arising from a density function $f(x; \theta)$. Let

$$\frac{\prod_{i=1}^{n} f(x_i; \theta_1)}{\prod_{i=1}^{n} f(x_i; \theta_0)} = t$$

Suppose that for some nonnegative constant k and critical region B_1 of size α, where $0 < \alpha < 1$, $t \geq k$ for $(x_1, x_2, \ldots, x_n)$ in B_1 and $t \leq k$ for $(x_1, x_2, \ldots, x_n)$ not in B_1. Then B_1 is the best critical region of size α.

Example 7.22

We shall find the best critical region for testing the hypothesis $H_0: \mu = \mu_0$ against the alternative hypothesis $H_1: \mu = \mu_1$, where $\mu_1 > \mu_0$. We assume that μ is the mean gray value for a class of images whose gray values are normally distributed with known variance σ^2. Applying the Neyman-Pearson Lemma, we have

$$t = \frac{e^{-1/2\sigma^2 \sum_{i=1}^{n}(x_i - \mu_1)^2}}{e^{-1/2\sigma^2 \sum_{i=1}^{n}(x_i - \mu_0)^2}}$$

$$= e^{[n(\mu_0^2 - \mu_1^2)/2\sigma^2 + (\mu_1 - \mu_0)n\bar{x}/\sigma^2]}$$

where

$$\bar{x} = \frac{\sum_{i=1}^{n} x_i}{n}$$

We need to find a region B_1 where $(x_1, x_2, \ldots, x_n) \in B_1$ implies that $t \geq k$. Setting $t \geq k$ and taking logs yields

$$\left(\frac{\mu_1 - \mu_0}{\sigma^2}\right) n\bar{x} \geq \ln k + \frac{n(\mu_1^2 - \mu_0^2)}{2\sigma^2}$$

Therefore, since $\mu_1 - \mu_0 > 0$,

$$\bar{x} \geq \left(\frac{\sigma^2}{\mu_1 - \mu_0}\right) \frac{\ln k}{n} + \frac{(\mu_1 + \mu_0)}{2}$$

Since k is nonnegative, $\ln k$ is real, and the right-hand side can take on any real number c. Therefore, $t \geq k$ for

$$\bar{x} = \frac{1}{n} \sum_{i=1}^{n} x_i \geq c$$

or

$$\sum_{i=1}^{n} x_i \geq nc$$

---

[9] Paul Hoel, Sidney Port, and Charles Stone, *Introduction to Statistical Theory* (Boston: Houghton Mifflin Company, 1971), p. 56.

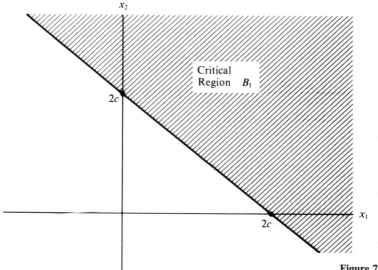

Figure 7.5 Critical Region

The case $n = 2$ is illustrated in Figure 7.5. For that case, an *a priori* specification of α will enable an appropriate value of c to be found. What is of more consequence, however, is that the critical region is always a half-plane to the right of the line $x_1 + x_2 = 2c$.

There is a result quite similar to the Neyman-Pearson Lemma for the case when θ is assumed to be a value attained from a random variable η. Here, it is assumed that the *a priori* distribution $f_\eta(\theta) = \pi_0 \delta(\theta - \theta_0) + \pi_1 \delta(\theta - \theta_1)$, for $0 < \pi_0 < 1$, is known. It is assumed that the probability of θ_0 occurring is π_0 and that of θ_1 occurring is π_1. Under these assumptions, we have the Bayesian version of Theorem 7.2.

Theorem 7.3.[10] Under the conditions of the preceding paragraph, the critical region associated with a random sample of size n of the random variable ξ with density $f_\xi(x \mid \theta)$ should be found by solving

$$\frac{\prod_{i=1}^{n} f(x_i \mid \theta_1)}{\prod_{i=1}^{n} f(x_i \mid \theta_0)} \geq \frac{\pi_0}{\pi_1}$$

---

[10] Ibid., p. 99.

The solution in Theorem 7.3 yields a decision function that minimizes

$$R = E(r(d,\eta)) = \int_{-\infty}^{\infty} f_\eta(\theta) \left[\int_{-\infty}^{\infty} \cdots \int_{-\infty}^{\infty} \ell(d(x_1, x_2, \ldots, x_n), \theta) \right.$$
$$\left. \times \prod_{i=1}^{n} f(x_i \mid \theta) \, dx_1 \, dx_2 \cdots dx_n \right] d\theta$$

The expression for $r(d, \theta)$ is exactly the same as has previously been given, except that $f(x \mid \theta)$ has been used instead of $f(x; \theta)$. (See Section 7.3.)

Using the 0–1 loss function previously described yields

$$R = \pi_0 P(a_1 \mid \theta_0) + \pi_1 P(a_0 \mid \theta_1) = \pi_0 \alpha + \pi_1 \beta$$

Example 7.23

Refer to Example 7.22 except this time suppose that the average gray value μ occurs as an observation of the random variable η having density $f_\eta(\mu) = \pi_0 \delta(\mu - \mu_0) + \pi_1 \delta(\mu - \mu_1)$, for $0 < \pi_0 < 1$. We seek the best critical region based on the Bayesian method given in the Theorem 7.3. As in Example 7.22, the ratio t of probabilities can be found and an inequality can be given involving $k = \pi_0/\pi_1$. We obtain

$$\bar{x} \geq \left(\frac{\sigma^2}{\mu_1 - \mu_0} \right) \frac{\ln(\pi_0/\pi_1)}{n} + \frac{(\mu_1 + \mu_0)}{2}$$

It is worthwhile discussing special cases of the result obtained in Example 7.23. In particular, if $\pi_0 = \pi_1 = \frac{1}{2}$, then we obtain $\bar{x} \geq (\mu_0 + \mu_1)/2$ (the median gray value of η). On the other hand, if $\pi_0 > \pi_1$, then $\bar{x}$ is greater than the median value, thereby decreasing the size of the critical region and making it more likely that μ_0 is chosen. In any case, as $n \to \infty$, $\bar{x} \geq (\mu_0 + \mu_1)/2$ results.

The Bayesian version of the Neyman-Pearson Lemma has natural extensions to multiple recognition problems. In these types of problems, the set of states of nature is partitioned into $k > 2$ subsets and, correspondingly, there are k parts to the action space. More specifically, it is assumed that the underlying random variable ξ has a density function depending on a parameter θ that can have, with nonzero probability, any one of the values $\theta_1, \theta_2, \ldots, \theta_k$. As before, the density is $f_\xi(x \mid \theta)$. Moreover, we assume that θ is a value of the random variable η having *a priori* density

$$f_\eta(\theta) = \sum_{i=1}^{k} \pi_i \delta(\theta - \theta_i)$$

If we now let $x_1, x_2, \ldots, x_n$ be the values of random sample $\xi_1, \xi_2, \ldots, \xi_n$,

$$\mathbf{v} = \begin{pmatrix} \xi_1 \\ \xi_2 \\ \vdots \\ \xi_n \end{pmatrix}$$

and $d(x_1, \ldots, x_n)$ denote the decision function, then the range space $\mathbf{v}(S^n)$ of the random vector $\mathbf{v}$ is partitioned into k sets $B_1, B_2, \ldots, B_k$. If $(x_1, \ldots, x_n) \in B_j$, then $d(x_1, \ldots, x_n) = a_j$ is the action indicative of choosing the hypothesis $H_j: \theta = \theta_j$. As was the case when $k = 2$, let the loss function ℓ be given by

$$\ell(a_i, \theta_j) = 1 - \delta_{ij}, \text{ where } \delta_{ij} = \begin{cases} 1 & \text{if } i = j \\ 0 & \text{if } i \neq j \end{cases}$$

Then zero loss occurs if a correct decision is made, and 100 percent loss occurs otherwise.

Theorem 7.4.[11] Under the conditions of the preceding paragraph, the best acceptance region B_i for the hypothesis $H_i: \theta = \theta_i$, for $i = 1, 2, \ldots, k$, is that part of the range space $\mathbf{v}(S^n)$ where

$$\pi_i \prod_{p=1}^{n} f(x_p \mid \theta_i) \geq \pi_j \prod_{p=1}^{n} f(x_p \mid \theta_j) \text{ for all } j \neq i$$

Points which can be assigned to more than one region should be placed in an arbitrary unique region.

Example 7.24

Let ξ be a random variable indicative of the background gray value appearing in a certain class of images. Suppose that

$$f_\xi(x \mid \theta) = \frac{1}{2\theta} I_{(-\theta, \theta)}(x)$$

where θ is a parameter satisfying the density function

$$f_\eta(\theta) = \sum_{i=1}^{3} \pi_i \delta(\theta - \theta_i)$$

The multidetection problem will be solved for

$$f_\eta(\theta) = \frac{1}{3}\delta(\theta - 1) + \frac{1}{6}\delta(\theta - 2) + \frac{1}{2}\delta(\theta - 3)$$

where only one sample will be taken. The acceptance region for $H_1: \theta = \theta_1 = 1$ is found by setting

$$\left(\frac{1}{3}\right)\left(\frac{1}{2}\right) I_{(-1,1)}(x) \geq \left(\frac{1}{6}\right)\left(\frac{1}{4}\right) I_{(-2,2)}(x)$$

and

$$\left(\frac{1}{3}\right)\left(\frac{1}{2}\right) I_{(-1,1)}(x) \geq \left(\frac{1}{2}\right)\left(\frac{1}{6}\right) I_{(-3,3)}(x)$$

[11] Ibid., p. 101.

The set of points in R simultaneously satisfying these inequalities is

$$A_1 = (-1, 1) \cup (-\infty, -3] \cup [3, \infty)$$

To find A_2, note that x is in A_2 if both of the following inequalities hold:

$$\left(\frac{1}{6}\right)\left(\frac{1}{4}\right) I_{(-2,2)}(x) \geq \left(\frac{1}{3}\right)\left(\frac{1}{2}\right) I_{(-1,1)}(x)$$

$$\left(\frac{1}{6}\right)\left(\frac{1}{4}\right) I_{(-2,2)}(x) \geq \left(\frac{1}{2}\right)\left(\frac{1}{6}\right) I_{(-3,3)}(x)$$

Thus, $A_2 = \emptyset$. Hence, if a single sample is taken, then we never decide in favor of H_2! Note that we did not let $A_2 = (-\infty, 3] \cup [3, \infty)$, since we used these sets in A_1. For A_3, we have

$$\left(\frac{1}{2}\right)\left(\frac{1}{6}\right) I_{(-3,3)}(x) \geq \left(\frac{1}{3}\right)\left(\frac{1}{2}\right) I_{(-1,1)}(x)$$

$$\left(\frac{1}{2}\right)\left(\frac{1}{6}\right) I_{(-3,3)}(x) \geq \left(\frac{1}{6}\right)\left(\frac{1}{4}\right) I_{(-2,2)}(x)$$

Therefore,

$$A_3 = (-3, -1] \cup [1, 3)$$

7.8 HOTELLING TRANSFORM

In Section 6.1, we briefly introduced the notion of a random field. In essence, a random field (not extended to include star values) is an array of random variables

$$F = (A_{ij})_{rt}$$

Each realization (sample observation) of the random field is a saturated bound matrix. For this reason, we refer to a random field as a *random image*.

The random image model is quite natural. For many reasons, such as sensor interference or garbled transmission, it is often unwarranted to assume that an image is any more than a random phenomenon. Each of its pixel gray values is then a random variable. For instance, one popular model takes the form of image-plus-noise: it is assumed that there is an image f that is deterministic and some noise random image N that is added to f. The result is a random image $F = \text{ADD}(f, N)$.

Given a random image $F = (A_{pq})_{rt}$, where each A_{pq} is a random variable, it is likely that the variables A_{pq} are correlated. The *Hotelling transform* provides a means of transforming the collection of random variables A_{pq} into a collection of uncorrelated variables B_{pq}. Once this transformation is accomplished, it is possible to perform judicious compression. Moreover, specialized versions of the transform are employed in image restoration and enhancement. In some of these applications, it is not the transformation procedure itself which provides the benefits; rather, it is manipulation in the transform world which is profitable.

The methodology of the Hotelling transform will be presented for the case of a random vector. By the use of VECTRAN, DOMAIN, and CREATE, the technique can be applied to random images. However, the Hotelling transform should not be viewed merely as a random form of an image-to-image transformation. Used properly, it has far-reaching application to the compression of feature vectors.

Let

$$\mathbf{v} = \begin{pmatrix} \xi_1 \\ \xi_2 \\ \vdots \\ \xi_n \end{pmatrix}$$

be a random vector, and suppose $\mathbf{v}_1, \mathbf{v}_2, \ldots, \mathbf{v}_K$ is an empirical sample of size K taken from the population described by $\mathbf{v}$. Then, for the input parameter $M \leq n$, the output $\mathbf{V}_M$ of the Hotelling transform is a random vector of M dimensions. The following sequence of steps produces $\mathbf{V}_M$:

1. Let $\bar{\mathbf{v}}$ be the empirical mean, i.e.,

$$\bar{\mathbf{v}} = \frac{1}{K} \sum_{i=1}^{K} \mathbf{v}_i$$

2. Let R denote the empirical covariance matrix associated with the $\bar{\mathbf{v}}_i$, i.e.,

$$R = \frac{1}{K} \sum_{i=1}^{K} (\mathbf{v}_i - \bar{\mathbf{v}})(\mathbf{v}_i - \bar{\mathbf{v}})'$$

3. It happens that the eigenvalues of R are real valued. These eigenvalues should be found together with the associated eigenvectors. The eigenvectors of distinct eigenvalues will be mutually orthogonal. Each individual eigenvalue possesses mutually orthogonal eigenvectors (Gram-Schmidt orthogonalization),[12] and these should be obtained. All eigenvectors must be normalized.

4. Form the matrix

$$Y = \begin{pmatrix} \cdots & \mathbf{Y}'_1 & \cdots \\ \cdots & \mathbf{Y}'_2 & \cdots \\ & \vdots & \\ \cdots & \mathbf{Y}'_M & \cdots \end{pmatrix}$$

where $\mathbf{Y}'_1, \mathbf{Y}'_2, \ldots, \mathbf{Y}'_M$ are the transposes of the eigenvectors $\mathbf{Y}_1, \mathbf{Y}_2, \ldots, \mathbf{Y}_M$ corresponding to the M largest eigenvalues, where an eigenvalue of mul-

---

[12] Paul R. Halmos, *Finite-Dimensional Vector Spaces*, 2d ed. (Princeton, NJ: D. Van Nostrand Co., Inc., 1958), p. 128.

Sec. 7.8   Hotelling Transform

tiplicity τ is counted τ times. The Hotelling transform for input parameter M is then given by

$$\mathbf{V}_M = Y(\mathbf{v} - \bar{\mathbf{v}}) = Y\mathbf{v} - Y\bar{\mathbf{v}}$$

Before proceeding to an example, let us consider the case of a random m by n image $F = (A_{pq})_{rt}$. Application of VECTRAN gives

$$\mathbf{F} = \text{VECTRAN}[(A_{pq})_{rt}] = \begin{pmatrix} A_{11} \\ A_{21} \\ \vdots \\ A_{m1} \\ A_{12} \\ \vdots \\ A_{mn} \end{pmatrix}$$

If $\mathbf{V}_M$ is the Hotelling transform of $\mathbf{F}$, with $M = mn$, then $\mathbf{V}_M$ is a random vector with components $B_{11}, B_{21}, \ldots, B_{mn}$, where each B_{pq} is a random variable and where the covariance matrix of $\mathbf{V}_M$ is diagonal. The latter condition is paramount. The Hotelling transform decorrelates the pixel random variables. Should one now desire that $\mathbf{V}_M$ be returned to image format, the operator CREATE can be applied to $\mathbf{V}_M$ and DOMAIN(F). The result is a random image $(B_{pq})_{rt}$ with the same domain as F, but with the B_{pq} uncorrelated.

Example 7.25

Let $F = (A \ B)_{0,0}$ be a random image. Then applying VECTRAN to F gives

$$\mathbf{v} = \begin{pmatrix} A \\ B \end{pmatrix}$$

Suppose the following images are obtained by sampling:

$$f_1 = (3 \ 1)_{0,0}$$
$$f_2 = (9 \ 5)_{0,0}$$

This sample will be used to find the Hotelling transform. Let

$$\mathbf{v}_1 = \text{VECTRAN}(f_1) = \begin{pmatrix} 3 \\ 1 \end{pmatrix}$$
$$\mathbf{v}_2 = \text{VECTRAN}(f_2) = \begin{pmatrix} 9 \\ 5 \end{pmatrix}$$

Now apply the four steps outlined previously for finding $\mathbf{V}_M$:

1.

$$\bar{\mathbf{v}} = \frac{1}{2}\left[\begin{pmatrix} 3 \\ 1 \end{pmatrix} + \begin{pmatrix} 9 \\ 5 \end{pmatrix}\right] = \begin{pmatrix} 6 \\ 3 \end{pmatrix}$$

2.
$$R = \frac{1}{2}\left\{\left[\binom{3}{1} - \binom{6}{3}\right]\left[\binom{3}{1} - \binom{6}{3}\right]' + \left[\binom{9}{5} - \binom{6}{3}\right]\left[\binom{9}{5} - \binom{6}{3}\right]'\right\}$$

$$= \frac{1}{2}\left\{\binom{-3}{-2}(-3 \quad -2) + \binom{3}{2}(3 \quad 2)\right\}$$

$$= \begin{pmatrix} 9 & 6 \\ 6 & 4 \end{pmatrix}$$

3. The eigenvalues of R are easily obtained by setting the determinant of $R - \lambda I$ equal to zero:

$$\det\begin{pmatrix} 9 - \lambda & 6 \\ 6 & 4 - \lambda \end{pmatrix} = \lambda^2 - 13\lambda = 0$$

Hence, $\lambda_1 = 13$ and $\lambda_2 = 0$. The eigenvectors of R are found by setting $(R - \lambda I)\binom{x}{y} = 0$ for each value of λ. For $\lambda_1 = 13$,

$$\begin{pmatrix} 9 - 13 & 6 \\ 6 & 4 - 13 \end{pmatrix}\binom{x}{y} = 0$$

Hence, $x = 3y/2$, and the first eigenvector is

$$\binom{3y/2}{y}, \text{ for } y \neq 0$$

After normalization,

$$\mathbf{Y}_1 = \frac{\binom{3y/2}{y}}{\sqrt{9/4y^2 + y^2}} = \frac{\binom{3}{2}}{\sqrt{13}}$$

The second normalized eigenvector is found in a similar fashion. For $\lambda_2 = 0$, we solve

$$\begin{pmatrix} 9 & 6 \\ 6 & 4 \end{pmatrix}\binom{x}{y} = 0$$

The resulting normalized eigenvector is

$$\mathbf{Y}_2 = \frac{\binom{-2}{3}}{\sqrt{13}}$$

Sec. 7.8   Hotelling Transform

4. The rows of Y are $\mathbf{Y}_1'$ and $\mathbf{Y}_2'$; hence,

$$Y = \frac{\begin{pmatrix} 3 & 2 \\ -2 & 3 \end{pmatrix}}{\sqrt{13}}$$

The Hotelling transform is

$$\mathbf{V}_2 = Y\mathbf{v} - Y\bar{\mathbf{v}}$$

$$= \begin{pmatrix} 3 & 2 \\ -2 & 3 \end{pmatrix} \frac{1}{\sqrt{13}} \left[\begin{pmatrix} A \\ B \end{pmatrix} - \begin{pmatrix} 6 \\ 3 \end{pmatrix} \right]$$

$$= \frac{1}{\sqrt{13}} \begin{pmatrix} 3A + 2B - 24 \\ -2A + 3B + 3 \end{pmatrix}$$

Finally, since $\text{DOMAIN}(f) = [(0, 0), (1, 0)]$, the output image is given by

$$G_2 = \text{CREATE}([(0, 0), (1, 0)], \mathbf{V}_2)$$

$$= \left(\frac{3A + 2B - 24}{\sqrt{13}} \quad \frac{-2A + 3B + 3}{\sqrt{13}} \right)_{0,0}$$

Had compression been desired in the preceding example, then only the eigenvector corresponding to the larger eigenvalue, $\lambda_1 = 13$, would have been used, and we would have had

$$Y = \left(\frac{3}{\sqrt{13}} \quad \frac{2}{\sqrt{13}} \right)$$

The compressed output for $M = 1$ would then have been

$$\mathbf{V}_1 = \left(\frac{3}{\sqrt{13}} \quad \frac{2}{\sqrt{13}} \right) \left[\begin{pmatrix} A \\ B \end{pmatrix} - \begin{pmatrix} 6 \\ 3 \end{pmatrix} \right] = \frac{1}{\sqrt{13}} (3A + 2B - 24)$$

It is important to realize that, whether or not there is compression, the meaning of the output random variables is not the same as that of the input variables. Compression will be discussed in detail following Example 7.27.

The Hotelling transform is considered to be an optimum transform technique and is used as a benchmark to which other transform methods are often compared. Unfortunately, because of the eigenvalue and eigenvector processing, it is computationally intense.

When employing the Hotelling transform, it is assumed that $E(\mathbf{v}) = \bar{\mathbf{v}}$, that is, the expected value of the random vector is equal to the empirical mean. But this implies that $E(\mathbf{V}_M) = 0$. Consequently, the covariance of the output $\mathbf{V}_M$ is given by

$$S = E(\mathbf{V}_M \mathbf{V}_M') = YE[(\mathbf{v} - E(\mathbf{v}))(\mathbf{v} - E(\mathbf{v}))']Y'$$

$$= YE[(\mathbf{v} - \bar{\mathbf{v}})(\mathbf{v} - \bar{\mathbf{v}})']Y'$$

It is also assumed that the empirical covariance R of $\mathbf{v}$ equals the true covariance of $\mathbf{v}$: $R = \text{Cov}[\mathbf{v}]$.

Example 7.26

Applying the preceding remarks to Example 7.25, we obtain

$$S = YRY' = \frac{\begin{pmatrix} 3 & 2 \\ -2 & 3 \end{pmatrix}}{\sqrt{13}} \begin{pmatrix} 9 & 6 \\ 6 & 4 \end{pmatrix} \frac{\begin{pmatrix} 3 & -2 \\ 2 & 3 \end{pmatrix}}{\sqrt{13}}$$

$$= \begin{pmatrix} 13 & 0 \\ 0 & 0 \end{pmatrix}$$

which is a diagonal matrix.

Example 7.27

Given the image

$$h = \begin{pmatrix} 2 & 4 & 3 \\ -2 & 0 & 1 \end{pmatrix}_{0,0}$$

assume that the Hotelling transform is to be applied to $(A \quad B)_{0,0} = F$, where samples of F arise from uncertainty in pixel location. Observations of F are equally likely and are the horizontal pixel subimages of length two in h. Specifically, the observations of F are obtained from h by the operations SELECT and TRAN, as depicted in Figure 7.6. Application of VECTRAN to each observed image gives

$$\mathbf{v}_1 = \begin{pmatrix} 2 \\ 4 \end{pmatrix} \quad \mathbf{v}_2 = \begin{pmatrix} 4 \\ 3 \end{pmatrix} \quad \mathbf{v}_3 = \begin{pmatrix} -2 \\ 0 \end{pmatrix} \quad \mathbf{v}_4 = \begin{pmatrix} 0 \\ 1 \end{pmatrix}$$

The average vector associated with the observed images is

$$\bar{\mathbf{v}} = \frac{1}{4}\left[\begin{pmatrix} 2 \\ 4 \end{pmatrix} + \begin{pmatrix} 4 \\ 3 \end{pmatrix} + \begin{pmatrix} -2 \\ 0 \end{pmatrix} + \begin{pmatrix} 0 \\ 1 \end{pmatrix}\right] = \begin{pmatrix} 1 \\ 2 \end{pmatrix}$$

The covariance matrix R is given by

$$R = \frac{1}{4}\left\{\begin{pmatrix} 1 & 2 \\ 2 & 4 \end{pmatrix} + \begin{pmatrix} 9 & 3 \\ 3 & 1 \end{pmatrix} + \begin{pmatrix} 9 & 6 \\ 6 & 4 \end{pmatrix} + \begin{pmatrix} 1 & 1 \\ 1 & 1 \end{pmatrix}\right\}$$

$$= \begin{pmatrix} 5 & 3 \\ 3 & 2.5 \end{pmatrix}$$

The eigenvalues of $2R$ are found by letting

$$\det \begin{pmatrix} 10 - \lambda & 6 \\ 6 & 5 - \lambda \end{pmatrix} = 0$$

This gives the characteristic equation $\lambda^2 - 15\lambda + 14 = 0$. Thus, the eigenvalues are $\lambda_1 = 14$ and $\lambda_2 = 1$. The eigenvectors are found by setting

$$\begin{pmatrix} 10 - \lambda & 6 \\ 6 & 5 - \lambda \end{pmatrix}\begin{pmatrix} x \\ y \end{pmatrix} = 0$$

Sec. 7.8 Hotelling Transform

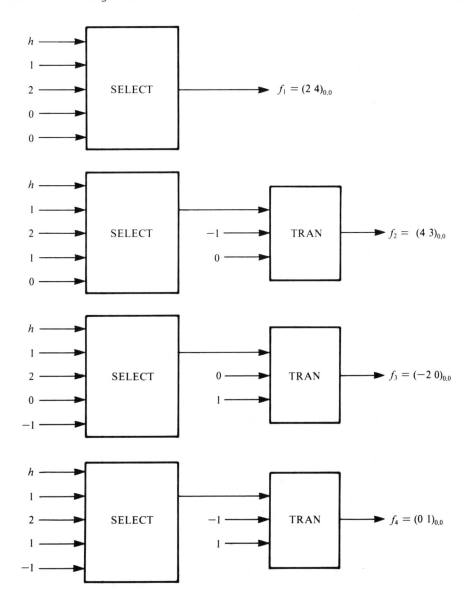

Figure 7.6 Observations Using Block Diagram

For $\lambda_1 = 14$,

$$\begin{pmatrix} -4 & 6 \\ 6 & -9 \end{pmatrix}\begin{pmatrix} x \\ y \end{pmatrix} = 0$$

and the first normalized eigenvector of R is

$$\mathbf{Y}_1 = \frac{\begin{pmatrix} 3 \\ 2 \end{pmatrix}}{\sqrt{13}}$$

For $\lambda_2 = 1$,

$$\begin{pmatrix} 9 & 6 \\ 6 & 4 \end{pmatrix}\begin{pmatrix} x \\ y \end{pmatrix} = 0$$

and the second normalized eigenvector is

$$\mathbf{Y}_2 = \frac{\begin{pmatrix} 2 \\ -3 \end{pmatrix}}{\sqrt{13}}$$

Using the transposes of the eigenvectors as rows of Y gives

$$Y = \frac{\begin{pmatrix} 3 & 2 \\ 2 & -3 \end{pmatrix}}{\sqrt{13}}$$

Hence,

$$\mathbf{V}_2 = \frac{\begin{pmatrix} 3 & 2 \\ 2 & -3 \end{pmatrix}}{\sqrt{13}}\left[\begin{pmatrix} A \\ B \end{pmatrix} - \begin{pmatrix} 1 \\ 2 \end{pmatrix}\right] = \frac{1}{\sqrt{13}}\begin{pmatrix} 3A + 2B - 7 \\ 2A - 3B - 4 \end{pmatrix}$$

Applying CREATE to $\mathbf{V}_2$ and DOMAIN(F) = [(0, 0), (1, 0)] gives the Hotelling image

$$G_2 = \left(\frac{3A + 2B - 7}{\sqrt{13}} \quad \frac{2A - 3B - 4}{\sqrt{13}}\right)_{0,0}$$

Under the assumptions discussed prior to this example, the diagonal covariance matrix for $\mathbf{V}_2$ is

$$S = \frac{\begin{pmatrix} 3 & 2 \\ 2 & -3 \end{pmatrix}}{\sqrt{13}} \frac{\begin{pmatrix} 10 & 6 \\ 6 & 5 \end{pmatrix}}{2} \frac{\begin{pmatrix} 3 & 2 \\ 2 & -3 \end{pmatrix}}{\sqrt{13}} = \begin{pmatrix} 7 & 0 \\ 0 & \frac{5}{26} \end{pmatrix}$$

We shall now focus on the compression characteristics of the Hotelling transform. For convenience, let $\mathbf{z} = \mathbf{v} - \bar{\mathbf{v}}$. Under the assumptions stated earlier regarding the equivalence of the expected value of $\mathbf{v}$ and the empirical mean, and the equivalence of the covariance and the empirical covariance, we have $E(\mathbf{z}) = \bar{\mathbf{z}} = 0$ and $E(\mathbf{z}\mathbf{z}') = R$. Letting $\mathbf{Z}_N$ denote the output of the Hotelling transform for input $\mathbf{z}$, with N equal to the dimension of the input random vector, it follows that $E(\mathbf{Z}_N) = 0$ and $E(\mathbf{Z}_N \mathbf{Z}'_N) = S = YRY'$.

Since the rows of Y are orthonormal, the inverse of Y is given by the transpose Y'. Consequently, inversion can take place and we have $\mathbf{z} = Y' \mathbf{Z}_N$. Hence, we see that the Hotelling transform acts like the matrix transforms examined in Section 5.9, except that in the present case the images being transformed are random images.

Let us look a bit deeper. Using matrix partitioning, we get

$$\mathbf{z} = (\mathbf{Y}_1 \mid \mathbf{Y}_2 \mid \cdots \mid \mathbf{Y}_N) \begin{pmatrix} Z^{(1)} \\ Z^{(2)} \\ \vdots \\ Z^{(N)} \end{pmatrix}$$

where the column matrix on the right represents $\mathbf{Z}_N$ and where Y_i denotes the ith eigenvector of the covariance matrix R. Therefore,

$$\mathbf{z} = \sum_{i=1}^{N} Z^{(i)} \mathbf{Y}_i$$

Suppose now that an estimate of $\mathbf{z}$ is desired which takes the form

$$\sum_{i=1}^{M} Z^{(i)} \mathbf{Y}_i + \sum_{i=M+1}^{N} C_i \mathbf{Y}_i$$

where $1 \leq M < N$. In other words, suppose we wish to estimate $\mathbf{z}$ by using M terms from the correct expansion and $N - M$ different terms, each of which is still a scalar multiple of $\mathbf{Y}_i$. Denote the estimate by $\mathbf{z}_M$, and let the error vector be

$$\mathbf{e}_M = \mathbf{z} - \mathbf{z}_M = \sum_{i=M+1}^{N} (Z^{(i)} - C_i) \mathbf{Y}_i$$

For a certain optimality criterion, it can be shown that $C_i = 0$ for $i = M + 1, \ldots, N$. Indeed, we wish to minimize $E(\mathbf{e}'_M \mathbf{e}_M)$. But

$$E(\mathbf{e}'_M \mathbf{e}_M) = \sum_{i=M+1}^{N} E(Z^{(i)} - C_i)^2$$

where the equality follows from the orthonormality of the $\mathbf{Y}_i$. Minimization of $E(\mathbf{e}'_M \mathbf{e}_M)$ is achieved by setting

$$C_i = E(Z^{(i)}) = 0$$

This means that the approximation of $\mathbf{z}$ is best accomplished by the *projection*

$$\mathbf{z}_M = \sum_{i=1}^{M} Z^{(i)} \mathbf{Y}_i$$

onto the subspace generated by the orthonormal eigenvectors $\mathbf{Y}_1, \mathbf{Y}_2, \ldots, \mathbf{Y}_M$. The situation is analogous to that encountered with orthonormal systems (see Chapter 5), the difference being that the $Z^{(i)}$ are random variables.

The question arises as to which eigenvectors should be kept for the expansion—that is, which should be kept to minimize the error

$$\sigma_M^2 = E(\mathbf{e}_M'\mathbf{e}_M) = \sum_{i=M+1}^{N} E([Z^{(i)}]^2)$$

By the manner in which the Hotelling transform is taken, $Z^{(i)} = \mathbf{Y}_i'\mathbf{z} = \mathbf{z}'\mathbf{Y}_i$. Substitution into σ_M^2 gives

$$\sigma_M^2 = \sum_{i=M+1}^{N} E(\mathbf{Y}_i'\mathbf{z}\mathbf{z}'\mathbf{Y}_i) = \sum_{i=M+1}^{N} \mathbf{Y}_i'R\mathbf{Y}_i$$

$$= \sum_{i=M+1}^{N} \lambda_i \mathbf{Y}_i'\mathbf{Y}_i = \sum_{i=M+1}^{N} \lambda_i$$

where λ_i is the ith eigenvalue of R. But by the construction of R, $\lambda_i \geq 0$ for all i. Consequently, the error in the approximation is minimized by keeping those eigenvectors which have the largest eigenvalues.

As previously noted, the expansion for $\mathbf{z}_M$ is similar to a projection onto a subspace. In fact, the analogy goes further. According to Theorem 5.4, an orthonormal projection is best when it utilizes the Fourier coefficients. Furthermore, part (ii) of the same theorem shows that the error is minimized by employing elements in the orthonormal system for which the Fourier coefficients of the element to be projected are maximal.

We close with some comments regarding the application of the Hotelling transform to feature vectors. If $\mathbf{v}$ happens to be a feature vector, it may very well be of benefit to compress it in order to reduce memory load and processing time. The Hotelling transform provides an optimality condition under which to choose the method of compression. Moreover, the expression of σ_M^2 as a sum of eigenvalues gives a quantitative measurement of the loss of information. It must, of course, be remembered that the output random variables in the Hotelling vector $\mathbf{V}_M$ have no particular meaning in terms of intuitive image parameters. But this is irrelevant for the comparison of feature vectors: as long as the knowledge-base feature vectors are stored in a (Hotelling) transformed mode, comparison of the compressed vectors can proceed.

EXERCISES

7.1. Show that if $\xi_1, \xi_2, \ldots, \xi_n$, $n > 1$ are independent identically distributed random variables on S; then

$$r = \frac{1}{n-1} \sum_{i=1}^{n} (\xi_i - \overline{\theta})^2$$

where

$$\bar{\theta} = \frac{\xi_1 + \xi_2 + \cdots \xi_n}{n}$$

is an unbiased estimator of the variance σ^2.

7.2. Let $\hat{\theta}_n = \varphi(\xi_1, \xi_2, \ldots, \xi_n)$ be an estimator of θ having the property that

$$E|\hat{\theta}_n - \theta|^2 \to 0 \text{ as } n \to \infty$$

Is $\hat{\theta}_n$ a consistent estimator of θ as defined in Section 7.2?

7.3. Use the Cramer-Rao inequality to investigate the efficiency of

$$\hat{\theta} = \frac{\xi_1 + \xi_2 + \cdots + \xi_n}{n}$$

as an estimator of the population mean, where $\xi_1, \xi_2, \ldots, \xi_n$ is a sample from a Gaussian distribution.

7.4. Use the random sample $\xi_1, \xi_2, \ldots, \xi_n$ from a Gaussian distribution of mean μ and variance 1 to investigate the sufficiency of the sample mean as an estimator of μ.

7.5. Refer to Example 7.7 and use the method of moments to estimate α and β for the gray value density of type gamma. Assume that the following five gray value runs have been observed:

$$s_1 = (0, 0)$$
$$s_2 = (0)$$
$$s_3 = (1, 2, 3)$$
$$s_4 = (1)$$
$$s_5 = (1, 2)$$

7.6. Let $\xi_1, \xi_2, \ldots, \xi_n$ constitute a random sample of background gray values whose density is exponential with parameter λ. Find the maximum likelihood estimator $\hat{\lambda}$ for λ.

7.7. Assume that

$$f_\xi(t) = \lambda \alpha t^{\alpha-1} e^{-\lambda t^\alpha} I_{(0,\infty)}(t), \text{ for } \lambda, \alpha > 0$$

is the density function of background gray values for a certain class of images. Determine the procedure for finding the maximum likelihood vector estimator for $\begin{pmatrix} \lambda \\ \alpha \end{pmatrix}$.

7.8. When a quadratic loss function is employed, the risk is minimized utilizing the conditional mean. When an absolute value loss function is utilized, what type of estimate minimizes the risk?

7.9. Verify the optimal estimation techniques, their biases, and the corresponding output covariance matrices given in Table 7.1.

7.10. Continue the recursive procedure from Example 7.16 by including a fifth equation, $3 = a + 2b$. Assume that

$$E(\mathbf{uu'}) = \begin{pmatrix} .1 & 0 & 0 & 0 & 0 \\ 0 & .1 & 0 & 0 & 0 \\ 0 & 0 & .2 & 0 & 0 \\ 0 & 0 & 0 & .4 & 0 \\ 0 & 0 & 0 & 0 & \sigma_5^2 \end{pmatrix}$$

and find $\hat{\mathbf{x}}_5$.

7.11. Relate the Kalman-Bucy procedure to the method used in solving Exercise 7.10.

7.12. Repeat Example 7.19 using the two observed images

$$f_1 = \begin{pmatrix} 0 & 1 \\ 0 & * \end{pmatrix}_{0,0}$$

and

$$f_2 = \begin{pmatrix} * & 3 \\ * & 4 \end{pmatrix}_{0,0}$$

7.13. Use the Neyman-Pearson criterion to determine the best critical region, as in Example 7.22. Assume, however, that the class of images possesses gray values that are exponentially distributed.

7.14. Repeat Example 7.24, this time using

$$f_\eta(\theta) = \frac{1}{4}\delta(\theta - 1) + \frac{1}{5}\delta(\theta - 2) + \frac{11}{20}\delta(\theta - 3)$$

7.15. Use the two random images

$$f_1 = (\text{②} \ 4)$$
$$f_2 = (\text{③} \ 5)$$

to find the Hotelling transform corresponding to the random image $F = (A \ B)_{0,0}$.

Appendix

Delta Functions

Generalized functions, in particular the delta function, are rigorously specified using the notion of a *good* (or *Schwarz*) function. A function $g(x)$ is said to be good if it is infinitely differentiable and if

$$\lim_{|x| \to \infty} \left| x^r \frac{d^k g(x)}{dx^k} \right| = 0$$

for all integers $r, k \geq 0$. (Such functions play a role in the inversion of the Radon transform. See Section 4.9.)

Example A.1

Functions such as e^{-x^2}, $x^3 e^{-x^2}$, and e^{-x^4} are good; whereas, e^{-x} is not good since it does not behave properly at $-\infty$. Similarly, the function $g(x) = xe^{-x^2} I_{(0,\infty)}(x)$ is not good since it is not infinitely differentiable. It does not even have a first derivative at the origin.

Sums and products of good functions are good, as are scalar multiples, derivatives, and translates of good functions. Good functions are always bounded. Indefinite integrals of good functions need not be good, as can be seen by observing the Gaussian distribution function.

A sequence $\{g_n(x)\}$ of good functions is said to be *regular* if

$$\lim_{n \to \infty} \int_{-\infty}^{\infty} g_n(x) g(x) \, dx$$

exists for every good function $g(x)$. Moreover, two regular sequences that give the same limit are said to be *equivalent*.

Example A.2

The sequence of "Gaussian density type" good functions

$$g_n(x) = \sqrt{\frac{n}{\pi}} e^{-nx^2}$$

is regular and has the property that

$$\lim_{n \to \infty} \int_{-\infty}^{\infty} g_n(x) g(x) \, dx = g(0)$$

for any good function $g(x)$. To see this, write

$$\int_{-\infty}^{\infty} g_n(x) g(x) \, dx = g(0) \int_{-\infty}^{\infty} g_n(x) \, dx + \int_{-\infty}^{\infty} (g(x) - g(0)) g_n(x) \, dx$$

$$= g(0) + \int_{-\infty}^{\infty} (g(x) - g(0)) g_n(x) \, dx$$

We will show that

$$\lim_{n \to \infty} \int_{-\infty}^{\infty} (g(x) - g(0)) g_n(x) \, dx = 0$$

Since derivatives of good functions are good, and good functions are bounded, there exists M such that $|g'(x)| \leq M$ for all x. It then follows that

$$|g(x) - g(0)| = \left| \int_0^x g'(t) \, dt \right| \leq M |x|$$

As a consequence

$$\left| \int_{-\infty}^{\infty} (g(x) - g(0)) g_n(x) \, dx \right| \leq 2M \sqrt{\frac{n}{\pi}} \int_0^{\infty} x e^{-nx^2} \, dx = \frac{M}{\sqrt{n\pi}}$$

and the result follows.

Example A.3

The sequence of good functions

$$h_n(x) = 1/2 \sqrt{\frac{n}{\pi}} e^{-nx^2/4}$$

is regular and

$$\lim_{n \to \infty} \int_{-\infty}^{\infty} h_n(x) g(x) \, dx = g(0)$$

for any good function $g(x)$. Consequently, the two regular sequences $\{h_n(x)\}$ and $\{g_n(x)\}$, from the previous example, are equivalent.

A *generalized function* is an equivalence class of regular sequences. More-

Appendix Delta Functions

over, if the regular sequence $\{g_n(x)\}$ has the limit

$$\lim_{n \to \infty} \int_{-\infty}^{\infty} g_n(x) \, g(x) \, dx = g(0)$$

for any good function $g(x)$, then the resulting generalized function is called the *delta function*, and it is denoted by $\delta(x)$. We write

$$\int_{-\infty}^{\infty} \delta(x) \, g(x) \, dx = g(0)$$

Example A.4

The regular sequences $\{g_n(x)\}$ and $\{h_n(x)\}$ defined in the previous examples are representatives (from the equivalence class) of the delta function.

Theorem A.1 Suppose $g(x)$ and $h(x)$ are two generalized functions defined, respectively, from the regular sequences $\{g_n(x)\}$ and $\{h_n(x)\}$. Then

(a) $g(x) + h(x)$ is defined by the regular sequence $\{h_n(x) + g_n(x)\}$.
(b) $cg(x)$ is defined by the regular sequence $\{cg_n(x)\}$.
(c) $g'(x)$ is defined by the regular sequence $\{g'_n(x)\}$.
(d) $g(ax + b)$ is defined by the regular sequence $\{g_n(ax + b)\}$.

Example A.5

If $f(x)$ is a good function, then from parts b and d of Theorem A1, it follows that

$$\int_{-\infty}^{\infty} c\delta(x - b) \, f(x) \, dx = cf(b)$$

It is important to identify the support region of a generalized function. The following definition is useful in this respect. Suppose the function $f(x)$ and the generalized function $g(x)$ have the property that, for any good function $h(x)$ with support (a, b),

$$\int_{-\infty}^{\infty} g(x) \, h(x) \, dx = \int_{-\infty}^{\infty} f(x) \, h(x) \, dx$$

Then we say that $f(x) = g(x)$ on (a, b), and conversely. From this definition it follows that, for the delta function, $\delta(x) = 0$ on $(-\infty, 0) \cup (0, \infty)$, and $\delta(x)$ is undefined at the origin. As a consequence,

$$\delta(x - a) = \begin{cases} 0, & \text{if } x < a \\ 0, & \text{if } x > a \\ \text{undefined}, & \text{if } x = a \end{cases}$$

If f is continuous at the origin, then the generalized function $\delta(x)f(x)$ is defined as $f(0) \cdot \delta(x)$. Similarly, if f is continuous at the point a, then $\delta(x - a) f(x)$

is defined as $f(a) \cdot \delta(x - a)$. Whenever $f(x)$ is continuous at the point a,

$$\int_c^b \delta(x - a) f(x) \, dx = \begin{cases} f(a), & \text{if } c < a < b \\ 0, & \text{if } a < c \\ 0, & \text{if } b < a \end{cases}$$

Delta functions are used extensively in Chapter 6 for describing the probability density function in the discrete case. Since

$$\int_c^x b\delta(t - a) \, dt = \begin{cases} b, & \text{if } c < a < x \\ 0, & \text{if } a < c \end{cases}$$

whenever a probability distribution function has a jump discontinuity at a point a with saltus b, the delta function $b\delta(x - a)$ is used as a generalized derivative (at that point).

Example A.6

If

$$F(x) = 1/2 \, I_{[0,2)}(x) + I_{[2,\infty)}(x)$$

then

$$f(x) = 1/2 \, \delta(x) + 1/2 \, \delta(x - 2)$$

is the generalized derivative of $F(x)$. Indeed,

$$\int_{-\infty}^x f(t) \, dt = \begin{cases} 0, & \text{if } x < 0 \\ 1/2, & \text{if } 0 < x < 2 \\ 1, & \text{if } 2 < x \end{cases}$$

Notice that $dF(x)/dx = f(x)$ everywhere except at $x = 0$ and $x = 2$.

The use of delta functions, as in Example A.6, can be employed to provide an operational calculus technique. Correct results are obtained, using this technique, for the case when the probability distribution function has no singular part, and the discontinuous part has no finite point as a limit point of discontinuities. Alternatively, Stieltjes integrals could be employed to rigorously obtain the same results.

Of special interest is the delta function in several variables. For convenience, we write

$$\delta(x_1 - a_1, \ldots, x_n - a_n, y_1 - b_1, \ldots, y_m - b_m)$$
$$= \delta(x_1 - a_1, \ldots, x_n - a_n)\delta(y_1 - b_1, \ldots, y_m - b_m)$$

and this function is zero everywhere except on the lines $x_i = a_i, i = 1, \ldots, n$, and $y_j = b_j, j = 1, \ldots, m$. Formal manipulation is carried out similarly to the one-dimensional case.

Example A.7

$$\int_{-\infty}^{\infty} \int_{-\infty}^{\infty} \delta(x - 3, y - 4) x^2 y^5 \, dx \, dy$$

$$= \int_{-\infty}^{\infty} \delta(x - 3) x^2 \, dx \int_{-\infty}^{\infty} \delta(y - 4) y^5 \, dy$$

$$= 3^2 \cdot 4^5$$

Example A.8

$$I = \int_{-\infty}^{\infty} \int_{-\infty}^{\infty} \delta(x - 3, y - 4) x \sin xy \, dx \, dy$$

$$= 3 \sin 12$$

Under certain conditions, the division of delta functions by other delta functions can be given meaning. Such is the case in Section 6.9, where a conditional discrete density is defined as the ratio of a joint density by a marginal density. In this application, nonzero conditional probabilities are obtained by utilizing those points for which the denominator is nonzero. More specifically, when delta functions are employed in the denominator, the points at which the delta functions "occur" are used. At these same locations, there will be delta functions in the numerator and, consequently, a cancellation should be performed.

Example A.9

Suppose

$$g(x, y) = \frac{1/2 \, \delta(x, y - 1) + 1/3 \, \delta(x, y - 2) + 1/6 \, \delta(x - 1, y - 1)}{5/6 \, \delta(x) + 1/6 \, \delta(x - 1)}$$

then

$$g(x, y) = \frac{1/2 \, \delta(x) \delta(y - 1) + 1/3 \, \delta(x) \delta(y - 2) + 1/6 \, \delta(x - 1) \delta(y - 1)}{5/6 \, \delta(x) + 1/6 \, \delta(x - 1)}$$

The denominator of g is zero everywhere except at $x = 0$ and $x = 1$. At $x = 0$, the terms $1/6 \, \delta(x - 1)$ and $1/6 \, \delta(x - 1) \delta(y - 1)$ are both zero; hence, at $x = 0$,

$$g(0, y) = \frac{1/2 \, \delta(x) \delta(y - 1) + 1/3 \, \delta(x) \delta(y - 2)}{5/6 \, \delta(x)}$$

Formally cancelling out $\delta(x)$ gives

$$g(0, y) = \frac{1/2 \, \delta(y - 1) + 1/3 \, \delta(y - 2)}{5/6}$$

$$= 3/5 \, \delta(y - 1) + 2/5 \, \delta(y - 2)$$

In a similar way

$$g(1, y) = \frac{1/6 \, \delta(x-1) \, \delta(y-1)}{1/6 \, \delta(x-1)}$$

$$= \delta(y-1)$$

These formal manipulations can be rigorously justified with the aid of Stieltjes-type integrals.

Bibliography

CHAPTER 2

AGGARWAL, J. K., R. O. DUDA, and A. ROSENFELD, eds. *Computer Methods in Image Analysis*. New York: IEEE Press, 1977.

AGIN, G. J., and T. I. BINFORD. (1973) "Computer Description of Curved Objects," *Proc. of the Intern. Joint Conf. on Artificial Intelligence*, Stanford, California (August 20–23, 1973): 629–640.

AHUJA, N., and B. J. SCHACHTER. *Pattern Models*. New York: John Wiley & Sons, 1983.

ANDREWS, H. C. *Computer Techniques in Image Processing*. New York: Academic Press, 1970.

ANDREWS, H. C., ed. *Digital Image Processing*. New York: IEEE Press, 1978.

ARCELLI, C. "Pattern Thinning by Contour Tracing," *Computer Graphics and Image Processing*, Vol. 17, No. 3 (October 1981): 130–144.

BAJCSY, R. "Computer Identification of Visual Surfaces," *Computer Graphics and Image Processing*, Vol. 2, No. 2 (October 1973): 118–130.

BAJCSY, R. "Three-Dimensional Scene Analysis," *Pro. Pattern Recognition Conf.*, Miami, Florida (December 1–4, 1980): 1064–1074.

BALLARD, D. H., and C. M. BROWN. *Computer Vision*. Englewood Cliffs, NJ: Prentice-Hall, Inc., 1982.

BALLARD, D. H., and D. SABBAH. "On Shapes," *Proc. of the Intern. Joint Conf. on Artificial Intelligence*, Vancouver, B.C. (August 24–28, 1981): 607–612.

BARROW, H. G., and J. M. TENENBAUM. "Recovering Intrinsic Scene Characteristics from Images," *Computer Vision Systems*, A. R. Hanson and E. M. Riseman, eds. New York: Academic Press, 1978.

BARROW, H. G., and J. M. TENENBAUM. "Computational Vision," *Proc. of the IEEE*, Vol. 69, No. 5 (May 1981): 572–595.

BERNSTEIN, R., ed. *Digital Image Processing for Remote Sensing*. New York: IEEE Press, 1978.

BERZINS, V. (1984) "Accuracy of Laplacian Edge Detectors," *Computer Vision, Graphics and Image Processing*, Vol. 27, No. 2 (August 1984): 195–210.

BINFORD, T. O. "Visual Perception by Computer," *Proc. IEEE Systems Science and Cybernetics Conf.*, Miami (December 1971).

BLAHUT, R. E. *Fast Algorithms for Digital Signal Processing*. Reading, MA: Addison-Wesley, 1985.

BRADY, J. M., ed. *Computer Vision*. Amsterdam: North-Holland Publishing Co., 1981.

BRADY, J. M., and R. PAUL, eds. *Robotics Research: The First International Symposium*. Cambridge, MA: MIT Press, 1984.

BRICE, C. R., and C. L. FENNEMA. "Scene Analysis Using Regions," *Artificial Intelligence*, Vol. 1, No. 3 (Fall 1970): 205–226.

CANNY, J. "Finding Edges and Lines in Images," *MIT AI Laboratory Technical Report* 720 (June 1983).

CASTLEMAN, K. R. *Digital Image Processing*. Englewood Cliffs, NJ: Prentice-Hall, Inc., 1979.

CORNSWEET, T. N. *Visual Perception*. New York: Academic Press, 1970.

DAVIS, L. S. "A Survey of Edge Detection Techniques," *Computer Graphics and Image Processing*, Vol. 4, No. 3 (September 1975): 248–270.

DODD, G. G., and L. ROSSOL, eds. *Computer Vision and Sensor-Based Robots*. New York: Plenum Press, 1979.

DOUGHERTY, E. R., and GIARDINA, C. R. *Matrix-Structured Image Processing*. Englewood Cliffs, NJ: Prentice-Hall, 1987.

DUDA, R. O., and P. E. HART. *Pattern Classification and Scene Analysis*. New York: John Wiley & Sons, 1973.

FAUGERAS, O. D., ed. *Fundamentals in Computer Vision*. Cambridge: Cambridge Univ. Press, 1983.

FRAM, J. R., and E. S. DEUTSCH. "On the Evaluation of Edge Detection Schemes and Their Comparison with Human Performance," *IEEE Trans. on Computers*, Vol. 24, No. 6 (June 1975): 616–628.

FREEMAN, H. "Techniques for the Digital Computer Analysis of Chain-Encoded Arbitrary Plane Curves," *Proc. National Electronics Conf.* Vol. 17 (Oct. 9–11, 1960): 421–432.

FREUDER, E. C. "On the Knowledge Required to Label a Picture Graph," *Artificial Intelligence*, Vol. 15, Nos. 1 & 2 (November 1980): 1–17.

GARDNER, W. E., ed. *Machine-Aided Image Analysis, 1978*. Bristol & London: The Institute of Physics, 1979.

GONZALEZ, R. C., and P. WINTZ. *Digital Image Processing*. Reading, MA: Addison-Wesley, 1977.

GREEN, W. B. *Digital Image Processing: A Systems Approach*. New York: Van Nostrand Reinhold Co., 1983.

GRIFFITH, A. K. "Edge Detection in Simple Scenes Using *A Priori* Information," *IEEE Trans. on Computers*, Vol. 20, No. 4 (April 1973): 371–381.

GRUEN, A. W. "Adaptive Least Squares Correlation—A Powerful Image Matching Technique," *Proc. ACSM-ASP Convention*, Washington, D.C. (March 1985).

GUPTA, J. N., and P. A. WINTZ. "A Boundary Finding Algorithm and Its Application," *IEEE Trans. on Circuits and Systems*, Vol. 22, No. 4 (April 1975): 351–362.

HABIBI, A. "Two Dimensional Bayesian Estimation of Images," *Proc. of the IEEE*, Vol. 60, No. 7 (July 1972): 878–883.

HALL, E. *Computer Image Processing and Recognition*. New York: Academic Press, 1979.

HANSON, A. R., and E. M. RISEMAN, eds. *Computer Vision Systems*. New York: Academic Press, 1978.

HARALICK, R. M. "Edge and Region Analysis for Digital Image Data," *Computer Graphics and Image Processing*, Vol. 12, No. 1 (January 1980): 60–73.

HERMAN, G. T., ed. *Image Reconstruction from Projections—Implementation and Applications*. New York: Springer-Verlag, 1979.

HILDRETH, E. C. *The Measurement of Visual Motion*. Cambridge, MA: MIT Press, 1983.

HORN, B. K. P., and E. J. WELDON. "Filtering Closed Curves," *Proc. Computer Vision and Pattern Recognition Conf.* San Francisco, California (June 19–23, 1985): 478–484.

HUANG, T. S., ed. *Image Sequence Processing and Dynamic Scene Analysis*. New York: Springer-Verlag, 1983.

HUANG, T. S., W. F. SCHREIBER, and O. J. TRETIAK. "Image Processing," *Proc. of the IEEE*, Vol. 59, No. 11 (November 1971): 1586–1609.

HUECKEL, M. "An Operator Which Locates Edges in Digital Pictures," *Journal of the ACM*, Vol. 18, No. 1 (January 1971): 113–125.

HUECKEL, M. "A Local Visual Operator Which Recognizes Edges and Lines," *Journal of the ACM*, Vol. 20, No. 4 (October 1973): 634–647.

JACOBUS, C. J., and R. T. CHIEN. "Two New Edge Detectors," *IEEE Trans. on Pattern Analysis and Machine Intelligence*, Vol. 2, No. 5 (September 1981): 581–592.

KANAL, L. N., ed. *Pattern Recognition*. Washington, D.C.: Thompson Book Co., 1980.

LEVINE, M. D. *Vision in Man and Machine*. New York: McGraw-Hill Book Co., 1985.

MARR, D., and E. HILDRETH. "Theory of Edge Detection," *Proc. of the Royal Society of London B*, Vol. 207 (1980): 187–217.

MODESTINO, J. W., and R. W. FRIES. "Edge Detection in Noisy Images Using Recursive Digital Filtering," *Computer Graphics and Image Processing*, Vol. 6, No. 5 (October 1977): 409–433.

NEVATIA, R. *Machine Perception*. Englewood Cliffs, NJ: Prentice-Hall, Inc., 1982.

NORTON, H. N. *Sensor and Analyzer Handbook*. Englewood Cliffs, NJ: Prentice-Hall, Inc., 1982.

OPPENHEIM, A. V., and A. S. WILLSKY. *Signals and Systems*. Englewood Cliffs, NJ: Prentice-Hall, Inc., 1983.

PAVLIDIS, T. "An Asynchronous Thinning Algorithm," *Computer Graphics and Image Processing*, Vol. 20, No. 2 (October 1982): 133–157.

PRATT, W. *Digital Image Processing*. New York: John Wiley & Sons, 1978.

ROSENFELD, A., ed. *Digital Picture Analysis*. New York: Springer-Verlag, 1976.

ROSENFELD, A., and A. C. KAK. *Digital Picture Processing*, Vols. 1 & 2, 2d ed. New York: Academic Press, 1982.

Rosenfeld, A. "Connectivity in Digital Pictures," *Journal of the ACM*, Vol. 17, No. 1 (January 1970): 146–160.

Schalkoff, R. J., and E. S. McVey. "A Model and Tracking Algorithm for a Class of Video Targets," *IEEE Trans. on Pattern Analysis and Machine Intelligence*, Vol. 4, No. 1 (January 1982): 2–10.

Scharf, D. *Magnifications—Photography with the Scanning Electron Microscope*. New York: Schocken Books, 1977.

Serra, J. *Image Analysis and Mathematical Morphology*. Boston, MA: Academic Press, 1985.

Shafer, S. A. *Shadows and Silhouettes in Computer Vision*. Boston, MA: Academic Press, 1985.

Stoffel, J. C., ed. *Graphical and Binary Image Processing and Applications*. Massachusetts: Artech House, Inc., 1982.

Stucki, P., ed. *Advances in Digital Image Processing: Theory, Application, Implementation*. New York: Plenum Press, 1979.

Sugihara, K. "Mathematical Structures of Line Drawings of Polyhedrons—Toward Man-Machine Communication by Means of Line Drawings," *IEEE Trans. on Pattern Analysis and Machine Intelligence*, Vol. 4, No. 5 (September 1982): 458–469.

Tanimoto, S., and A. Klinger, eds. *Structured Computer Vision: Machine Perception through Hierarchical Computation Structures*. New York: Academic Press, 1980.

Ullman, S. *The Interpretation of Visual Motion*. Cambridge, MA: MIT Press, 1979.

Ullman, S., and W. Richards, eds. *Image Understanding 1984*. Norwood, NJ: Ablex Publishing Corp., 1984.

Wojcik, Z. M. "An Approach to the Recognition of Contours and Line-Shaped Objects," *Computer Vision, Graphics and Image Processing*, Vol. 25, No. 2 (February 1984): 184–204.

Woodham, R. J. "Analysing Images of Curved Surfaces," *Artificial Intelligence*, Vol. 17, Nos. 1–3 (August 1981): 117–140.

Yakimovsky, Y. "Boundary and Object Detection in Real World Images," *Proc. of the Intern. Joint Conf. on Artificial Intelligence*, Tbilisi, Georgia, U.S.S.R. (September 3–8, 1975): 695–704.

CHAPTER 3

Beucher, S. "Random Processes Simulations on the Texture Analyser," *Lecture Notes in Biomathematics*, No. 23 (1977): Springer-Verlag.

Beucher, S., and Lantuejoul, Ch. "Use of Watersheds in Contour Detection," *Int. Workshop on Image Processing*, CCETT. Rennes, France (1979).

Blaschke, W. V. "Vorlesungen über Integral Geometrie," Teubner, Leipzig (1936).

Bookstein, F. L., ed. "The Measurement of Biological Shape and Shape Change," *Lecture Notes in Biomathematics*, No. 24, Springer-Verlag, Berlin, Heidelber, New York (1978).

Calabi, L. "A Study of the Skeleton of Plane Figures," *Parke Math. Lab, Inc.* (1965).

COLEMAN, R. "An Introduction to Mathematical Stereology," *Memoirs n° 3*, Department of theoretical statistics, Univ. of Aarhus (1979).

CROFTON, M. W. "On the Theory of Local Probability, Applied to Straight Lines Drawn at Random in a Plane, the Method Used Being Also Extended to the Proof of Certain New Theorems in the Integral Calculus," *Philos. Trans. Roy. Soc., London*, 158 (1868): 181–189.

CRUZ-ORIVE, L. M. "Particle Size-Shape Distribution: The General Spheroid Problem II, Stochastic Model and Practice Guide," *J. of Micro.*, 112, part 2 (1978): 153–168.

DOUGHERTY, E. R., and GIARDINA, C. R. *Matrix-Structured Image Processing*. Englewood Cliffs, N.J: Prentice-Hall, Inc., 1987.

DOUGHERTY, E. R., and GIARDINA, C. R. "Error Bounds for Morphologically Derived Feature Measurements," *SIAM Journal on Applied Mathematics*. In Press.

DOUGHERTY, E. R., and GIARDINA, C. R. "Binary Euclidean Images and Convergence," *1986 Conference on Intelligent Systems and Machines*, Published in Conference Proceedings, Oakland Univ., Rochester, Michigan.

DOUGHERTY, E. R., and GIARDINA, C. R. "Bound Matrix Structured Image Processing," *1986 Conference on Intelligent Systems and Machines*, Published in Conference Proceedings, Oakland Univ., Rochester, Michigan.

DOUGHERTY, E. R., and GIARDINA, C. R. "Sampling Criteria for Euclidean Images," *39th Annual Conference, SPSE*, Minnesota (May 1986).

DOUGHERTY, E. R., and GIARDINA, C. R. "A Digital Version of Matheron's Theorem for Increasing Mappings in Terms of a Basis for the Kernel," *IEEE Computer Vision and Pattern Recognition*, Miami (June 1986).

DUFF, M. J. B., and WATSON, D. M. "The Cellular Logic Array Image Processor," *Dept. of Physics and Astronomy*, Univ. College, London (1974).

ELIAS, H., and WEIBEL, E. R. *Quantitative Methods in Morphology*, Springer-Verlag, Berlin (1967).

FEDERER, H. *Geometric Measure Theory*. Berlin: Springer-Verlag, 1969.

FLOOK, A. G. "The Use of Dialtion Logic on the Quantimet to Achieve Fractal Dimension Characterisation of Textured and Structured Profiles," *Powder Technology*, 21 (1978): 295–298.

GOETCHARIAN, V. "From Binary to Grey Level Tone Image Processing by Using Fuzzy Logic Concepts," *Pattern Recognition*, Vol. 12 (1980): 7–15.

GOETCHERIAN, V. "Parallel Image Processes and Real-Time Texture Analysis," Thesis Doctor of Philosophy, Univ. College, London (1980).

GOLAY, M. J. E. "Hexagonal Parallel Pattern Transformation," *IEEE Trans. Comput.*, C-18 (1969): 733–740.

HADWIGER, H. *Vorslesungen Über Inhalt, Oberfläche and Isoperimetrie*. Berlin: Springer-Verlag, 1957.

KENDALL, M. G. G., and MORAN, P. A. P. *Geometrical Probability*. London: Griffin, 1963.

KENDALL, D. G. "Foundations of a Theory of Random Sets," in *Stochastic Geometry*, E. F. Harding and D. G. Kendall, eds. New York: John Wiley & Sons, 1974.

KLEIN, J. C., and SERRA, J. "The Texture Analyser," *J. of Micr.*, 95, part 2 (April 1973): 349–356.

Lay, Steven. *Convex Sets and Their Applications.* New York: John Wiley & Sons, 1982.

Mandelbrot, B. B. *Fractals: Form, Chance, Dimension.* San Francisco & London: W. H. Freeman & Co., 1977.

Matheron, G. "Eléments pour une Théorie des Milieux Poreux," Masson, Paris (1967).

Matheron, G. "Random Sets Theory and Its Applications to Stereology," *J. of Micr.*, 95, part 1 (Feb 1972): 15–23.

Matheron, G. *Random Sets and Integral Geometry.* New York: John Wiley & Sons, 1975.

Matheron, G. "La Formule de Crofton pour les Sections Épaisses," *J. Appl. Prob.*, 13 (1976): 707–713.

Miles, R. E., and Serra, J., eds. "Geometrical Probabilities and Biological Structures," *Lecture Notes in Biomathematics*, No. 23 (1978): Springer-Verlag.

Minkowski, H. "Volumen and Oberfläche," *Math. Ann.*, Vol. 57 (1903): 447–495.

Moran, P. A. P. "The Probabilistic Basis of Stereology," *Suppl. Adv. Appl. Prob.* 69–91.

Santalo, L. A. *Introduction to Integral Geometry.* Paris: Hermann, 1953.

Santalo, L. A. *Integral Geometry and Geometric Probability.* Reading, MA: Addison-Wesley, 1976.

Serra, J. *Image Analysis and Mathematical Morphology.* New York: Academic Press, 1983.

Underwood, E. E. *Quantitative Stereology.* Reading, MA: Addison-Wesley, 1970.

Valentine, F. A. *Convex Sets.* New York: McGraw-Hill Book Co. 1964.

Watson, G. "Mathematical Morphology," Tech. Report No. 21, Dept. of Stat., Princeton Univ., New Jersey (1973)

Weibel, E. R. "Practical Methods for Biological Morphometry, Vol. 1: Stereological Methods," New York: Academic Press, 1980.

CHAPTER 4

Akhiezer, N. I. *Theory of Approximation.* New York: Frederick Ungar Publishing Co., 1956.

Ash, J. M., ed. *Studies in Harmonic Analysis.* New York: The Mathematical Association of America, 1976.

Bachman, G. *Elements of Abstract Harmonic Analysis.* New York: Academic Press, 1964.

Bochner, S., and K. Chandrasekharan. *Fourier Transforms.* Princeton, NJ: Princeton Univ. Press, 1949.

Doetsch, G. *Guide to the Applications of the Laplace and Z-Transforms.* New York: Van Nostrand Reinhold Co., 1971.

Goldberg, R. R. *Fourier Transforms.* Cambridge: Cambridge Univ. Press, 1965.

Goldberg, S. *Unbounded Linear Operators: Theory and Applications.* New York: McGraw-Hill Book Co., 1966.

Halmos, P. R. *Measure Theory.* Princeton, NJ: Van Nostrand Reinhold Co., 1966.

Helgason, S. *The Radon Transform.* Boston, MA: Birkhausser, 1980.

KOOPMANS, L. H. *The Spectral Analysis of Time Series*. New York: Academic Press, 1974.

LIGHTHILL, M. J. *Introduction to Fourier Analysis and Generalised Functions*. Cambridge: Cambridge Univ. Press, 1962.

LOGAN, B. "Hillert Transform of a Function Having a Bounded Integral and a Bounded Derivative," *SIAM Jour. Math Analysis*, Vol. 14, No. 2 (1983).

MUNROE, M. E. *Measure & Integration*, 2d ed. Reading, MA: Addison-Wesley, 1971.

NATANSON, I. P. *Constructive Function Theory*. New York: Frederick Ungar Publishing Co., 1964.

NATANSON, I. P. *Constructive Function Theory*, Vol. 2. New York: Frederick Ungar Publishing Co., 1965.

NATANSON, I. P. *Constructive Function Theory*, Vol. 3. New York: Frederick Ungar Publishing Co., 1965.

NATANSON, I. P. *Theory of Functions of a Real Variable*, Vol. I. New York: Frederick Ungar Publishing Co., 1964.

NATANSON, I. P. *Theory of Functions of a Real Variable*, Vol. II. New York: Frederick Ungar Publishing Co., 1967.

RIVLIN, T. J. *An Introduction to the Approximation of Functions*. New York: Dover Publications, Inc., 1969.

ROYDEN, H. L. *Real Analysis*, 2d ed. New York: Macmillan Co., 1968.

RUDIN, W. *Real and Complex Analysis*. New York: McGraw-Hill Book Co., 1966.

TITCHMARSH, E. C. *Introduction to the Theory of Fourier Integrals*. London: Clarendon Press, 1962.

WIDDER, D. V. *The Laplace Transform*. Princeton, NJ: Princeton Univ. Press, 1946.

CHAPTER 5

AKHIEZER, N. I., and I. M. GLAZMAN. *Theory of Operators in Hilbert Space*, Vol. 1. New York: Frederick Ungar Publishing Co., 1966.

AKHIEZER, N. I., and I. M. GLAZMAN. *Theory of Operators in Hilbert Space*, Vol. 2. New York: Frederick Ungar Publishing Co., 1963.

ALEXITS, G. *Convergence Problems of Orthogonal Series*. New York: Pergamon Press, 1961.

COURANT, R., and D. HILBERT. *Methods of Mathematical Physics*, Vol. 1. New York: Interscience Publishers, Inc., 1953.

DOUGHERTY, E., and C. R. GIARDINA. *Matrix Structured Image Processes*. Englewood Cliffs, NJ: Prentice-Hall, Inc., 1986.

DUNFORD, N., and J. T. SCHWARTZ. *Linear Operators: Part I: General Theory*. New York: John Wiley & Sons, 1976.

DYM, H., and H. P. MCKEAN. *Fourier Series and Integrals*. New York: Academic Press, 1972.

FINE, N. J. "On the Walsh Functions," *American Math. Society* 65 (1949)

GIARDINA, C. R. "Bounds on the Truncation Error for Walsh Expansions," *Notice AMS*, Vol. 25, No. 2 (1978).

GOFFMAN, C., and G. PEDRICK. *First Course in Functional Analysis*. Englewood Cliffs, NJ: Prentice-Hall, Inc. 1965.

HALMOS, P. R. *Introduction to Hilbert Space and the Theory of Spectral Multiplicity*. New York: Chelsea Publishing Co., 1957.

HOBSON, E. W. *The Theory of Functions of a Real Variable and the Theory of Fourier's Series*. New York: Dover Publications, Inc., 1957.

KREYSZIG, E. *Introductory Functional Analysis with Applications*. New York: John Wiley & Sons, 1978.

KUHL, F., and C. R. GIARDINA. "Elliptic Fourier Features of a Closed Contour," *Computer Graphics and Image Processing*, 18 (1982).

ZYGMUND, A. *Trigonometrical Series*. New York: Dover Publications, 1955.

CHAPTER 6

ASH, R. B. *Basic Probability Theory*. New York: John Wiley & Sons, 1970.

BHARUCHA-REID, A. T. *Elements of the Theory of Markov Processes and Their Applications*. New York: McGraw-Hill Book Co., 1960.

BHATTACHARYA, R. N., and R. R. RAO. *Normal Approximation and Asymptotic Expansions*. New York: John Wiley & Sons, 1976.

COX, D. R., and H. D. MILLER. *The Theory of Stochastic Processes*. New York: John Wiley & Sons, 1968.

CRAMER, J., and M. R. LEADBETTER. *Stationary and Related Stochastic Processes*. New York: John Wiley & Sons, 1967.

DAVID, H. A. *Order Statistics*. New York: John Wiley & Sons, 1970.

DOOB, J. L. *Stochastic Processes*. New York: John Wiley & Sons, 1967.

FELLER, W. *An Introduction to Probability Theory and Its Applications*, Vol. 1, 3d ed. New York: John Wiley & Sons, 1957.

FELLER, W. *An Introduction to Probability Theory and Its Applications*, Vol. 2. New York: John Wiley & Sons, 1966.

FREUND, J. E. *Mathematical Statistics*, 2d ed. Englewood Cliffs, NJ: Prentice-Hall, Inc., 1971.

HOEL, P. G., PORT, S. C., and C. J. STONE. *Introduction to Probability Theory*. Boston, MA: Houghton Mifflin Co., 1971.

HOEL, P. G., PORT, S. C., and C. J. STONE. *Introduction to Stochastic Processes*. Boston, MA: Houghton Mifflin Co., 1972.

KHINCHIN, A. I. *Mathematical Foundations of Information Theory*. New York: Dover Publications, Inc., 1957.

LOEVE, M. *Probability Theory I*, 4th ed. New York: Springer-Verlag, 1977.

LOEVE, M. *Probability Theory II*, 4th ed. New York: Springer-Verlag, 1978.

NEUTS, M. F. *Probability*. Boston, MA: Allyn & Bacon, Inc., 1973.

PARZEN, E. *Modern Probability Theory and Its Applications*. New York: John Wiley & Sons, 1960.

PARZEN, E. *Stochastic Processes*. San Francisco: Holden-Day, Inc., 1962.

ROZANOV, Y. A. *Introductory Probability Theory*. Englewood Cliffs, NJ: Prentice-Hall, Inc., 1969.

YAGLOM, A. M. *Stationary Random Functions*. New York: Dover Publications, Inc. 1962.

CHAPTER 7

BELLMAN, R. *Introduction to Matrix Analysis*. New York: McGraw-Hill Book Co., 1970.

BOULLION, T. L., and P. L. ODELL. *Generalized Inverse Matrices*. New York: Wiley-Interscience, 1971

BRUNK, H. D. *An Introduction to Mathematical Statistics*. Waltham, MA: Blaisdell Publishing Co., 1965.

BUCY, R. S., and P. D. JOSEPH. *Filtering for Stochastic Processes with Applications to Guidance*. New York: Interscience Publishers, 1968.

CONOVER, W. J. *Practical Nonparametric Statistics*. New York: John Wiley & Sons, 1971.

CRAMER, H. *Mathematical Methods of Statistics*. Princeton, NJ: Princeton Univ. Press, 1946.

DEGROOT, M. H. *Optimal Statistical Decisions*. New York: McGraw-Hill Book Co., 1970.

FISZ, M. *Probability Theory and Mathematical Statistics*. New York: John Wiley & Sons, 1963.

GELB, A., ed. *Applied Optimal Estimation*. Cambridge, MA: MIT Press, 1974.

GHOSH, B. K. *Sequential Tests of Statistical Hypothesis*. Reading, MA: Addison-Wesley, 1970.

HALMOS, P. R. *Finite-Dimensional Vector Spaces*. Princeton, NJ: Van Nostrand Reinhold Co., 1958.

HOEL, P. G. *Introduction to Mathematical Statistics*. New York: John Wiley & Sons, 1962.

HOGG, R., and A. CRAIG. *Introduction to Mathematical Statistics*. London: Macmillan Co., 1970.

KENDALL, M., and A. STUART. *The Advanced Theory of Statistics, Vol. 1*, 4th ed. New York: Macmillan Co., 1977.

KENDALL, M., and A. STUART. *The Advanced Theory of Statistics, Vol. 3*, 3d ed. New York: Hafner Press, 1976.

LEWIS, T. O., and P. L. ODELL. *Estimation in Linear Models*. Englewood Cliffs, NJ: Prentice-Hall, Inc., 1971.

MOOD, A. M., GRAYBILL, F. A., and D. C. BOES. *Introduction to the Theory of Statistics*, 3rd ed. New York: McGraw-Hill Book Co., 1963.

ROZANOV, Y. A. *Innovation Processes*. New York: John Wiley & Sons, 1977.

WALD, A. *Sequential Analysis*. New York: Dover Publications, Inc., 1973.

WASAN, M. T. *Parametric Estimation*. New York: McGraw-Hill Book Co., 1970.

ZEHNA, P. W. *Probability Distributions and Statistics*. Boston, MA: Allyn & Bacon, Inc., 1970.

Index

Abel transform, 213
ABS, 49
Absolutely continuous, 316
Absolutely convergent, 246, 320
Absolute value loss function, 388
Accept (hypothesis), 411
Acceptable estimator, 376
Acceptance region, 412
ADD, 19, 48
Additive inverse, 245
Additivity, 126, 306
Affine function, 330
Algebraic closing, 159
Algebraic opening, 159
Almost everywhere, 193
Alphabet, 277
Alternative hypothesis, 411
Analog image, 183
Antiextensive, 108
A posteriori probability, 310
A priori probability, 310
Arc length, 279
Area, 120, 124
Area vector, 269
Associativity, 36, 94, 193, 200, 245
Asymptotically unbiased, 377
AVER, 35
Average, 320
Averaging mask, 40

Back projection, 217
Baire function, 321, 328
Bandlimited, 235
Barycentic coordinates, 115

Basis, 158, 160
Bayesian estimation, 389
Bayes risk, 389
Bayes theorem, 309
B-closed, 110
Best approximation, 242
Best critical region, 414
Best-fit plane, 70
Best-fit quadric surface, 81
Best least-squares approximation, 242
Best linear unbiased estimator (BLUE), 378
Best test, 414
Beta density, 329
Beta function, 212, 328
Better estimator, 378
BETWEEN, 34
Bias, 377
Biased estimator, 377
Bilateral Laplace transform, 210
Bimodal, 325
Binary operation, 20
Binomial density, 328
Bivariate normal density, 361
BLUE, 378, 394
B-open, 110
Borel set, 312, 335
Borel field, 312
Bound matrix, 11
Bounded set, 115
Bounded variation, 197
Bromwich integral formula, 209, 210

C-additivity, 126
Cantor intersection theorem, 129

Caratheodory's theorem, 114
CARD, 30
Cauchy density function, 317
Cauchy distribution function, 317
Cauchy principle value (CPV), 197
Cauchy projection theorem, 132
Cauchy-Schwarz inequality, 38, 247
Cauchy sequence, 248
Center of gravity, 320
Central moments, 347
Central n^{th} moment, 321
Chain code, 277
Characteristic function, 322
Chebyshev inequality, 323
Circular symmetry, 201
Clipping circuit, 333
Closed set, 115
Closed with respect to set B, 109
Closing, 105, 145
Closure, 115
Commutativity, 20, 94, 193, 200, 245
Commuting diagram, 233
Compact set, 115
Compatible with translation, 152
Complement, 96
Complete, 256
Complex trigonometric system, 251
Compression, 10, 242, 423
Computer tomography, 217
Computer vision system, 2
Concatenation, 277
Conditional density function, 341
Conditional distribution, 341
Conditional mean, 390
Conditional moments, 344
Conditional probability, 309
Consistent, 377
CONST, 30
Constant image, 139
Constant random variable, 312
Continuous from above, 129
Continuous functional, 127
Continous random variable, 316
Convex, 113
Convex combination, 113
Convex hull, 113
Convex ring, 133
Convolution, 193, 199, 206
Convolutional formula, 355
Convolutional method of inversion, 234
Convolution theorem, 209
Co-occurrence matrix, 79
Coordinate system, 241

Correlation coefficient, 348
Countable additivity, 306
Covariance, 137, 348
Covariance matrix, 350
Covariance update, 401
Cramer-Rao inequality, 379
CREATE, 29
Critical region, 412

Decision rule, 411
Decreasing, 100
Delta function, 316, 433
Density function, 318, 338
Dependent, 311, 345
Design matrix, 393, 403
Detection theory, 410
Deterministic image, 307
Difference operator, 52
Digital linear granulometric size distribution, 149
Digitization, 8, 139, 171
Digitized image, 4
Dilation, 99, 141
Dini's test, 264
DIRECT, 61
Directional derivative, 61
Discrete amplitude spectrum, 262
Discrete density, 317
Discrete Fourier transform, 275
Discrete spectrum, 262
Distribution-free estimation, 371
Distribution function, 313, 342
Distributivity, 103, 245
DIV, 22
DOMAIN, 29
Dominant image, 27
DOT, 35
Dot image, 292
Duality, 96
Dual mapping, 155
DX, 52
DY, 52

Edge image, 69
Eigenfunction, 192
Eigenvalue, 420
Eigenvector, 420
Ellipse of concentration, 363
Elliptical radii, 366
Empirical mean, 375
Energy signal, 198
ε-net, 292
Enhancement, 5
EQUAL, 34

Index

Erosion, 99, 143
Error, 139, 378, 412
Error covariance extrapolation, 404
Error covariance update equation, 405
Error of nondetection, 412
Estimate, 372
Estimation, 371, 372
Estimator, 372
Euclidean granulometry, 166
Euler formula, 258
Even signal, 206
Event, 305
Expected value, 321, 346
Exponential density, 329
EXTADD, 28, 48
EXTEND, 27
Extensive, 109
EXTMAX, 28
EXTMIN, 28
EXTMULT, 28
Extrapolation equation, 404
EXTSCALAR, 48

Factorial moment, 323
False-alarm-type error, 412
Feature vector, 6, 410
FILTER, 41
First-order central moments, 347
FLIP, 25
Fourier coefficient, 241
Fourier cosine transform, 207
Fourier series, 261
Fourier sine transform, 207
Fourier transform, 187, 322
Fractional integration, 213
Frequency, 187
Frequency spectrum, 187
Fuzzy measure, 133

Gamma density, 327, 329
Gamma function, 210, 212, 327
Gaussian density, 319, 329
Gauss inequality, 326
Generalized Fourier transform, 208, 210
Generalized function, 432
Generator, 169
Geometric density, 322, 328
Good function, 431
GRAD, 53
GRADEDGE0, 65
GRADEDGE1, 65
GRADEDGE2, 65
Gradient, 53

Gradient filter, 53
GRADMAG0, 55
GRADMAG1, 54
GRADMAG2, 54
Granulometry, 164
Gray level density function, 333
Gray value distribution, 314
GREATER, 34

Hadamard matrix, 269
Hadwiger's theorem, 133
Hankel transform, 202
Harmonic ellipsi, 289
Hausdorff metric, 127
Higher-order moments, 323
High-pass filter, 43
Hilbert space, 204, 249
Hilbert transform, 213
HIST, 75
Histogram, 76
Histogram equalization, 77
Holder condition, 264, 294
Homogeneity, 101, 125
Homogenous, 188
Horizontal digital covariance function, 151
Hotelling transform, 419
Hough transform, 236
Hyperplane, 227

Idempotent, 109
Identically distributed, 373
Illumination gradient, 62
Image convergence, 127
Image functional, 123
Image-to-image transformation, 8
Image-to-parameter transformation, 8
Image representation, 8
Image vector, 48
Increasing, 100, 109, 126
Independent, 310, 345
Inferential statistics, 410
Infinitely differentiable, 226
Information density, 16
Inner product, 246
Inner product space, 246
Input noise, 393
Instrumental errors, 307
Integrable image, 198
Integrable signal, 187
Integral geometry, 131
Interior, 115
Interquantile range, 325
Inverse discrete Fourier transform, 276

Inverse Fourier transform, 196
Inverse Hilbert transform, 214
Inverse Laplace transform, 209
Inversion theorem, 204

Jacobian determinant, 354
Joint distribution function, 336
Joint probability distribution, 336
Jordan's test, 264

Kalman-Bucy filtering, 402
Kalman-Bucy gain equation, 405
KBF, 402
Kernel, 152

L_2 convergence, 257
L_1 norm, 54
L_2 norm, 54
L_∞ norm, 54
Laplace transform, 208
Laplacian, 227
Least-squares criterion, 70
Least-squares solution, 393, 408
Linear, 188, 199, 219, 376, 392
Linear digital granulometry, 149
Linear space, 245
Line-drawing-type image, 185
Lipshitz condition, 216, 294
Lognormal density, 329
Loss function, 388
Low-pass filter, 42

Magnification, 289
MAG1, 49
MAG2, 49
MAG0, 49
Marginal density function, 341
Marginal distribution, 340
Matheron Representation Theorem, 155, 156, 163, 169
Matrix image transform, 276
MAX, 21
Maximum likelihood estimation, 384
Maximum likelihood estimator, 385, 394
Maximum likelihood filter, 394
Maximum likelihood value, 327
Mean, 320
Mean deviation, 322
Mean risk, 389
Mean square error, 378
Mean value of random vector, 249
Mean-value theorem, 272
Measure, 120

Medical imaging, 217
Mellin transform, 208, 211
Mellin-type convolution, 359
Method of moments, 383
MIN, 21
Minimal bound matrix, 13, 16
Minkowski addition, 92, 140
Minkowski algebra, 91
Minkowski functionals, 134
Minkowski subtraction, 95, 142
Mixed moments, 347
MLF, 385, 394
Mode, 325
Model equation, 407
Moment factorial generating function, 322
Moment generating function, 322
Moments, 320, 347
Monotonicity, 133
Morphology, 91
Most efficient estimator, 379
Moving-average filter, 40
MULT, 20
Multiple recognition, 417
Multiplication identity element, 245
Multivariable normal density, 360
Multivariate Gaussian density, 360
Mutually exclusive, 306

Negation, 22
Negative binomial density, 328
Neighborhood, 40
Nested sets, 129
Neyman criterion for sufficiency, 382
Neyman-Pearson lemma, 414
NINETY, 24
NINETY^2, 25
NINETY^3, 25
Noise, 5, 33, 307
Noise vector, 403
Nondestructive testing of materials, 217
Norm, 247
NORM, 38
Normal density function, 319
Normal-form equation of the line, 218
Normalization, 289, 292
n^{th} moment, 320
Null hypothesis, 411

Observation, 371
Observation equation, 403
Odd signal, 206
One norm, 199
Opening, 105, 144

Open with respect to set B, 109
Orthogonal, 215, 241
Orthogonal projection, 242
Orthogonal set, 249
Outcome, 305
Output covariance matrix, 379
Overdetermined, 393

Parameter space, 371
Parameter-to-decision transformation, 8
Parametric estimation, 371
Pareto density, 329
Parseval's theorem, 259
Partition, 308
Pearsonian skewness, 326
Periodic extension, 261
Perpendicular, 241
Perpendicular distance, 243
Piecewise-linear line image, 278
Pixel, 11
PIXSUM, 35
Plancherel's theorem, 204
Pointwise convergence, 256
Poisson density, 328
Polytope, 115
PREWDX, 57
PREWEDGE0, 65
PREWEDGE1, 65
PREWEDGE2, 65
Prewitt compass gradient, 67
Prewitt difference operators, 57
Prewitt mask, 58
PREWMAG1, 58
PREWMAG2, 58
PREWMAG0, 58
Probability, 306
Probability density function, 316
Probability frequency function, 316
Probability measure, 305
Probability space, 305
Projection, 124, 217, 218, 241
Pseudoinverse, 393
p^{th} quantile, 325
Pure energy signal, 203

Quadratic loss function, 388
Quantization, 8

Rademacher functions, 266
Radial image, 201
Radon transform, 217, 227
Random field, 307
Random image, 419

Random sampling, 373
Random vector, 335
Random (vector) sample of size M, 373
RANGE, 29
Rapidly decreasing function, 226
Real linear space, 246
Reciprocation, 22
Recognition scheme, 366
Recursive algorithm, 399
Reflection, 96
Regularity criteria, 380
Regular sequence, 432
Rejection region, 412
REST, 39
Restoration, 4, 407
Rieman-Lebesque lemma, 188, 199, 263
Riesz-Fischer theorem, 257
Riesz potential, 228
Risk function, 388
ROBEDGE0, 65
ROBEDGE1, 65
ROBEDGE2, 65
Roberts gradient, 60
ROBGRAD, 60
ROBMAG1, 61
ROB1, 60
ROB2, 60
Rotation, 289, 366
Rotational invariance, 124
Rotation operator, 24

Sample mean, 374, 375
Sample median, 376
Sample moment, 383
Sample space, 305
Sample value, 371
Sample variance s^2, 377
Sampling, 139
Sampling error, 173
Sampling problem, 4
Saturated bound matrix, 16
SCALAR, 21, 48
Scalar multiplication, 21, 96
Schwarz function, 431
Second moment matrix, 350
Segment, 5
SELECT, 26
Sensor bias, 5
Sequency constants, 269
Serial product, 356
Sieving, 163
Sigma field, 305
Signature, 362

Sign function, 280
Singular part, 316
Size distribution, 120
Slack variables, 397
SMOOTH, 42
SOBEDGE0, 65
SOBEDGE1, 65
SOBEDGE2, 65
Sobel gradient, 59
Sobel masks, 59
SOBMAG0, 59
SOBMAG1, 59
SOBMAG2, 59
Space-invariant filter, 40
Space-variant filter, 40
Spectral amplitude, 187
SQROOT, 49
SQUARE, 40, 49
Square summable, 246
Standard deviation, 321
State equation, 401
State estimate update, 405
States of nature, 410
State variables, 403
State vector, 402
Statistic, 376
Statistical inference, 371
Stereology, 131
Strong neighbor, 40
Structuring element, 91, 120, 141
SUB, 22
Subimage, 100
Subordinate image, 27
Subtraction, 22
Sufficiency, 380
Summable, 246
Summation operator, 35
Support, 116
SYMDX, 55
SYMDY, 55
SYMGRAD, 56
SYMMAG0, 57
SYMMAG1, 57
SYMMAG2, 57
Symmetric difference operators, 55
Symmetric gradient, 56
Synthetic aperture radar (SAR), 1
System model, 402

τ-closing, 160
τ-opening, 159, 160
Tempered distribution, 226

Template, 141
Test, 411
THRESH, 30
Thresholding, 332
Threshold operator, 30
Total variation, 197, 282
t-oversize, 164
TRAN, 23
Transition matrix, 403
Translation, 92, 289, 366
Translational invariance, 95, 124
Triangle inequality, 248
Trigonometric polynomial, 252
Trigonometric system, 251
TRUNC, 33
Truncation, 333
Type I error, 412
Type II error, 412

Unbiased estimator, 377
Uncorrelated (random) variables, 348
Uniform bound, 176
Uniform density, 319, 329
Uniformly continuous, 188, 199
Uniformly distributed, 365
Unimodal, 325
Uniqueness theorem, 195, 200
Unit normal, 134
Unit step function, 188
Unit tangent, 134

Variance, 321, 348
Vertical digital covariance function, 151
Vertices, 115

Walsh column vector, 266
Walsh functions, 266
Walsh image, 292
Walsh row vector, 267
Weak neighbor, 40
Weiball density, 329
Weyl transform, 213
White noise, 403
Window, 26

x projection, 280

y projection, 280

Zero-order Bessel function, 201, 210
Zero vector, 245